FROM DNA TO PROTEIN

From DNA to Protein

The Transfer of Genetic Information

Maria Szekely

Department of Biochemistry,
Imperial College, London, UK

First edition 1980
Reprinted (with corrections) 1981

Published by
THE MACMILLAN PRESS LTD
London and Basingstoke
Companies and representatives throughout the world

Published in the USA by
Halsted Press, a Division of
John Wiley & Sons, Inc, New York

LCCCN 79-11894

Printed in Hong Kong

British Library Cataloguing in Publication Data

Szekely, Maria
From DNA to protein.
1. Molecular biology
I. Title
575.2′1 QH430

ISBN 0-333-21836-1
ISBN 0-333-21838-8 Pbk

Contents

Acknowledgements

I am greatly indebted to scientists, colleagues and friends from all over the world for the kind help that they gave me throughout the task of writing this book.

I wish to express my gratitude to Dr F. Sanger and to colleagues in his Division at the MRC Laboratory of Molecular Biology, Cambridge, in particular to Dr G. G. Brownlee, Dr B. G. Barrell, Dr C. Milstein, Dr P. H. Hamlyn and Dr F. E. Baralle; also to Professor H. G. Wittmann and to colleagues in his Division at the Max-Planck-Institut für Molekulare Genetik, Berlin, in particular to Dr R. Brimacombe, Dr R. A. Garrett, Dr A. R. Subramanian, Dr V. A. Erdmann, Dr K. H. Nierhaus, Dr K. Isono and Dr S. Isono. They all not only provided me with valuable material for this book in the form of reprints, preprints, photographs, etc. but also gave me the opportunity of having very fruitful discussions on recent progress in their field of research.

It is a pleasure to thank all those who kindly helped by sending me photographs for illustrations, preprints and other miscellaneous information on their research work. In this connection I wish to thank Dr G. Stöffler, Dr P. Leder, Dr M. Wu, Dr R. H. Cohn, Dr P. Sloof, Dr B. E. Griffin, Dr A. Jacobson, Dr J. A. Lake, Dr S. L. McKnight, Dr K. E. Koths, Dr D. Dressler, Professor O. L. Miller Jr, Dr D. F. Klessig, Dr B. Johnson, Mr M. J. F. Fowler, Dr A. Klug, Dr R. A. Cox, Dr A. J. Shatkin and Professor A. R. Fersht.

I am grateful for the valuable comments and advice of those who read some chapters of the manuscript, in particular to Dr F. Sanger, Dr H. Gould, Dr B. Johnson and Mrs T. Anderton.

I am also indebted to Miss M. Gordon for her help in compiling and correcting reference lists.

Abbreviations used throughout the text

MW, molecular weight

ss and ds (in relation to DNA or RNA), single stranded and double stranded, respectively.

A, C, G, T and U stand for nucleotides in RNA or DNA, N for an unspecified nucleotide. Only where the text leaves ambiguities in this respect are d and r used to distinguish between deoxyribonucleotides and ribonucleotides. Similarly, the position of the phosphate group (pN for 5′ phosphorylated nucleotides and Np for 3′ phosphorylated nucleotides) is indicated only if there is special emphasis on this structure or if the text would otherwise allow ambiguous interpretation.

bp, base pairs

kb, kilobases

cDNA, complementary DNA

r-protein, ribosomal protein

rRNA; ribosomal RNA

In the lists of references:

PNAS, *Proc. natl. Acad. Sci. U.S.A.*

Nature NB, *Nature New Biology*

Introduction

The last twenty-five years saw the birth and rapid development of a new discipline, molecular biology, which unites modern trends of biochemistry, biophysics and genetics. We can place its origin at the discovery of the double-helical structure of DNA, the molecular structure which gave us the first insight into the exact nature of genetic information. Development followed along the lines of research into the conservation and transfer of genetic information. Our present knowledge in this field has evolved through distinct stages landmarked by the introduction of new concepts, those of the double helix, the messenger, the Central Dogma and the genetic code, and with them new attitudes in approaching these problems. It may be added that recent results on the structure of overlapping genes and split genes have revealed new possibilities of how genetic information can be built into the DNA molecule. These results, which may well lead to a new concept of gene structure, will be discussed later, in Chapters 1 and 3.

THE DNA DOUBLE HELIX

At the time of Watson and Crick's discovery[1], DNA had already been recognised as the genetic material which comprises, encoded in its structure, the information for all genetically determined characteristics of any living organism. This also implied that such information is passed on from one generation to the next by producing exact replicas of these molecules. The Watson–Crick model of DNA revealed the way in which the genetic information can be built into the molecular structure and already pointed to the biological mechanism by which faithful copying of the structure can be achieved, ensuring that this information will be conserved over the generations. All the required information can be encoded into the sequence of nucleotides in the two strands of DNA. We can calculate in how many different ways the four bases can be arranged in a DNA stretch of known length; the number of different nucleotide sequences which in theory may form a 1000 nucleotide-long stretch (this is a reasonable size for one gene) is $4^{1000} = 10^{600}$. This enormous variability of nucleotide sequences in the two polynucleotide chains is more than sufficient to account for the number of different genes in different chromosomes. It follows that exact reproduction of the nucleotide sequence is the key to the conservation of genetic information.

The structure of the double helix was based on the recognition of the rules of base pairing: hydrogen bonding occurs between complementary bases, between A and T and between G and C. The strict complementarity of the two strands in the double-helical molecule means that the nucleotide sequence of one strand is unambiguously determined by that of the other. It also follows from the rules of base pairing that each DNA strand can serve as a template for the synthesis of a new strand of exactly defined nucleotide sequence. The mechanism of DNA replication that emerges, based on the rules of complementarity, explains how the two strands of DNA direct the synthesis of complementary strands, resulting in the production of two daughter molecules identical to the parental DNA (*figure 0.1*).

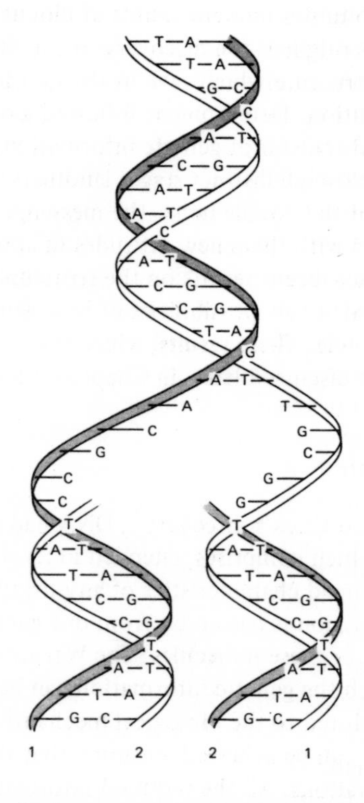

Figure 0.1 DNA double helix in the process of replication. 1: Parental strands; 2: newly synthesised strands. Shaded strands are complementary to white strands. Note that the base sequences of shaded strands 1 and 2 are identical, as are the base sequences of white strands 1 and 2.

The formation of pairs between complementary bases also proved to be the basis of the transfer of information from one molecular species to another. The synthesis of RNA on a DNA template follows the above mechanism as far as the copying of nucleotide sequences is concerned. In the synthesis of proteins the translation of nucleotide sequences into amino acid sequences is achieved via interaction of complementary trinucleotides. Base pairing has even more widespread significance: it is probably the basis of recognition of different control signals and plays a role in every nucleic acid–nucleic acid interaction.

THE MESSENGER CONCEPT

The messenger concept evolved from the necessity of finding a mediator between the biochemically stable, invariant DNA and the much more flexible protein pattern of the cell which, in prokaryotes, adjusts rapidly to the altered environment and to changes in the needs of the cell for different proteins. In eukaryotes the compartmentalisation of the cell which separates the site of protein synthesis from the site of genetic material localisation emphasised the need for a more mobile mediator which could establish direct contact with both the DNA which carries the information and the protein in which the information is expressed. In their classical paper, Jacob and Monod[2] describe the molecule which fulfils this function as a polynucleotide with a high rate of turnover which can temporarily associate with ribosomes and which is synthesised on, and reflects the base composition of, the DNA of the structural genes.

The obvious candidate for the function of messenger was RNA, a smaller molecule, biochemically less stable, present in the nucleus as well as in the cytoplasm, which can carry, encoded in its nucleotide sequence, the same messages as those present in the genetic material. Experimental evidence that RNAs indeed act as messengers was obtained soon after the launching of the messenger hypothesis, but it was years before the first mRNA was in fact isolated from animal cells. The difficulties which caused this delay were partly inherent in the problem—the huge number of different mRNA species in the cell, the low amount in which any individual species is present, the lability of bacterial mRNAs—and were partly due to technical shortcomings—the inadequacy of current techniques for separation and identification of mRNAs. Improvement in the methods of fractionation by gel electrophoresis and development of efficient cell-free protein-synthesising systems led eventually to the isolation of a pure RNA fraction from reticulocytes which directed the synthesis of globin in a reticulocyte lysate[3]. Isolation of a number of other mRNAs followed when more information had been collected on some structural characteristics of these molecules which facilitated their isolation, and when, in the knowledge of the genetic code and in possession of new methods for nucleic acid sequence analysis, identification of mRNAs by their nucleotide sequences became possible. Progress in the physical mapping of genomes and in the isolation of single genes also opened the way to the isolation of the corresponding mRNAs from bacterial cells. Today, quite a few cellular and viral mRNAs can be obtained in pure form; their number is increasing rapidly, as is the information collected on the structure of these molecules.

THE CENTRAL DOGMA

The Central Dogma was formulated by Crick to fit into a clear pattern the different observations available in the late 'fifties on the flow of information from one molecular species to another[4]. The rules laid down in the Central Dogma govern the transfer of genetic information in all living organisms. In their original, simpler form these rules described the ways generally used for information transfer in cellular processes:

$$\text{DNA} \xrightarrow[\text{transcription}]{} \text{RNA} \xrightarrow[\text{translation}]{} \text{protein}$$

which means that information can be transferred from DNA to RNA and from RNA to protein, but information cannot leave the protein and be passed to another molecule. This is strictly true for the mechanisms operating within the cell's own system of information transfer: RNA is synthesised on a DNA template, a process called transcription; and protein is made on an mRNA template by the process called translation. Transcription produces an RNA copy of one or a few messages encoded in the genetic material: the mRNA molecule. Translation leads to the synthesis of a protein the amino acid sequence of which corresponds to the same information content as that encoded into the nucleotide sequence of mRNA. The position of RNA in the diagram corresponds to the function of the messenger discussed above: it constitutes the link between the genetic material and the protein. It is the aim of this book to describe the mechanisms by which faithful transfer of information occurs between these molecules and to explore the manifold interactions between nucleic acids and proteins which ensure the high accuracy achieved in these processes.

The above diagram applies to the general mechanisms functioning in the cells. It does not account for the special processes observed in certain viruses which also make use of other ways to store, transfer and express genetic information. In RNA viruses the genetic information is originally encoded into RNA and not DNA. Some viruses of this group can replicate and produce messenger RNAs by copying their RNA molecules, thus producing new RNA strands. Others, the RNA tumour viruses, follow a different mechanism. They contain an enzyme, reverse transcriptase, which copies the viral RNA into a DNA molecule. The discovery of this enzyme[5] caused considerable excitement, as it seemed to bring about a flow of information in a direction opposite to that defined in the Central Dogma. In 1971, Crick[6] described the complete range of 'permitted' routes of information transfer, incorporating these special viral processes (*figure 0.2*).

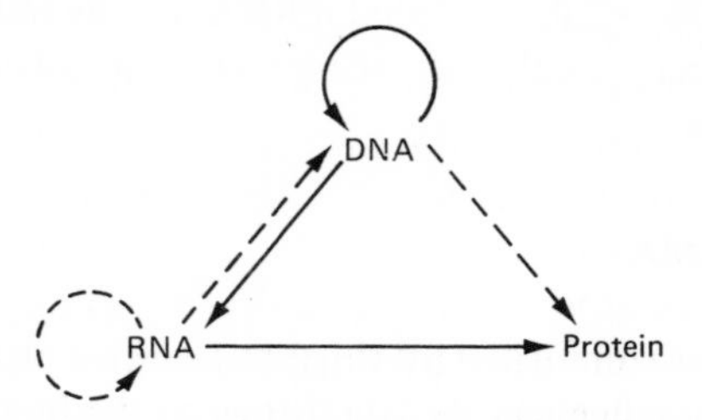

Figure 0.2 The routes of information transfer permitted by the Central Dogma.

The diagram in *figure 0.2* includes both the usual (solid lines) and special (broken lines) routes. It also shows (circular arrows) that DNA is used as a template for self-replication and that a similar self-replication process of RNA may exist in some viruses.

THE GENETIC CODE

If we only consider the processes of transcription and translation with respect to the ways in which genetic information is transferred from one molecule to another and disregard for the present the actual enzymic mechanisms by which these molecules are synthesised, we find that the first step involves simple copying of a nucleotide sequence, while the second requires a specific code system to relate the nucleotide sequences in RNA to the amino acid sequences in protein. Establishing the exact nature of this code and disclosing the actual code alphabet were the problems of molecular biology which probably attracted the greatest interest in the early 'sixties.

Some characteristics of the genetic code could be predicted on theoretical considerations, before the code alphabet had been determined. The discrepancy between the number of nucleotides, 4, and the number of amino acids, 20, suggested the necessity of a triplet code: only a combination of 3 nucleotides could yield more than 20 different codewords. However, the total number of trinucleotides, 64, is higher than that required to provide one codeword for each amino acid. This indicated either a degenerate code, in which several codons are used for one amino acid, or the existence of a number of 'nonsense codons', codewords which are not used in the message because they do not correspond to any amino acid. The nature of the code is a decisive factor with regard to the possible structures of mRNAs: a great number of nonsense codons would cause strong restrictions on the primary structure of any mRNA, while a degenerate code would allow a degree of flexibility: different primary structures could still comprise the same information. With the determination of the code alphabet and later with the nucleotide sequence analysis of different messages, it has been clearly established that the code is degenerate, and the corresponding flexibility in mRNA structure has also become apparent. Only three nonsense codons have been detected, all of which serve as termination signals of protein synthesis. Further characteristics, such as the lack of punctuation, i.e. the lack of any additional signals which separate the codon for one amino acid from that for the next, and the one ambiguity in the code, involving the initiation codon, were revealed in the course of investigations which led to the deciphering of the code alphabet and were confirmed when nucleotide sequences of natural mRNAs had been established. The lack of punctuation has the consequence that any nucleotide sequence can be interpreted in three different 'reading frames' (*figure 0.3*).

This is not only a theoretical possibility, for it has been found recently that the same nucleic acid sequence can comprise two messages: it can direct the synthesis of two different proteins which are encoded in the same sequence but in

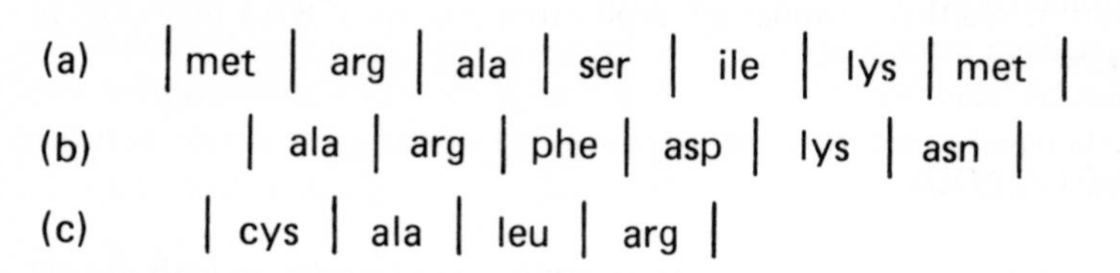

Figure 0.3 Decoding of a message in three different reading frames. The amino acid sequences (a) and (b) actually exist in proteins of ΦX174 phage. In reading frame (c) termination should occur after the fourth amino acid, as a nonsense codon, UAA, is present in the next position.

different reading frames[7]. Two of the amino acid sequences shown in *figure 0.3* actually exist in two viral proteins. It follows that faithful decoding of any message depends on recognition of the proper reading frame, which requires recognition of the specific site where decoding starts. There are two codons which can define initiation sites: AUG, which is generally used as initiation codon, and GUG, which occurs only infrequently at the initiation sites of mRNAs. Considering the important role of initiation codons in ensuring the accuracy of translation, the discovery that both triplets AUG and GUG are ambiguous in their interpretation was rather unexpected: in addition to functioning as initiation codons they can also code for amino acids at internal positions, AUG for methionine, GUG for valine. These represent the only ambiguity of the genetic code.

The code alphabet, shown in *figure 0.4*, has been worked out over a period of 4 years, by the combined efforts of several research groups using different approaches. The first codon, UUU, specifying phenylalanine, was determined by Nirenberg and Matthaei[8], who discovered that in a cell-free system the synthetic polynucleotide poly(U) directed the synthesis of polyphenylalanine. By similar experiments, some more codewords were revealed, but the method was restricted to nucleotide triplets present in synthetic polynucleotides that could function as artificial messengers. It had a further shortcoming in that it provided only the composition of the triplet and not the actual sequence of the three nucleotides. Khorana's group applied a more direct approach: they synthesised oligonucleotides of known structure and translated them into oligopeptides[9]. These experiments yielded the exact structure of a number of codewords and confirmed in a direct way the triplet nature of the code. Nirenberg and Leder[10] eventually developed a new assay technique which led to the elucidation of all trinucleotides coding for the 20 amino acids. This was based on the property of aminoacyl-tRNAs (see below) that they can bind to the ribosome in the presence of their cognate codon. Phenylalanyl-tRNA, for example, binds to the ribosome in the presence of the trinucleotide UUU. By testing the binding of all available aminoacyl-tRNAs with all possible trinucleotides, Nirenberg and coworkers were able to assign each codon to an amino acid, with the exception of the three nonsense codons.

Concomitantly with the investigation into the nature of the genetic code, studies have been carried out to reveal the mechanism by which the coded message can be 'read' by the protein-synthesising system. Considering the very different chemical structures involved, direct and specific interaction between trinucleotides and amino acids could be ruled out. The participation of an

Second letter

First letter	U	C	A	G	Third letter
U	UUU Phe	UCU Ser	UAU Tyr	UGU Cys	U
U	UUC Phe	UCC Ser	UAC Tyr	UGC Cys	C
U	UUA Leu	UCA Ser	UAA Ter*	UGA Ter*	A
U	UUG Leu	UCG Ser	UAG Ter*	UGG Trp	G
C	CUU Leu	CCU Pro	CAU His	CGU Arg	U
C	CUC Leu	CCC Pro	CAC His	CGC Arg	C
C	CUA Leu	CCA Pro	CAA Gln	CGA Arg	A
C	CUG Leu	CCG Pro	CAG Gln	CGG Arg	G
A	AUU Ile	ACU Thr	AAU Asn	AGU Ser	U
A	AUC Ile	ACC Thr	AAC Asn	AGC Ser	C
A	AUA Ile	ACA Thr	AAA Lys	AGA Arg	A
A	AUG Met	ACG Thr	AAG Lys	AGG Arg	G
G	GUU Val	GCU Ala	GAU Asp	GGU Gly	U
G	GUC Val	GCC Ala	GAC Asp	GGC Gly	C
G	GUA Val	GCA Ala	GAA Glu	GGA Gly	A
G	GUG Val	GCG Ala	GAG Glu	GGG Gly	G

Figure 0.4 The genetic code.* The three triplets UAA, UAG, UGA with no amino acid allocated to them, are nonsense codons which lead to the termination of the polypeptide chain.

adaptor molecule was indicated in the decoding process. This proved to be an RNA molecule, and by a lucky coincidence, it was an RNA species which could be isolated in pure form and was therefore accessible to functional studies at that time impossible with other RNAs. The specific adaptors were found to be transfer RNAs (tRNAs), small RNA molecules which can fulfil this function because they possess a specific site which recognises the codon (by complementary base pairing) and another specific site to which the corresponding amino acid can be bound. These molecules thus provide the link between the codon and the amino acid specified by it. Another function of tRNAs is that they carry the amino acids to the site of protein synthesis. The aminoacylated tRNA can bind to the ribosome if its cognate codon is present. (This property has been exploited to decipher the genetic code.) In the process of decoding the message, transfer RNAs thus play a central role.

The mechanism of these processes, as well as of other steps in translation, will be described in later chapters of this book. It may be useful, however, to present here a brief outline of the gross events in translation, as we know them today, so as to enable the reader to fit the different components of the system and their functions into a proper pattern.

THE MAIN EVENTS IN TRANSLATION

The ribosomal particle provides the basic machinery for the synthesis of polypeptide chains, for all components of the translation system attach to it: the mRNA which carries the genetic message, the aminoacyl-tRNA which reads the message and delivers the corresponding amino acid to the ribosome, and the growing peptide chain, also bound to a tRNA molecule, in the form of peptidyl-tRNA.

The first step in translation is the formation of an *initiation complex*, which consists of the mRNA, the ribosome which binds to the specific initiation site in this mRNA and the initiator tRNA which interacts with the initiation codon (AUG or GUG) and carries a free or formylated methionine residue. This methionine is required for initiation; it is removed from most proteins before or shortly after synthesis is terminated. Some protein factors, the initiation factors, also participate in the formation of the initiation complex; they are released after completion of this first stage. At the next stage, elongation, the ribosome moves along the mRNA in the direction from the 5′ to the 3′ end. To each codon which subsequently comes into contact with the ribosome, a specific aminoacyl-tRNA binds and then attaches also to the ribosome, in a position which places the amino acid at the functional centre where the peptide bond is formed. Protein factors, the elongation factors, also take part in these processes. The actual chemical reaction which leads to peptide bond formation is carried out by components of the ribosomal particle itself. It involves a transfer of the existing peptidyl group from tRNA to the α-amino group of the incoming aminoacyl-tRNA.

In this way, growth of the peptide chain takes place from the N-terminus to the C-terminus. At each step of the elongation of the polypeptide chain, there are two tRNA molecules attached to two specific sites on the ribosome: one tRNA carries the existing peptide chain and the other carries the new amino acid. When a termination (nonsense) codon is reached which has no cognate tRNA, only the peptidyl-tRNA remains on the ribosome, and it undergoes hydrolysis instead of peptide transfer, yielding a free complete polypeptide chain. With the aid of protein factors, the release factors, all components are released from the ribosome and the whole process can start anew.

In the following chapters, the different mechanisms will be described by which the genetic information is transferred from one molecule to another: from the parental DNA to the daughter molecules in replication, from DNA to RNA in transcription and from RNA to protein in translation. The structure of DNA and RNA will be discussed in relation to the ways in which information is encoded into these molecules. The structural features which function as control signals, safeguarding the accuracy or ensuring the flexibility of these processes, will be emphasised. As the main subject of this book is the mechanism of information transfer, less emphasis will be placed on the purely enzymological aspects of the synthesis of nucleic acids and proteins. The many rather sophisticated techniques which are applied today to study these procedures will be described only as far as the principle of the methods is concerned, with the exception of the determination of nucleotide sequences. As the nucleotide sequence of a DNA or mRNA molecule constitutes the genetic information, it is of special interest to describe the current techniques which enable us to establish these sequences and to learn in which form this information is built into the genome or into the messenger molecule.

REFERENCES

1 Watson, J. D. and Crick, F. H. C. *Nature,* **171**, 737, 964 (1953).
2 Jacob, F. and Monod, J. *J. Mol. Biol.,* **3**, 381 (1961).
3 Laycock, D. G. and Hunt, J. A. *Nature,* **221**, 1118 (1969).
Lockard, R. E. and Lingrel, J. B. *Biochem. Biophys. Res. Comm.*, **37**, 204 (1969).
4 Crick, F. H. C. *Symp. Soc. Exptl. Biol.,* **12**, 138 (1958).
5 Temin, H. M. and Mizutani, S. *Nature,* **226**, 1211 (1970).
Baltimore, D. *Nature,* **226**, 1209 (1970).
6 Crick, F. *Nature,* **227**, 561 (1970).
7 Barrell, B. G., Air, G. M. and Hutchison, C. A. III *Nature,* **264**, 34 (1976).
Smith, M., Brown, N. L., Air, G. M., Barrell, B. G., Coulson, A. R., Hutchison, C. A. III and Sanger, F. *Nature,* **265**, 702 (1977).
Shaw, D. C., Walker, J. E., Northrop, F. D., Barrell, B. G., Godson, G. N. and Fiddes, J. C. *Nature,* **272**, 510 (1978).
8 Nirenberg, M. W. and Matthaei, J. H. *PNAS,* **47**, 1588 (1961).
9 Khorana, H. G. *et al. Cold Spring Harbor Symp.,* **31**, 39 (1966).
10 Nirenberg, M. and Leder, P. *Science,* **145**, 1399 (1964).
Nirenberg, M. *et al. Cold Spring Harbor Symp.,* **31**, 39 (1966).

PART 1

The Genetic Material

1 The Structure of DNA and the Structure and Organisation of Genes

SIZE AND CODING POTENTIAL

The genetic information for all biological functions of an organism is encoded into the nucleotide sequence of its DNA. The questions which arise are: how much information can be carried by a DNA molecule; how large must a chromosome be to provide all the necessary information for the organism?

Obviously, the number of different proteins synthesised in the cells varies from one organism to another, and with their number the number of genes varies also, increasing greatly from the genomes of the simplest viruses to those of complicated animal chromosomes. The size of the chromosome also shows an increase, from the viral DNAs which fall into the molecular weight range of 1 million to over 100 million, and the bacterial DNAs which have molecular weights of about a billion, to the 1000 times bigger DNA of the human chromosomes, with a molecular weight of the order of magnitude of 10^{12} (*see table 1.1*).

Table 1.1 The size of DNA in different organisms

	Molecular weight	Base pairs	Kilobases
Viruses			
SV40	3×10^6	5×10^3	5
Adenovirus	1.4×10^7	2.1×10^4	21
λ	3.3×10^7	5×10^4	50
T_4	1.3×10^8	2×10^5	200
Bacteria			
E. coli	2.2×10^9	4.6×10^6	4600
Animal cells			
Mouse	1.5×10^{12}	2.3×10^9	2.3 million
Human	1.8×10^{12}	2.8×10^9	2.8 million

It is, however, difficult to correlate the size of the DNA with the actual number of functional genes, as in most cases we have no exact knowledge of the latter. Nevertheless, it is possible to calculate the total coding potential of the different DNAs and to compare these with approximate estimates of the information content required in different organisms. As we are interested in information content, it is more to the point to define the size of DNA in terms of base pairs. As three nucleotides determine one amino acid, about one thousand base pairs can code for a protein of 30 000 daltons, which can be considered an average molecular weight. In *table 1.1*, the sizes of different DNAs are also shown in base pairs (bp). Large DNA molecules or long DNA stretches are often described in units of kilobases (kb), i.e. in units of 1000 base pairs. A 1 kb-long DNA stretch can thus be accepted as an average length for a gene. Calculating on this basis, the genomes of the smallest viruses may contain enough information for about 4 to 5 genes, the chromosome of *Escherichia coli* may contain 3000 to 4000 genes, and the giant chromosomes of higher organisms could provide sufficient information for over a million genes.

The coding potential could be exactly determined in some small viruses where it was possible to measure the length of each gene or to establish directly the nucleotide sequence of the DNA. Considering the size of these molecules, it is understandable that exact data are available only at the small end of the scale; the data shown in *table 1.1* for large chromosomes are only approximations.

In small viruses it is also possible to determine the number of genes which are actually present. Genetic maps and physical maps can be constructed and the number of genes can be compared with the coding potential. In larger genomes, it has to be taken into account that not all genes may have been detected, that there may still be gaps in the genetic map. In some cases the calculated coding capacity and the actual number of genes are in good agreement. The map of the genome of monkey virus SV40 (Simian virus 40) is shown in *figure 1.1*. This DNA codes for five proteins. Two pairs of proteins (VP2/VP3 and small t/large T antigens) contain parts of identical amino acid sequences; these are coded for by the same DNA stretch. The messenger RNAs specifying the five proteins account for almost the whole length of the genome. In bacteriophage λ, about

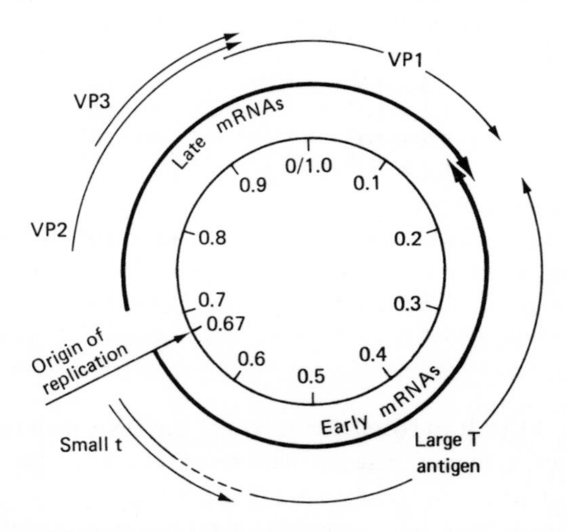

Figure 1.1 Map of SV40 DNA. Inner circle: DNA genome; thick lines: mRNAs produced early and late in the infection cycle; thin lines: regions coding for viral proteins. Arrows show the 5′ → 3′ direction of mRNAs and the N-terminal → C-terminal direction of proteins.

40 genes have been mapped so far, which is not much less than its expected total coding capacity. The genome of bacteriophage T_4 is much bigger, its total coding capacity amounting to about 150 genes, 100 of which have already been identified.

If we look, however, at very small or very large DNAs, we find different situations. At one extreme, some very small bacteriophages (ΦX174, G4) contain genomes which seem too small to code for all the phage proteins–we will discuss later how an excess of information can be compressed in their DNA. At the other extreme, in the giant eukaryotic chromosomes, the estimated number of genes is one or two orders of magnitude lower than the calculated coding potential of their DNA. The DNA of the human chromosomes, with its 2.8×10^9 base pairs, could contain over 2 million genes, but even the highest estimates for genes actually functioning are well below 10^5.

We have taken a rather simplified view so far of the information content of DNAs. The number of genes is, in fact, not equal to the number of different proteins. DNA does not only code for proteins, and the messenger is not the only RNA species which is copied from the nucleotide sequence of DNA. All the other RNAs of the cell, ribosomal RNAs and transfer RNAs, are also synthesised on the DNA template. *Figure 1.2* shows a part of the genetic map of *E. coli* DNA which contains genes for ribosomal and transfer RNAs as well as mRNAs.

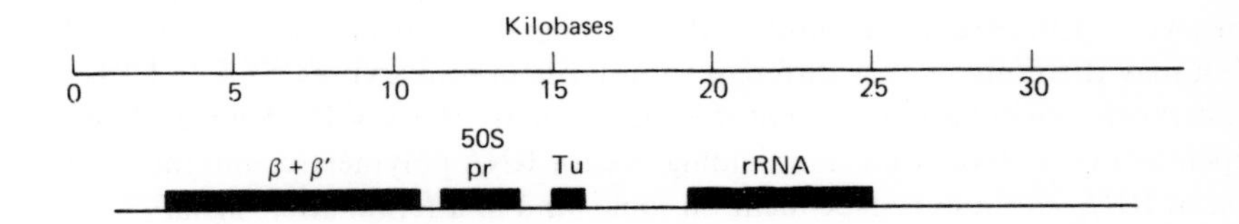

Figure 1.2 Map of a part of the *E. coli* genome, which contains genes for ribosomal components (based on data from Nomura[3]). β, β′: genes for two subunits of *E. coli* RNA polymerase; 50S pr: genes for ribosomal proteins; Tu: gene for translation elongation factor EF-Tu; rRNA: ribosomal RNA transcription unit, comprising genes for 5S, 23S and 16S rRNAs and for some tRNAs. The exact structure of this transcription unit is shown in Figure 1.16.

In addition to these structural genes, there are also regions in the DNA which are not transcribed but which, nevertheless, have very important functions. These are control regions, which regulate all synthetic processes: the replication of the DNA and the expression of the structural genes. Their total length makes up only a very small part of the DNA molecule, but their importance is great. Many control sites are present in every genome: these are the sites of interaction with different regulatory proteins which initiate or terminate, activate or prevent a synthetic process. One such control site is shown on the genetic map of the SV40 DNA in position 0.67: this is the origin of DNA replication. Another, more complex control region, part of the *E. coli* genome, can be seen in *figure 1.3*. In prokaryotes, the genes for some functionally related enzymes are organised into a structural and functional unit, the operon. The figure shows the organisation of the lac

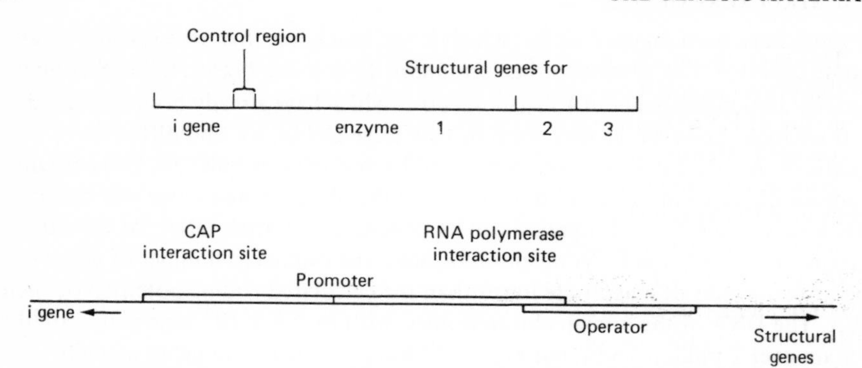

Figure 1.3 The lac operon. Above: structure of the complete operon, drawn in proportion to the actual lengths of genes and control regions. Below: structure of the control region. For explanation, see text.

operon which produces three enzymes required for the metabolism of lactose (*see* p. 95). In addition to the three structural genes coding for these three enzymes, the lac operon contains: a site of interaction with RNA polymerase, the enzyme synthesising mRNA (promoter site); a site of interaction with the repressor, the protein which can prevent the synthesis of mRNA (operator site); a site of interaction with the catabolite activator protein (CAP site) which activates the synthesis of mRNA; and a gene coding for the synthesis of the repressor protein (i gene). Eukaryotic genes are not organised into operons; the control regions in eukaryotic DNA may therefore be very different from those in prokaryotes. Still, in both prokaryotic and eukaryotic organisms, the synthesis of every RNA molecule is dependent on a recognition and binding site for RNA polymerase, and the termination of RNA synthesis is dependent on some kind of a termination signal.

A further discrepancy between total information content of a genome and the number of different proteins and RNAs it codes for, arises because of the presence of reiterated genes. Very efficient synthesis of some RNAs and proteins is made possible by the presence of several copies of the corresponding genes in the DNA. In prokaryotes, the number of copies is still relatively low: the *E. coli* genome contains seven genes for each ribosomal RNA. In eukaryotes, reiterated genes are more redundant, the number of ribosomal RNA genes varies in different organisms – there are 140 copies in yeast DNA. Transfer RNA genes are also redundant; in the DNA of yeast there are on average 10 copies of each gene coding for the different tRNA species. With the exception of histone genes, the genes coding for proteins are almost always unique. In rapidly dividing nuclei, fast production of histones is required and this is made possible by the presence of several hundred copies of histone genes.

The great discrepancy between the size of eukaryotic DNA and the required information content is by no means abolished if all the above additional information is included in our calculations. Reiterated genes, control regions, etc. account only for a few per cent of the total coding capacity of the eukaryotic genome. The problem is still unsolved: it is not known whether any function can be ascribed to the major part of eukaryotic DNA molecules.

However, the recent discovery of split genes brings us closer to an explanation of this phenomenon. Comparison of the structure of eukaryotic DNA with the structure of RNAs synthesised on this template produced unexpected results which encouraged intensive research into the exact mapping of the sites in the eukaryotic genome corresponding to the nucleotide sequences found in the mature mRNAs, ribosomal RNAs and tRNAs. These investigations revealed that our ideas about the structure of the genes, which we envisaged as linear, uninterrupted stretches of DNA, are true only for prokaryotes and have to be completely re-assessed if we are to obtain a true picture of the organisation of the eukaryotic genome. In the DNA of eukaryotes, one gene, the carrier of one well defined genetic message, is often comprised of several non-contiguous stretches; it is interrupted by one or more inserts which may be a few hundred or even a thousand base pairs long. These inserts are not expressed in the protein molecule or in the mature RNA. Such inserts of 'silent DNA' (Gilbert introduced the name 'intron') have been detected in viral messages as well as in the genes coding for globin, ovalbumin, immunoglobulin and in some organisms in rRNA and the tRNA genes also[1]. In *figure 1.1* a dotted line shows one such interruption in the DNA sequence coding for the large T antigen. Many genes have a mosaic structure, built of two or more remote stretches.

These findings certainly necessitate a basic change in our image of gene structure and of the organisation of the genome. This problem will be discussed in greater detail later, but with respect to the size of eukaryotic DNA and its coding capacity, we can draw some conclusions here from this new concept of 'expressed' and 'silent' DNA regions. The data available are not yet sufficient for us to calculate what proportions of the total DNA of higher organisms fall into one or the other category. It seems, however, that the interruptions may be as long (e.g. in the globin gene) or even longer (e.g. in the ovalbumin gene) than the mRNA molecule. It should also be taken into account that mRNA molecules also contain untranslated sequences in addition to those which are actually coding for the structure of the protein. The ovalbumin gene is over 7000 base pairs long; the length of the mature ovalbumin mRNA is 1859 nucleotides, of which 1158 comprise the actual code for the ovalbumin molecule. If the length of introns and other redundant sequences turned out to be in excess over the functional regions of the eukaryotic genome, this might at least partly account for the discrepancy between the size and information content of these DNA molecules.

ISOLATION OF GENES. PHYSICAL MAPPING

The isolation of well defined identifiable fragments containing individual genes was made possible by the discovery of a very specific group of enzymes, the *restriction endonucleases.* Other nucleases which can break down DNA are not specific for the site where they introduce a split. Endonucleases, enzymes which split DNA somewhere at the inside of the polynucleotide chain, are often specific for the substrate, e.g. they cleave only DNA not RNA, or they cleave only single-stranded or only double-stranded nucleic acids, but they usually cleave at any one site inside the chain irrespective of what bases are present at that site. There is, however, one group of endonucleases which show a very high degree of specificity for the cleavage site: they are specific, not for a base, but for a longer nucleotide sequence in the DNA molecule. These endonucleases are called restriction enzymes. They are produced mainly by bacterial cells and their function is to degrade foreign DNA which enters the cell. The nucleotide sequences recognised by a few restriction enzymes are shown in *table 1.2*. It can be seen that enzymes

Table 1.2 Recognition and cleavage sites of restriction endonucleases

Enzyme	Sequence recognised (cleavage sites marked by arrows)
*Eco*RI	CTTAA↓G G↑AATTC
*Hin*dIII	TTCGA↓A A↑AGCTT
*Hae*III	CC↓GG GG↑CC
*Hpa*I	CAA↓TTG GTT↑AAC
*Hpa*II	GGC↓C C↑CGG

prepared from different bacteria split the DNA at different sites. A common characteristic of the different sequences recognised by different restriction endonucleases is a twofold symmetry at all these sites. (This kind of symmetrical sequence is called a *palindromic* sequence, and will be described in more detail on p. 28.) The names of the restriction enzymes (and their abbreviations) refer to the microorganism from which they have been obtained. As the nucleotide sequences recognised are 4 to 6 base pairs long and occur therefore only with limited frequency in any DNA molecule, it follows that restriction enzymes split DNA molecules only at a limited number of sites. The restriction enzyme of *E. coli*, *Eco*RI, causes only one cleavage in the over 5000-bp-long DNA of SV40 virus and five cleavages in bacteriophage λ DNA.

The fragments produced by the restriction enzymes can be identified in different ways; their nucleotide sequences can be established, and the proteins or RNAs for which they are coding can be identified. By choosing suitable restriction enzymes, a DNA stretch carrying only one gene or part of a gene can be isolated. The position of restriction fragments in the chromosome can be determined by the successive and combined use of different restriction endonucleases. This leads to the construction of restriction maps or physical maps.

The following example shows a relatively simple and elegant way of constructing a restriction map by measuring the exact length of different restriction fragments.

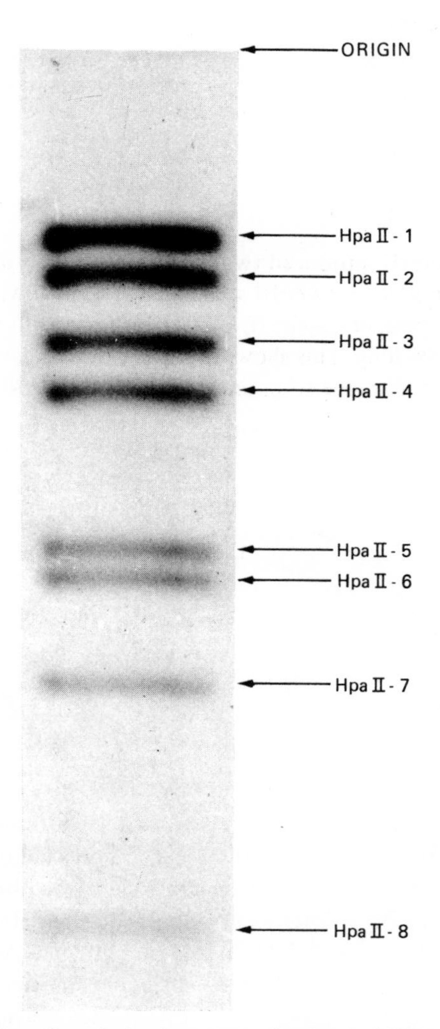

Figure 1.4 Separation of restriction fragments of polyoma DNA by electrophoresis in polyacrylamide–agarose gel. (By courtesy of Griffin *et al.*[2]. Photograph by courtesy of Dr B. E. Griffin.)

Griffin *et al.*[2] digested DNA from polyoma virus with three restriction enzymes: *Eco*RI, *Hpa*II and *Hin*dIII. This virus contains a circular double-stranded DNA genome, the total length of which is about 5200 bp. *Eco*RI splits the DNA at only one site, thereby converting the circular molecule into a linear form. This cleavage site was chosen as the startpoint of the map, 0%, and the total length of the DNA was defined as 100%. *Hpa*II produced eight fragments which were separated according to size by gel electrophoresis (*figure 1.4*); the length of the fragments, expressed as percentage of the total length, was the following:

Fragment *Hpa*II –	
1	27.3%
2	21.4
3	16.8
4	13.2
5	7.7
6	6.7
7	5.2
8	1.8

The third enzyme, *Hin*dIII, produced two fragments, of 56% and 44% length. If the linear molecule obtained by *Eco*RI splitting was treated with *Hpa*II, all the above fragments were present except *Hpa*II-2, which was split into two smaller stretches, 20% and 1.5% long. This shows that the *Eco*RI cleavage site is in the *Hpa*II-2 fragment, 1.5% distance from one end. If the two *Hin*dIII fragments were

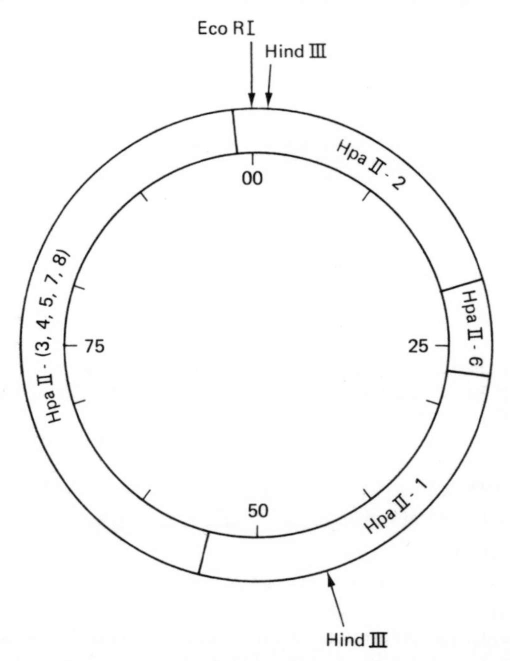

Figure 1.5 Partial map of polyoma virus.

digested with *Hpa*II, the following fragments were obtained: from *Hin*dIII 56%; *Hpa*II-3, 4, 5, 7, 8 and two new fragments, of 2.5% and 8.5% length; from *Hin*dIII 44%: *Hpa*II-6 only and two new fragments of 18% and 19.5% length. The two cleavage sites of *Hin*dIII thus fall into the fragments *Hpa*II-1 and *Hpa*II-2, and the map can be partially constructed as shown in *figure 1.5.*

To find the exact position of *Hpa*II fragments 3, 4, 5, 7 and 8, partial digestion was carried out with this enzyme, and led to the isolation of larger products which could be further split into the already identified final restriction fragments. Such partial digestion products were found to comprise fragments 1 and 3, fragments 3 and 5, fragments 4 and 5, fragments 7 and 8, and fragments 2 and 7. It could thus be concluded that of the *Hpa*II fragments, the following pairs were adjacent in the genome: fragments 1 and 3; 3 and 5; 5 and 4; 8 and 7; 7 and 2 (*figure 1.6*). This unambiguously determines the position of these fragments in the map.

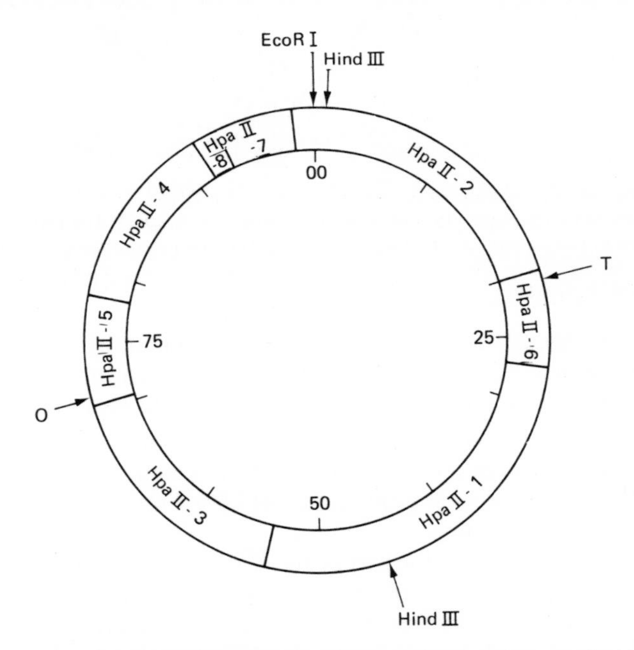

Figure 1.6 Restriction map of polyoma DNA (by courtesy of Griffin *et al.*[2]). The cleavage sites of *Eco*RI, *Hin*dIII and *Hpa*II are shown. O and T: origin and termination points of replication, respectively.

In a similar but often more complicated way, physical maps have been obtained from a number of viral genomes and from parts of some bacterial and some eukaryotic chromcsomes.

Restriction maps can be constructed of genomes of mutants defective in different genes, and the genetic loci of these mutations and thus of the genes, can be assigned to specific restriction fragments. Another way of allocating functional sites to definite positions is to identify the RNAs or proteins for which the individual restriction fragments code.

Some of the methods which allow such identification will be discussed later in this chapter. An interesting example of the location of a functional site is described in the work of Griffin *et al.*[2]. They identified the origin of replication in polyoma DNA with the aid of electron microscopy. At the first stage of replication a small bubble appeared at the site of the origin. This was seen to be growing in both directions as replication proceeded (*see* Chapter 2). The electron micrograph of a replicating polyoma DNA molecule is shown in *figure 1.7*. If the molecule is split

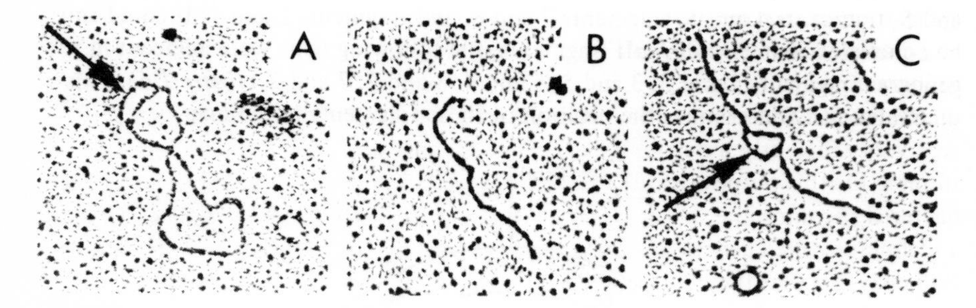

Figure 1.7 Electron micrographs of A: replicating polyoma DNA; B and C: two fragments of this DNA, obtained by cleavage with restriction endonuclease *Hin*dIII. The replication bubble (marked by arrow) can be seen on the larger fragment. Its position was determined as 25% from one end and 30% from the other. (By courtesy of Griffin *et al.*[2]. Photograph by courtesy of Dr B. E. Griffin.)

into two parts by *Hin*dIII, the replication bubble is found in the larger restriction fragment; it must therefore be at the left side of the map. The distance of the mid-point of the bubble from the ends of the fragment could be measured in the electron microscope and was found to be 30% from one end and 25% from the other. As the two ends cannot be distinguished by this method, there is still some ambiguity: the origin is either at position 71% or 76%. Further studies supported the former as the exact location.

Nomura's group fragmented the part of the *E. coli* genome which contained clustered genes for ribosomal components[3]. They first used *Eco*RI, then dissected the fragments thus obtained by treatment with other restriction enzymes. The genes these fragments carried could be identified by the corresponding gene products. One of the methods used to locate individual genes involved the transcription and translation of these DNA fragments in an *in vitro* system, and identification of the RNAs and proteins synthesised. The map so obtained from a 72.3-kb-long DNA fragment is shown in *figure 1.2*. Another DNA fragment, located at a different part of the chromosome, was found to carry the genes for a great number of ribosomal proteins and in addition, those for two elongation factors required for protein synthesis and that for the α subunit of RNA polymerase. It is interesting that while on one hand the genes for ribosomal components seem to be clustered together in a few well defined regions of the *E. coli* genome, on the other hand, these genes are interspersed by genes for tRNAs, different subunits of RNA polymerase and translational protein factors. Note the presence of such genes also on the map in *figure 1.2*.

GENETIC ENGINEERING

Characterisation of different restriction fragments, whether it means the identification of the genes they contain or the determination of their exact primary structure, often requires larger amounts of material than can be obtained directly by restriction enzyme digestion. The modern techniques of genetic engineering and cloning make it possible to isolate a restriction fragment or a single gene in milligramme quantities.

Restriction fragments can be inserted into foreign DNA molecules which replicate very efficiently in bacterial cells. Phage DNAs or plasmid DNAs are most suitable for this purpose as they both replicate at a high rate, independently of the bacterial chromosome. Insertion of a stretch of foreign DNA into these molecules, a technique called genetic manipulation, is based on the property of some restriction enzymes that they cut the two strands of DNA a few nucleotides apart, leaving short, single-stranded tails ('sticky ends') on both fragments (*see table 1.2*). If two DNA molecules are cleaved by the same enzyme, the same sticky ends will be produced. As these short tails have complementary nucleotide sequences in both molecules, annealing can occur between the sticky ends of different DNA molecules, as shown below.

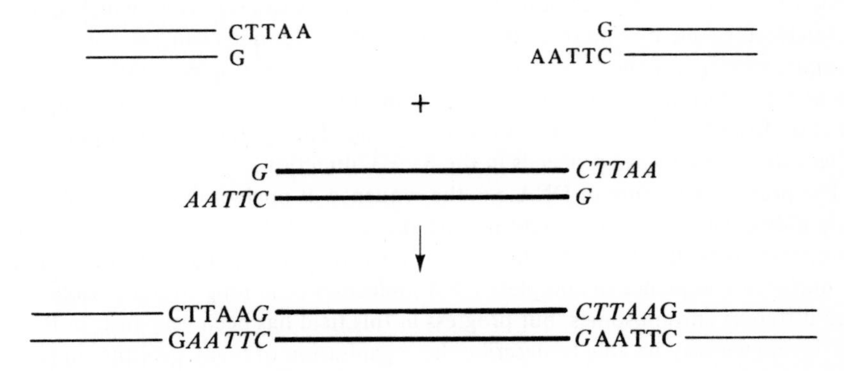

The missing phosphodiester bond between the adjacent nucleotides of the two DNAs can be formed by a specific enzyme, DNA ligase, which seals the two fragments together in covalent linkage*.

This artificially constructed DNA molecule is introduced into bacterial cells, where it is cloned to produce by efficient replication a great number of identical DNA molecules. These can be isolated and treated again with the same restriction

*Specific techniques have been worked out also for producing artificial 'tails' which make it possible to integrate practically any DNA fragment into a plasmid or phage DNA.

enzyme. The same DNA fragment which had been originally inserted can thus be excised and recovered in vastly increased amount.

In recent years, an increasing number of prokaryotic and eukaryotic genes have been cloned in this way, as well as control regions of DNA, e.g. DNA fragments containing the operator of the lac operon, or the origins of replication of the *E. coli* chromosome and of some viral genomes.

Transducing phages and plasmids carrying a well defined part of bacterial DNA have been widely used to study the organisation of the genome. An elegant example of the use of this technique for the mapping of the genes for ribosomal components in the *E. coli* chromosome has been described by Nomura (ref. 3, 1976).

THE PRIMARY STRUCTURE OF DNA

In the DNA molecule, the four mononucleotides dAMP, dCMP, dGMP and dTMP are linked by 3′, 5′-phosphodiester bonds, as seen in *figure 1.8*. This gives a chemically determined direction to the polynucleotide chain which is recognised by enzyme systems. The two ends of the polynucleotide chain can be defined as the 5′ end where the 5′ hydroxyl is not linked to another nucleotide, and the 3′ end where the 3′ hydroxyl is not involved in an internucleotide bond. The 5′ and 3′ hydroxyls need not be in free form; they may be phosphorylated but only in a phosphomonoester linkage, not in a phosphodiester bond.

Many enzyme reactions are specific for proceeding either only in the direction from the 5′ to the 3′ end (5′ → 3′) or in the opposite direction from the 3′ to the 5′ end (3′ → 5′). The degradation of DNA by exonucleases (enzymes which split off nucleotides one by one from the end of a polynucleotide chain) is, for example, specific for the 5′ or 3′ end: the exonuclease from spleen starts degradation at the 5′ terminus, while the enzyme from snake venom works in the opposite direction from the 3′ terminus towards the 5′ end. The synthesis of new polynucleotide chains always proceeds in the 5′ → 3′ direction.

The primary structure of DNA, i.e. the sequence of nucleotides in the polynucleotide chains, is of paramount importance, as this is the form in which genetic information is built into DNA. Only recently has it become possible to determine the nucleotide sequence in complete DNA molecules or in long stretches isolated from different chromosomes, but progress in this field has been very fast. In the near future we may be able to describe the organisation of many genomes in terms of exact chemical structure and to explain the control mechanisms of gene expression on a strictly structural basis.

The complete nucleotide sequence of the genomes of coliphages ΦX174 (ref. 4) and G4 (ref. 38) have been determined by Sanger's group in Cambridge. The genome of monkey virus SV40 has been completely sequenced by Fiers' group in Belgium and by Weissmann's group in the USA[5]. The sequence of polyoma virus DNA has been established by Griffin and coworkers[5a]. In bacteriophage λ DNA, a sequence about 1 kb long has been determined which contains control signals as well as coding sequences[6], and the complete sequence of the gene for the λ repressor protein has also been established[39].

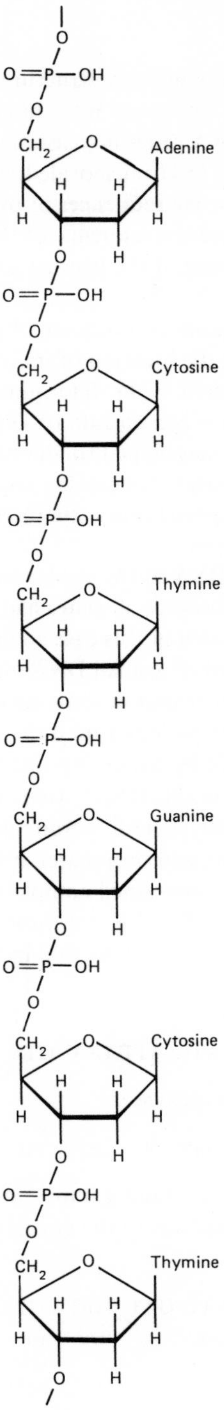

O
O=P—OH
O
CH$_2$ O Adenine
H H
H H
O H
O=P—OH
O
CH$_2$ O Cytosine
H H
H H
O H
O=P—OH
O
CH$_2$ O Thymine
H H
H H
O H
O=P—OH
O
CH$_2$ O Guanine
H H
H H
O H
O=P—OH
O
CH$_2$ O Cytosine
H H
H H
O H
O=P—OH
O
CH$_2$ O Thymine
H H
H H
O H

Figure 1.8 Structure of a part of a polydeoxyribonucleotide chain.

Fewer data are available on the primary structure of non-viral DNAs. A DNA sequence in yeast coding for 5S ribosomal RNA has been established by Valenzuela *et al.* and by Maxam *et al.*[7]. tRNA genes in yeast have also been sequenced[40]. Brosius *et al.*[40a] determined the 1541 nucleotide-long sequence of a 16S ribosomal RNA gene in *E. coli*. Nucleotide sequences of protein genes are also known: Farabough[41] recently determined the sequence of the gene coding for the lac repressor in *E. coli*, and sequencing of the histone gene repeat unit in sea urchin DNA is also well under way[42].

The sequencing of coding regions is obviously of great interest for the chemical mapping of the genome, i.e. for the location of individual genes within an exact chemical structure. A good example is the determination of the structure of the histone gene repeat unit. Here the comparison of DNA sequences with the amino acid sequences of different histones helped to identify which genes were coding for the H2a and H3 histone species[8]. Nucleotide sequence determination led in some cases to the detection of special gene structures, thus revealing new aspects of the organisation of the genome.

The single-stranded circular DNA of the small coliphage ΦX174 was the first DNA genome in which the positions of all genes had been located in the nucleotide sequence[4]. The results revealed in this case an unexpected and amazing way in which the information content of a small DNA molecule can be efficiently used. It has been mentioned earlier that in some small viruses the coding potential of the genome seems insufficient to code for all the viral proteins; ΦX174 is one such virus. It has been discovered by Sanger's group that two pairs of proteins are coded for by overlapping genes in this DNA[9]. Gene D of ΦX174 extends from nucleotide 390 to nucleotide 846. Gene E is contained completely within this region: it starts at nucleotide 568 and terminates at nucleotide 840. The same is true for genes A and B: gene B is contained completely in the gene A region. The amino acid sequences of proteins A and B and those of proteins D and E are not identical; the nucleotide sequences are translated in two different reading frames. The sequence

ATG CGC GCT TCG ATA AAA ATG ATT GGC GTA TCC AAC ATG CAG AGT

corresponds to the amino acid sequence

met arg ala ser ile lys met ile gly val ser asn leu gln ser

in protein A, while, by shifting the reading frame one nucleotide to the left, the same nucleotide sequence corresponds to the entirely different amino acid sequence of protein B:

AT GCG CGC TTC GAT AAA AAT GAT TGG CGT ATC CAA CCT GCA GAG
ala arg phe asp lys asn asp trp asn ile gln pro ala glu

The arrangement of genes in ΦX174 DNA is shown below in a linear sequence:

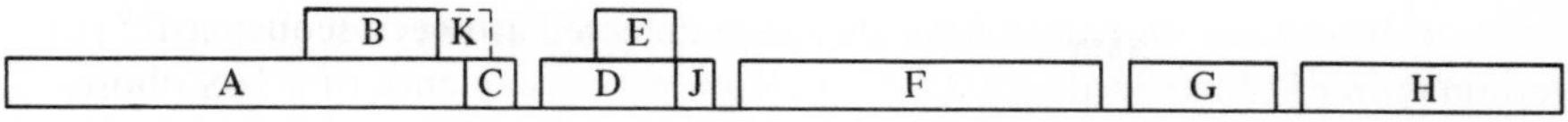

B K E
A C D J F G H

In the closely related bacteriophage G4 a third overlap has been detected: gene K overlaps with genes B, A and C. A few nucleotides in the gene K region overlap briefly with two other genes simultaneously, so that all three reading frames of the DNA are used[10]. Gene K is probably also present in the ΦX174 genome.

It cannot be judged today whether this is a very special way of compressing information into a small DNA molecule, and is used only by some small viruses, or if overlapping genes have a more general significance. The phenomenon is not restricted to bacteriophages; in the genome of monkey virus SV40, the regions coding for the C-terminal part of VP2 protein and for the N-terminal part of VP1 protein overlap over a distance of 122 nucleotides[5,11]. These proteins are also translated in different reading frames.

Knowing the complete primary structure of SV40 DNA[5], it is now possible to map all genes exactly and to locate the overlaps as well as the interruptions which occur in this genome. The work of Weissmann's group on the comparison of nucleotide sequences in SV40 DNA and in the mRNA coding for the major protein of this virus, VP1, was one of the earliest investigations which revealed that protein genes are interrupted in this genome[12]. They found that the nucleotide sequences of the mRNA corresponded to two distant parts of the SV40 DNA which were about 1000 base pairs apart in the genome. Interruptions of different lengths have since been located in the sequence coding for every mRNA of this virus. Comparison of the nucleotide sequences of tRNA genes and mature tRNA molecules in yeast led to the detection of short interruptions in these genes also[40].

The primary structure of non-coding regions of different genomes has also been the subject of extensive investigations. These regions are of interest because they may have regulatory functions and contain control signals. In SV40 DNA, a 600 base pair-long region between the start-points of the early and late mRNAs (*figure 1.1*) contains information for the initiation of replication and for early and late transcription. The nucleotide sequence around the origin of replication shows a very high degree of symmetry[13].

3′ 5′
G G C T T T T G C A A A A A G C T T T G C A A A C A T G
C C G A A A A C G T T T T T C G A A A C G T T T C T A C
5′ 3′

Symmetrical sequences are often found in regions of DNA which have a regulatory function. Such sequences may influence the secondary structure (*see figure 1.11*) and thereby the binding of specific proteins. Symmetrical sequences may therefore be of great importance for protein-DNA interactions. Sequences with a two-fold rotational symmetry have been detected in the control regions of the lac operon[14] as well as of the gal operon[15]. Symmetrical structures which may give rise to hairpin loop formation have also been detected at sites which control the termination of transcription.

A characteristic feature of non-coding regions in eukaryotic DNAs is that they often contain highly repetitive sequences. These may consist of very simple, short nucleotide sequences repeated hundreds of times. They may be present in spacer regions, between genes, or may form quite long DNA stretches. Such DNA regions, because of their different base composition, have a different buoyant density from the bulk of the DNA and can be separated from the rest of the chromosome on this basis. They can be isolated by density gradient centrifugation in the form of small, well defined DNA molecules, called *satellite DNAs.* Because of their smaller size and simple structure, satellite DNAs have been accessible to sequence studies for several years, and the nucleotide sequences of a number of satellite DNAs from different organisms have been established[16]. The repetitive rRNA genes of the yeast chromosome could also be isolated in the form of a satellite DNA.

An interesting structure of symmetrical repeated sequences which often occurs in DNA molecules is the *palindrome.* Here the sequences in the two strands are the same, if read in the same polarity, e.g.

$$
\begin{array}{c}
5' \longrightarrow 3' \\
\text{CATTATATAATG} \\
\text{GTAATATATTAC} \\
3' \longleftarrow 5'
\end{array}
$$

Such a region has a two-fold rotational symmetry and can form hairpin loops. Short palindromes occur with great frequency in different DNAs, but in some molecules several hundred base pair-long palindromic sequences are also present. The function of these long palindromes is not known. If such structures occur at the ends of linear DNA molecules, they may, however, play an important role in replication[17].

SECONDARY STRUCTURE OF DNA

The discovery of the double-helical structure of DNA is considered the greatest biological discovery of the century. It was achieved in the early 'fifties as the result of brilliant work in two laboratories: at King's College, London, Wilkins and Franklin obtained X-ray diffraction data which suggested that DNA molecules may have a helical structure; and at the Cavendish Laboratory, Cambridge, Watson and Crick constructed models which fitted the crystallography data as well as the size and three-dimensional structure of the nucleotides.

The X-ray diffraction patterns indicated that the helical molecules were built of more than one polynucleotide chain. After several attempts to arrange two or three polynucleotide chains in a helical conformation, a model emerged which fulfilled the stereochemical requirements and was compatible with the X-ray diffraction patterns: a double helix with the sugar–phosphate backbone at the outside and the bases pointing to the inside, in a plane perpendicular to the axis. Watson's ingenious idea that the two helical strands are held together by hydrogen bonds between complementary base pairs was based partly on stereochemical studies on hydrogen bond formation between the different bases, and partly on earlier data of Chargaff, who found that the base composition of DNAs from different organisms showed a certain regularity: they all contained an equal number of A and T residues as well as an equal number of G and C residues. This rather curious fact found a logical explanation when hydrogen bond formation between different bases was studied: both the complementary base pair A and T and the complementary base pair G and C can be held together by a stable hydrogen bonded structure, the former forming two, the latter three, hydrogen bonds. The base pairs thus formed are almost equal in shape and size. The space requirement of the A–T base pair is 11 Å, that of the G–C base pair 10.8 Å. They can thus fit into a regular double helix filling in the space between the two sugar–phosphate backbones.

Hydrogen bonds can also be formed between other nucleotides. We will see later that other types of base pairs contribute, for example, to the tertiary structure of

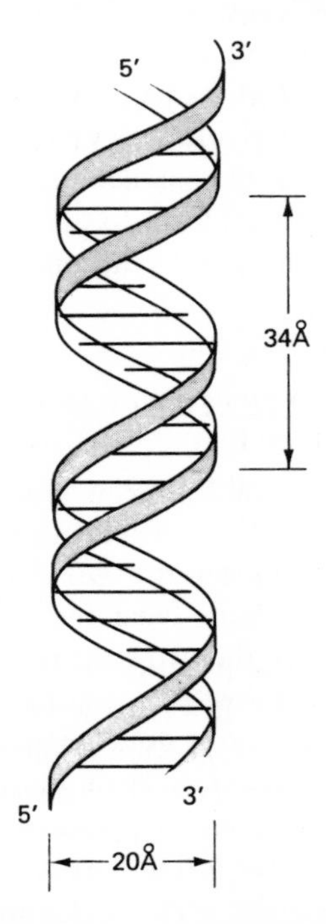

Figure 1.9 Diagrammatic representation of the double-helical structure of DNA. The sugar–phosphate backbones of the two strands are shown in the form of helical bands at the outside. The base pairs which are holding the two strands together are perpendicular to the axis of the helix; they are represented by horizontal lines.

RNA molecules (Chapter 7). However, these 'non-Watson–Crick-type' pairs are less stable and do not fit into the stereochemically specified space in the double-helical structure. Such base pairs do not occur in a double-helical DNA.

As, in the double helix, for each base A in one strand there is a complementary base T present in the opposite strand, and for each base G in one strand a complementary base C in the opposite strand, the number of A residues equals that of T residues and the number of C residues equals that of G residues in any double-stranded DNA molecule.

The strict complementarity of the strands implies that the nucleotide sequence of one strand determines the nucleotide sequence of the other. The significance of this fact for the replication of DNA is obvious: each DNA strand can serve as a template to direct the synthesis of a complementary strand, thus producing two double-stranded DNA molecules identical to the parental DNA. Watson and Crick, in their now historical Letter to *Nature*[18], in which they first described the double-helical model of DNA, had already called attention to the implications of this structure which have proved to be of fundamental biological importance. Their paper ends with the remark: 'It has not escaped our notice that the specific pairing we have postulated immediately suggests a possible copying mechanism for the genetic material.'

The model of the double helix is shown on *figure 1.9*. The direction of the two strands is antiparallel: the 3′ end of one strand is paired with the 5′ end of the complementary strand. Each turn of the helix comprises 10 base pairs, which are at a distance of 3.4 Å. The length of a full turn is 34 Å. Interaction occurs between bases in the vertical direction also: the parallel position of the apolar heterocyclic structures within a polar environment leads to the 'stacking' of the bases: they are held together by apolar forces which give additional stability to the double-helical structure.

DISSOCIATION AND ASSOCIATION OF COMPLEMENTARY STRANDS

The two strands of the double-helical DNA can be separated by physical or chemical methods. Heating a solution of DNA results in denaturation, i.e. the hydrogen bonds between the bases are broken and the two strands dissociate. Denaturation can be partial, when the two strands are separated only in some regions of the DNA molecule, or complete, when the two strands fall apart.

Denaturation of DNA can be followed by recording the optical density of the solution at 260 nm. The absorbance of a double-stranded DNA is always lower than that of a single-stranded DNA, which again shows a lower absorbance than an equimolar amount of free nucleotides in solution. This '*hypochromic effect*' is caused by the suppression of the free rotation of the bases in a polynucleotide structure. Some stacking of bases also occurs in a single polynucleotide chain, but the effect is much stronger in a double-stranded molecule. Separation of the two strands is therefore accompanied by an increase in optical density (*hyperchromic effect*). This increase is proportional to the degree of dissociation and amounts to

20–30% when complete strand separation is attained. When registering the increase in the optical density as a function of temperature, a 'melting curve' is obtained which is characteristic of certain features in the primary and secondary structure of a nucleic acid (*figure 1.10*).

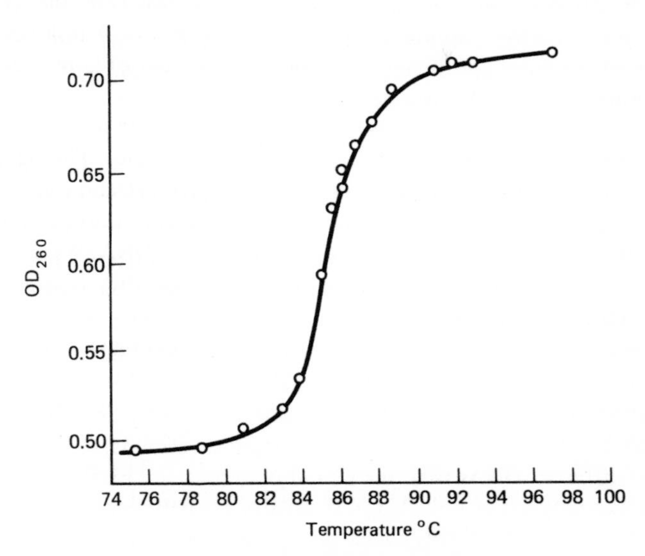

Figure 1.10 'Melting curve' of DNA.

Increasing the temperature has no effect on the optical density until the transition point, T_m (melting point), is reached, where the hydrogen bonds are broken. If the primary and secondary structure is homogeneous all along the molecule, all base pairs dissociate at about the same temperature: a sudden sharp increase in the optical density shows that strand separation has occurred. In molecules which do not have a completely double-helical structure (*see* Chapter 5 on the secondary structure of RNA), the transition occurs over a wider temperature range, as some areas, held together more tightly by more hydrogen bonded pairs, require a higher temperature for dissociation than areas held together by fewer hydrogen bonds. The primary structure also has an influence on the transition temperature and on the shape of the melting curve. For the above reason, A–T base pairs, held together by two hydrogen bonds, are easier to disrupt than G–C base pairs, which form three hydrogen bonds. DNA molecules with high G–C content therefore have a higher transition point than those with low G–C content. If G–C-rich and A–T-rich regions are not homogeneously distributed along the DNA molecule, the transition will either be less sharp, or more than one transition point will be found.

Mild denaturation, which separates only A–T base pairs but not G–C base pairs, can also be used for mapping DNAs in the electron microscope. Where long A-T-rich stretches are present, single-stranded 'bubbles' will be seen on the electron micrograph, while the G–C-rich regions will appear as linear double-stranded stretches. These 'partial denaturation maps' can be characteristic of different DNA molecules.

There are also chemical reagents which bring about the separation of the two DNA strands. Alkaline solutions, and some organic solvents like formaldehyde, formamide and dimethyl sulphoxide, denature DNA. Formaldehyde also causes substitution of the bases and thereby prevents any re-annealing of the two strands. If no substitution occurs, denaturation is reversible, and complementary strands may re-anneal.

In spite of the many easy ways in which their dissociation can be achieved, the isolation of the separate strands is a rather difficult task. The structures of the two strands are so similar that most isolation techniques prove inadequate for their separation. Besides, complementary sequences in the two strands may re-anneal and thus prevent separation. Specific techniques have been successfully used to isolate the two strands of some DNA molecules. If the base composition of the strands differs strongly and if the proportion of purines is high in one and low in the other strand, this may cause a sufficient difference in the density of the two strands to allow separation on this basis. In this case, centrifugation in a CsCl gradient may be sufficient to isolate the two separate strands. It has been found in a number of different DNAs that one strand preferentially binds synthetic polynucleotides like poly(G) or poly(U,G). If density gradient centrifugation is carried out in the presence of these polynucleotides, the strand which combines with them sediments much faster and can thus be well separated from the free DNA strand. Maxam and Gilbert recently described a simple electrophoretic technique for the separation of the strands of alkali-denatured restriction fragments[19]. This is based on an earlier observation that complementary strands show different electrophoretic mobilities in gels.

Although dissociation is reversible, special conditions are required for the perfect re-annealing of the two separated DNA strands. Slow cooling to a moderately low temperature can ensure that the two heat-denatured strands meet and anneal to re-form the original double-stranded structure. If these conditions are not met, mis-matching may occur. Relatively short complementary stretches in two molecules are sufficient for annealing and if such sequences are present at several different sites in the two strands, the 'wrong' parts may anneal to each other. Sometimes, one strand may fold back upon itself if it contains complementary stretches, so that a short double-stranded region and a single-stranded loop are formed. Such 'looped-out' structures may arise not only by denaturation and re-annealing; they are supposed to exist even in native DNA molecules at regions which show symmetrical sequences (*see* p. 27). Different control regions, e.g. the operator regions of the lac and gal operons, could easily form such hairpin loops; two possible looped-out structures for the origin of replication in SV40 virus are shown in ***figure 1.11***. It has been suggested that such special secondary structures may have a role in the interaction of DNA with different proteins.

The ability of even short complementary stretches to form base-paired struc-

```
        A—G                                   G—C            G—C
       /   \                                 /   \          /   \
      A     C                               T     A        T     A
       \   /                                 \   /          \   /
        A T                                   T A            T A
        | |                                   | |            | |
        A T                                   T A            T A
        | |                                   | |            | |
        A T                   5'---G—G—C—T A—A—G—C G—A---3'
        | |                   3'---C—C—G—A T—T—C—G C—T---5'
        C G                                   | |            | |
        | |                                   A T            A T
        G C                                   | |            | |
        | |                                   A T            A T
        T A                                  /   \          /   \
        | |                                 A     T        A     T
        T A                                  \   /          \   /
        | |                                   C—G            C—G
        T A
        | |
5'---G—G—C—T G—A—T—G---3'
3'---C—C—G—A C—T—A—C---5'
        | |
        A T
        | |
        A T
        | |
        A T
        | |
        C G
        | |
        G C
        | |
        T A
        | |
        T A
        | |
        T A
       /   \
      T     G
       \   /
        T—C
```

Figure 1.11 Possible hairpin-loop structures at the origin of replication of SV40 DNA. (After Jay *et al.*[13]. Reproduced with permission from *Biochem. Biophys. Res. Comm.*, 69, 678 (1976).)

tures provides a technique for detecting such complementary sequences in two different molecules. The technique is made very versatile by the fact that annealing may also occur between complementary regions of a DNA and an RNA molecule. Annealing between different molecules is called hybridisation; the double-stranded molecule formed in this latter case is a DNA-RNA hybrid.

APPLICATIONS OF THE HYBRIDISATION TECHNIQUE

Hybridisation has become a very widely used technique for the location of genes in the chromosome and for the determination of the number of copies of individual genes. It follows from the mechanism of transcription (Chapter 3) that mRNAs and other RNA transcripts are comprised of nucleotide sequences complementary to the gene. The transcript which can often be easily isolated can therefore hybridise to each copy of the gene in the chromosome. This provides a convenient way of identifying individual genes. Labelled transcripts can be used as probes to anneal to the DNA region which contains the corresponding genes. By determining the amount of RNA annealed to the DNA, which can be done by estimating the radioactivity in the DNA-RNA hybrid, the number of genes present can be calculated. Schweizer *et al.* hybridised labelled ribosomal RNA to denatured yeast DNA[20]. From the total radioactivity found annealed to DNA, they calculated that 140 rRNA molecules hybridised to one molecule of yeast DNA, i.e. that the number of rRNA genes in the yeast chromosome was 140. Using a basically similar method, the number of rRNA genes in *E. coli* was found to be 7 and in human liver cells 180. The genes coding for proteins can be identified by using messenger RNA as probe. Applying this principle to a mixture of mRNAs from HeLa cells, Klein *et al.* have shown that the majority of protein genes are unique[21]. However, difficulty in this latter method arises with mRNAs which cannot be easily labelled. In these cases an indirect method can be used: a labelled copy (complementary DNA, or cDNA) can be prepared from the RNA *in vitro* (*see* p. 72) and this can be hybridised to the chromosomal DNA.

It has been mentioned previously that when physical maps of genomes are constructed, the position of individual genes can be determined by identifying the corresponding gene products. This can be done either by transcribing and translating different restriction fragments *in vitro* and identifying the RNAs or proteins which are synthesised (*see* p. 22), or by hybridising the transcripts to the DNA fragments. Hybridisation will occur only with the restriction fragments which contain sequences complementary to the RNA probe, i.e. only with the fragments which contain at least part of the corresponding genes.

Many different variants of the hybridisation technique are used today for physical mapping of different DNA regions. Southern[22] worked out a very useful technique to detect which restriction fragments of a DNA molecule contain the specific coding sequences of an individual gene. The DNA fragments are separated by electrophoresis in agarose gel. The DNA bands thus obtained are transferred (blotted) onto nitrocellulose filters and are treated with the radioactive RNA transcript. The RNA probe hybridises only to DNA fragments with complementary sequences, and consequently only the DNA bands comprising the specific coding sequences will become radioactive. (For mapping eukaryotic protein genes, usually a labelled cDNA probe is used instead of the mRNA.) The method has very wide applications. To mention but a few interesting results obtained in this way, the interrupted structures of a number of eukaryotic genes (e.g. of the globin and ovalbumin genes) have been established by mapping the gene region with the aid of this 'blotting' technique.

As fragments of the chromosomal DNA as well as cDNA probes can be inserted into plasmids and cloned, larger amounts of the reactants can be produced. This greatly extends the field of application of hybridisation techniques.

The formation of molecular hybrids can also be observed under the electron microscope. The work of Kedes and coworkers on the mapping of histone genes in sea urchin DNA shows an interesting example of this latter method[23].

Histone genes are reiterated in most organisms; in sea urchin DNA they are present in several hundred copies. By fragmentation with restriction enzymes a repeat unit about 7 kb long could be isolated from this DNA which contained one gene each for the five histone species H1, H2A, H2B, H3 and H4. In order to determine the organisation of this histone gene repeat unit, Kedes and coworkers prepared subfragments by digestion with other restriction nucleases and hybridised these subfragments to purified histone mRNAs. Of seven subfragments, two did not hybridise to any of the histone mRNAs, while the others were found to hybridise to only one (in one case, to two) mRNA species. The map in *figure 1.12* shows the position of the restriction fragments and of the five coding regions.

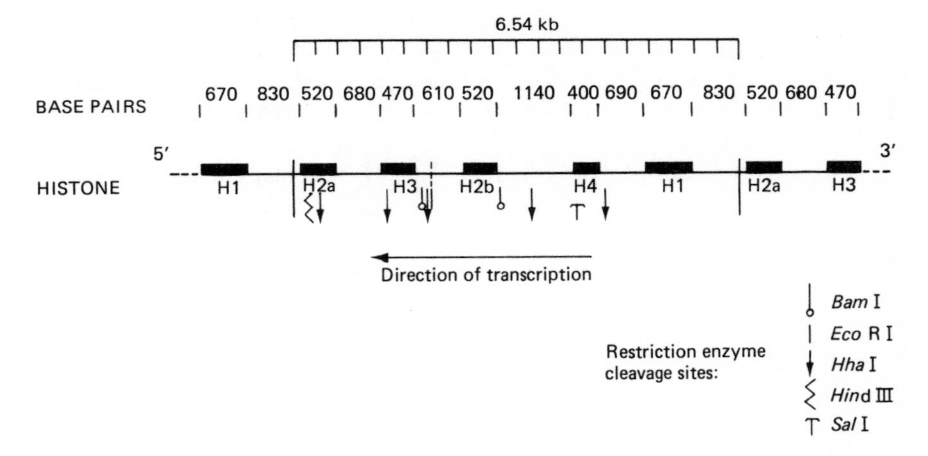

Figure 1.12 Structure of the histone gene repeat unit in sea urchin DNA. (According to Cohn *et al.*[23]. Reproduced with permission from *Cell*, 9, 147 (1976).)

For the construction of the exact map, an elegant new technique was introduced by Wu *et al.*[23] to detect RNA-DNA hybrids in the electron microscope. Long DNA fragments were denatured and hybridised with a mixture of mRNAs. This treatment yielded a DNA preparation which was single stranded wherever there were no coding regions present, and carried an annealed mRNA at each of the coding regions. A protein which binds only to single-stranded nucleic acids and which shows up very strongly in the electron microscope was added to this preparation, and the picture shown in *figure 1.13* was thus obtained. The dark regions represent the single-stranded, non-hybridised stretches of the DNA while the thin lines correspond to the histone genes to which mRNAs are annealed. The lengths of the genes can be measured exactly (this serves also as a way of identification), as can the lengths of the stretches between two genes, the *spacer* regions. As a result of these studies, the exact map, as seen in *figure 1.12*, giving the total

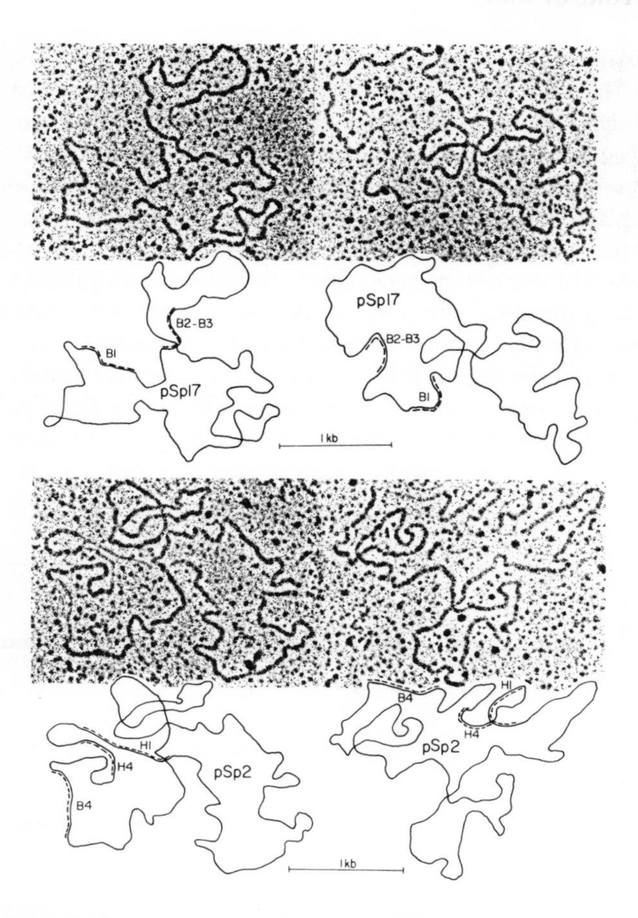

Figure 1.13 Hybridisation of histone mRNAs to the DNA of a histone gene repeat unit, investigated by the electron microscopic technique of Wu *et al.*[23]. The thin, faintly visible stretches are the DNA–RNA hybrids which represent the positions of the genes. The darker stretches are the single-stranded DNA regions, made visible by the binding of protein. These represent the spacer regions. Genes designated in the figure as B1, B2–B3 and B4 have been identified as genes coding for histones H3, H2A and H2B, respectively. (Reproduced with permission from *Cell*, 9, 163 (1976). Photograph by courtesy of Dr M. Wu.)

length of the repeat unit as 6540 bp, could be constructed.

Hybridisation of some eukaryotic mRNAs to the corresponding DNA revealed the existence of interrupted, split-gene structures in eukaryotic chromosomes. mRNAs of several eukaryotic viruses (adenovirus type 2, SV40, polyoma) as well as some host mRNAs (globin, ovalbumin, immunoglobulin), were found to anneal to several distant stretches in the genome. Using electron microscopy, Leder *et al.* studied the formation of molecular hybrids between mouse globin mRNA and the segment of mouse DNA which carries the globin gene[24]. *Figure 1.14* shows the result of such an experiment. This hybridisation technique produces stretches of DNA-RNA hybrids where complementary sequences are present; the other strand of the DNA in these stretches remains single stranded. The resulting structure is a

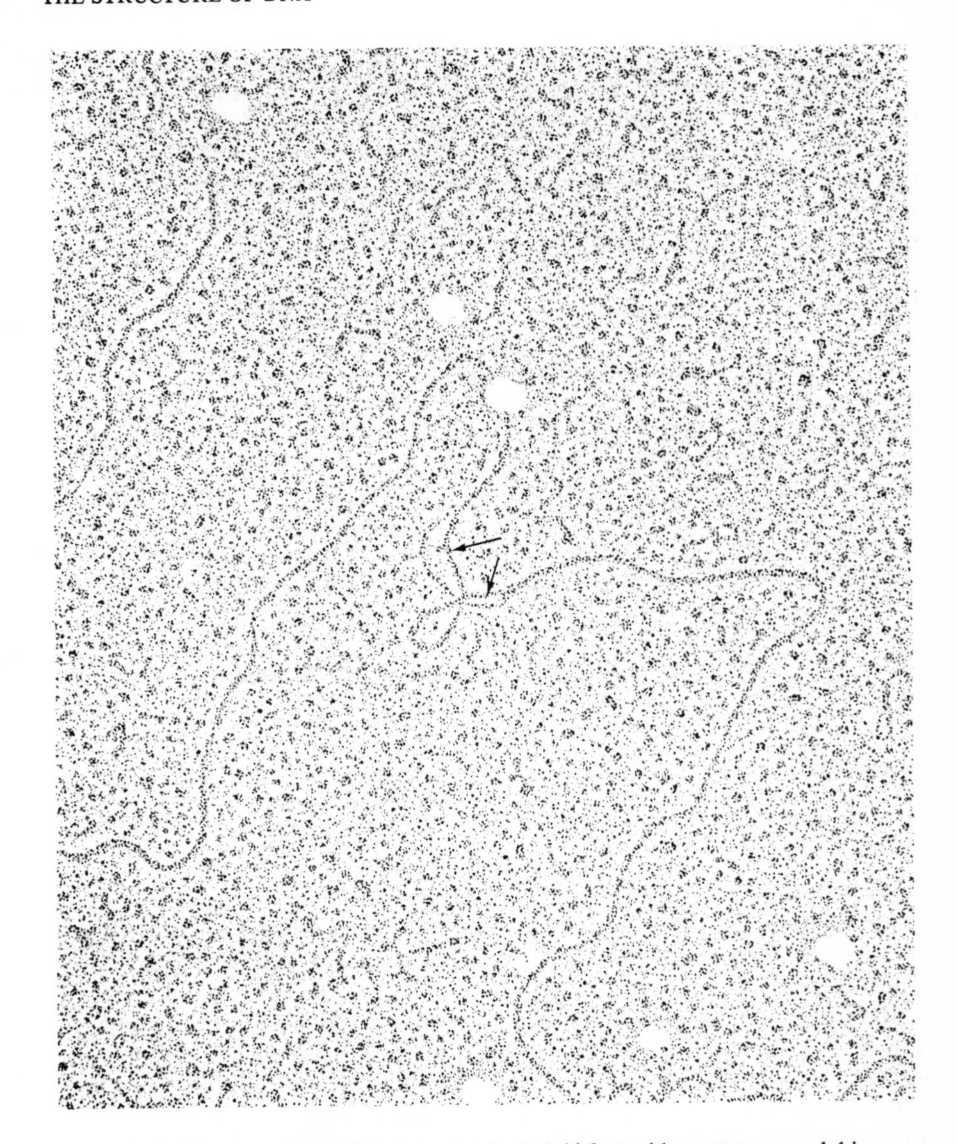

Figure 1.14 Electron micrograph of a molecular hybrid formed between mouse globin mRNA and mouse DNA (Leder *et al.*[24]). Annealed structures are seen as 'R-loops' (marked by arrows), where one strand of the DNA hybridises to the RNA, the other DNA strand remaining single stranded. Two such loops are formed with globin mRNA, which shows that the RNA contains complementary sequences to two distant stretches in the DNA. Between these two stretches, a double-stranded DNA segment which does not hybridise to mRNA forms a separate looped-out structure. This DNA stretch is not complementary to globin mRNA, it is an interrupting sequence in the middle of the globin gene. For further explanation, see text. (By courtesy of Dr P. Leder.)

loop (R loop) which can be observed in the electron microscope. It can be seen in the figure that two such loops are formed with globin mRNA (marked by arrows), because the mRNA hybridises to two distant stretches in the DNA. Between these stretches an unchanged, double-stranded DNA segment is seen; this is the DNA stretch which separates the two parts of the globin gene, and is an interruption in the middle of the globin coding region. No nucleotide sequence corresponding to this DNA region is present in the mature mRNA molecule. The interpretation of this result is that the genetic message is not present in a continuous nucleotide sequence in the DNA, that the gene has a split structure. A second, short interruption has also been detected in the globin gene by this technique.

Hybridisation of ovalbumin mRNA to the corresponding region in chicken DNA revealed seven interruptions in this gene[43]. O'Malley and coworkers located these intervening sequences by constructing restriction maps of the relevant region of chicken DNA and hybridising each restriction fragment not to the mRNA itself, but to a labelled copy of it, to a cDNA probe (*see* p. 72). In this way, the map of expressed and intervening DNA sequences in the natural ovalbumin gene could be constructed, as shown in *figure 1.15*.

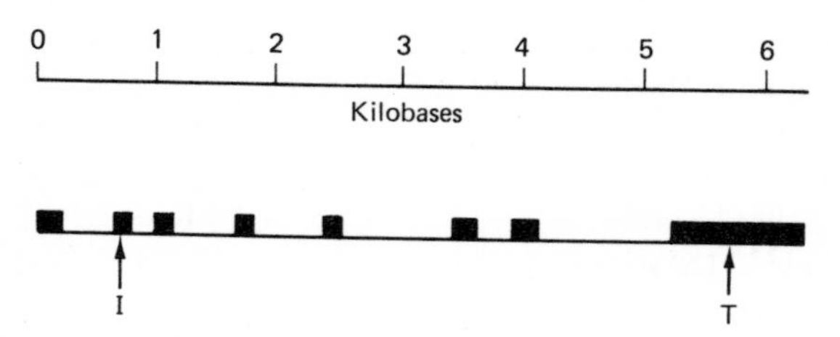

Figure 1.15 Map of the interrupted ovalbumin gene in chick DNA. Black areas represent expressed, structural gene sequences, white areas intervening sequences. The initiation and termination sites of the ovalbumin message are marked I and T, respectively. (After Dugaiczyk *et al.*[43].)

The increasing number of genes in which split structures have been detected emphasises the significance of this discovery. The phenomenon is not restricted to mRNAs, as hybridisation of rRNAs to some eukaryotic genomes have yielded similar results[25].

Hybridisation techniques work both ways: they can be used for the identification of DNA stretches which contain complementary sequences to a well defined RNA molecule, or they can be used for the identification or isolation of an RNA molecule which is complementary to one strand of a well defined DNA. As it is possible today to produce substantial quantities of isolated genes, these can be used as 'probes' for isolating the corresponding mRNA. Yanofsky's group applied this technique for the isolation of the mRNA of the tryptophan operon (the operon containing genes for five enzymes taking part in the synthesis of tryptophan). The DNA probes were immobilised on a membrane filter and an extract of *E. coli* cells containing trp mRNA was passed through it. The RNA which hybridised to the DNA probe was retained on the filter and could be subsequently released from DNA and obtained in pure form. By using as probes restriction fragments containing different parts of the operon, they isolated different functional parts of the trp mRNA molecule[26].

SOME ASPECTS OF THE ORGANISATION OF THE GENOME

(1) Transcription units

Techniques of genetic mapping, physical mapping, hybridisation and nucleotide sequence determination make it possible today to determine exactly the position of a number of genes in different genomes. *In vitro* transcription and/or translation of well defined DNA stretches, as well as studies on the structure of *in vivo* transcripts, provided in some cases exact information not only on the structure of single genes but also on that of structural-functional units comprised of a number of genes. The function and organisation of these transcription units and operons will be discussed later in relation to the mechanism and control of their transcription, but a few data relevant to the topics discussed earlier in this Chapter may be better mentioned here.

From the physical maps shown in *figures 1.2* and *1.12*, it can be seen that genes determining functionally related RNAs or proteins are often clustered together in one or a few specific parts of the genome. In some cases, such clustered genes form a functional unit: they are expressed together. In *E. coli*, ribosomal RNA genes are present in such *transcription units*, which contain the nucleotide sequences coding for the three ribosomal RNAs, separated from each other by spacer regions. It has been shown recently that transfer RNA genes are present in the spacer regions[27]. The exact structure of a ribosomal RNA transcription unit is shown in *figure 1.16*. This unit comprises the genes for 16S RNA, $tRNA^{glu}$, 23S RNA, 5S RNA and $tRNA^{trp}$ and $tRNA^{asp}$ at the distal end. Ribosomal RNA

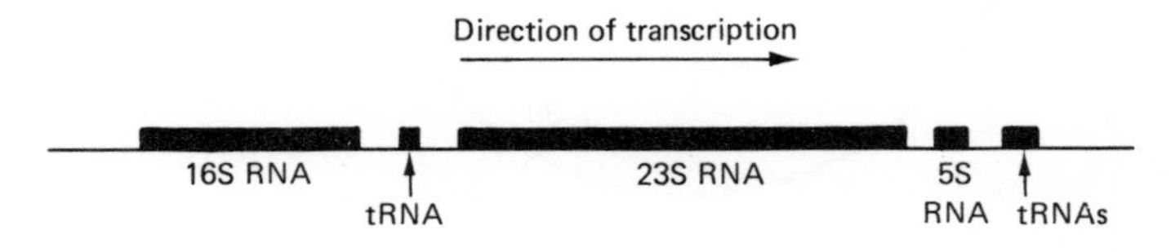

Figure 1.16 Structure of a ribosomal RNA transcription unit in *E. coli.* (According to the data of Lund *et al.* and Morgan *et al.*[27].) Black regions: genes for rRNAs and tRNAs. White regions: spacers.

genes are repetitive: in *E. coli*, seven transcription units have been detected, four of which contain $tRNA^{glu}$ gene in the spacer region between the 16S RNA and 23S RNA genes, while the three others have both $tRNA^{ala}$ and $tRNA^{ile}$ genes in their spacers. Further tRNA genes have been detected at the distal end of ribosomal transcription units. The expression of these rRNA and tRNA genes is under common control; the whole unit is transcribed into one long RNA molecule. The redundant sequences are eliminated by a post-transcriptional process, which eventually produces the mature RNA molecules.

In eukaryotes, the genes for three out of the four ribosomal RNAs (Chapter 6), i.e. the genes for 18S, 5.8S and 28S RNA are present in one transcription unit. The 5S rRNA gene does not belong to this unit and is transcribed independently.

In some higher organisms in which the ribosomal genes are highly repetitive, a very great number of such transcription units follow each other in tandem, separated by spacer regions.

Transfer RNA genes are dispersed in the chromosome. Nevertheless, there is evidence that in prokaryotes they often do not occur as single copies but rather tend to form small clusters[28]. At different positions in the *E. coli* genome, two or three tRNA genes are not only clustered but are transcribed together. Such groups or pairs may consist of copies of the same gene, e.g. three or four genes for $tRNA^{gly}$ form such a group, two genes for $tRNA^{tyr}$ form a pair. Alternatively, two different genes may be present in tandem, e.g. the pair of genes for $tRNA^{gly}$ and $tRNA^{thr}$ (*figure 1.17*). These tandem genes are separated by short (about 100 bp) spacer regions.

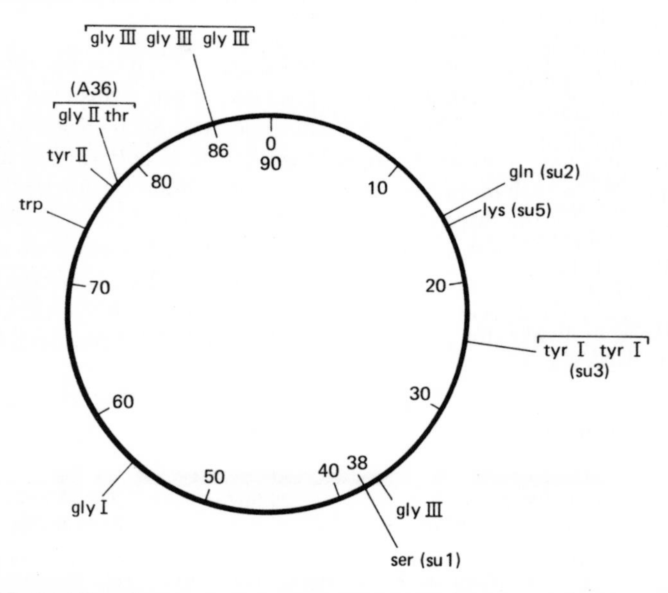

Figure 1.17 Position of some tRNA genes in the *E. coli* genome. (According to Smith[28]. Reproduced with permission from *Progr. Nucl. Acid Res. Mol. Biol.*, **16**, 25 (1976).)

In eukaryotes, the tRNA genes are also partly clustered and partly dispersed in the chromosome. In *Xenopus laevis*, the arrangement of tRNA genes has been studied in detail. *Xenopus* DNA contains on average 200 copies of each tRNA gene, making a total of 8000 genes. Most of the tRNA genes are grouped together with long (about 800 bp) spacer regions between them, covering 100 kb-long stretches of the DNA. No evidence has been found so far, however, for several genes being co-transcribed.

The organisation of protein genes is more complex than that of the genes coding for stable RNAs, and is very different in prokaryotes and eukaryotes. In the *E. coli* chromosome, near the ribosomal RNA genes, the genes for ribosomal proteins are also clustered. Genes for about 50 r-proteins are organised into several transcription units, some of which also contain other protein genes. The structures of these transcription units have been extensively studied by Nomura's group[29] (*see* p. 213). Within each transcription unit the expression of all genes is regulated by a common control mechanism, i.e. the genes are under coordinate control.

In prokaryotes many, though not all, genes for functionally related enzymes are organised into operons. These transcription units contain, in addition to the structural genes, i.e. the nucleotide sequences coding for these proteins, a control region (*figure 1.3*) which interacts with different regulatory proteins and thus exercises a flexible control over the transcription of all structural genes of the operon. The expression of these genes can be switched on or off coordinately (Chapter 3). It is due to these sophisticated control mechanisms that the bacterial cell can rapidly adapt to changes in the environment which cause changes in the requirement of the cell for different enzymes.

In higher organisms in which the enzyme patterns of the different cells are fairly constant, the protein genes are not part of operons. They are dispersed in the chromosome, transcribed separately, and their expression is in most cases individually controlled. Even in the case of histone genes which are present in structurally organised repeat units, each structural gene is expressed separately[29a]. Nevertheless, in some cases a functional control coordinating the expression of a number of related genes may also be at work in eukaryotes. Gorenstein and Warner[30] studied the controlled synthesis of ribosomal proteins in yeast and found that mutations in one or two regulatory genes cause coordinated inhibition of the synthesis of at least 40 ribosomal proteins, although the genes for these proteins are dispersed in the chromosome.

(2) 'Unusual' gene structures

Our classical idea of the gene as a continuous stretch in the DNA genome which carries in its base sequence the information for one protein molecule, has to be revised in view of the discovery of 'unusual' gene structures described earlier in this chapter.

(a) The presence of overlapping genes in bacteriophages reveals that the same DNA stretch can carry information for two different proteins, corresponding to two different reading frames[9,10,38]; short nucleotide stretches may even have a third interpretation. In animal viruses, two kinds of overlap have been detected A 122 nucleotide-long region in SV40 DNA codes for two proteins in two different reading frames: for the C-terminal part of VP2 protein and for the N-terminal part of the major viral protein, VP1 (*figure 1.1* and ref. 11). In the same viral genome, some DNA stretches direct the synthesis of two different protein molecules, although the DNA sequence is read in the same reading frame. This results in the production of different protein molecules which have in part identical amino acid sequences. Such overlaps can also be seen in *figure 1.1*. Part of protein VP2 is coded for by the same DNA region as that for VP3. As the reading frame is the same, the amino acid sequence of VP3 is identical to that of the C-terminal part of VP2. In the case of the small t and large T antigens, the synthesis of their N-terminal parts is directed by the same DNA stretch; consequently, these two proteins have identical N-terminal amino acid sequences.

(b) Interruptions have been detected in a number of eukaryotic genes. As far as protein genes are concerned, the positions of expressed and of intervening sequences have been exactly mapped in the globin gene[24], the ovalbumin gene[43] and in protein genes of adenovirus type 2 (ref. 30a) and of SV40 (ref. 5). Interruptions have also been detected in the genes for immunoglobulin, for lysozyme and for proteins of polyoma virus. The idea that a natural gene carries information for one protein in the form of a continuous nucleotide sequence must thus be abandoned. The organisation and expression of these genes resemble in some respects a transcription unit: the DNA region comprising the gene contains structural and intervening sequences which are all transcribed into a giant RNA molecule. The primary transcript contains a large number of redundant sequences which will be eliminated in a post-transcriptional process, the *splicing* of mRNA. In some viruses the situation is even more complex: some of the expressed DNA stretches may belong to more than one protein gene. A more detailed description of the structure of some split genes and of their spliced mRNAs will be given in Chapter 3.

THE STRUCTURE OF CHROMATIN

The DNA of eukaryotes is present in the cell nucleus in the form of chromatin. This is a tight nucleoprotein complex containing basic proteins, histones, and acidic, non-histone proteins. (Non-histone chromatin protein is abbreviated to NHCP.) The histones constitute a much larger proportion of the total protein in chromatin, and although only five different species, H1, H2A, H2B, H3 and H4, are present, a very great number of copies of these molecules can cover practically the whole length of the DNA. The NHCPs show a much greater variability, but each of them is present in very small amounts only. The histones show a remarkable similarity in different cells and organisms. With very few exceptions the same histone species are present in every cell and the primary structure of histones shows an unusual degree of conservation: for example, the amino acid sequences of histones H3 and H4 are identical in plants and animals. In contrast to this, a great number of different NHCPs are present in every cell and, although little is known about their actual structure, it is certain that they are tissue and species specific. These two groups of proteins play different roles in forming the structure and controlling the function of the chromatin.

The arrangement of DNA and proteins in chromatin has been studied by electron microscopy, X-ray crystallography, analysis of digestion products, and reconstitution of chromatin from its constituents. In spite of the complexity of chromatin structure, an amazing uniformity in the basic pattern has emerged from all these investigations: an almost invariant repeating pattern in every cell, built of very similar repeat units, 100 Å in diameter, called nucleosomes[31]. Mild digestion of chromatin by nucleases produces monomers and oligomers of these nucleosomes which can be separated by gel electrophoresis. Analysis of the main components of nucleosomes has revealed the presence of a DNA stretch about 200 bp long and a complete set of all five histone species. Further digestion removes histone H1 and a length of DNA, slightly variable in different cells. The remaining nucleoprotein particle, the core, contains a DNA stretch consistently 140 bp long and two molecules each of the H2A, H2B, H3 and H4 histone species. The nucleosome is thus built of a core particle (the sedimentation coefficient of which is 11S) and a linker which connects one particle to the next. The linker also has a compact structure with a strongly folded DNA. Stoichiometry of the proteins shows that there is only one molecule of histone H1 attached to this DNA stretch. It is assumed that histone H1 influences the structure of the linker region and thus determines the spacing of nucleosomes.

Attachment of proteins to the DNA causes a strong deformation which results in sevenfold shortening in the chromatin fibre. The length of a 200 bp stretch in free DNA is 680 Å, whereas the dimensions of the nucleosome are around 100 Å. The tight packing in the particle is brought about by winding the DNA double helix around the eight proteins of the core particle, as shown in *figure 1.18a.* From X-ray diffraction analysis and electron microscopic studies of crystals of nucleosome core particles, Klug's group deduced a model for this superhelical DNA, wound in $1\frac{3}{4}$ turns around a flat, wedge-shaped protein particle of the dimensions 110 × 110 × 57 Å (ref. 32). The shape of the superhelical DNA is shown in *figure 1.18b.* The superhelical structure requires a distortion in the

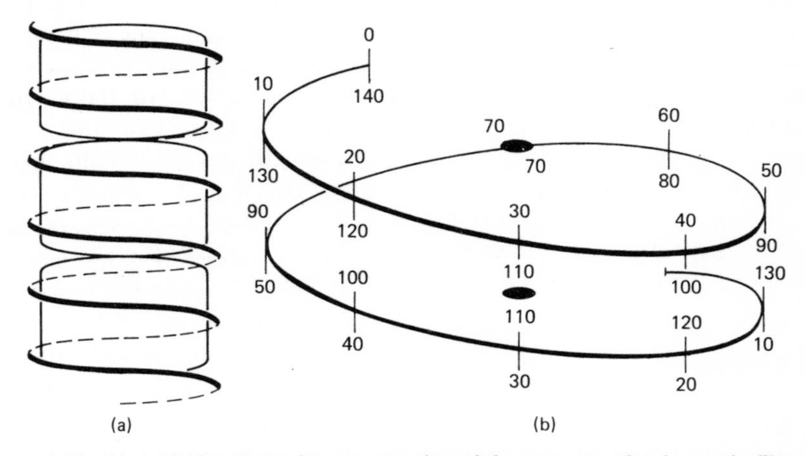

Figure 1.18 (a) A highly schematic representation of the structure of a chromatin fibre. (b) Diagram of a 140-bp-long superhelical DNA wound in 1¾ turns around the proteins of the nucleosome core particle. (According to Finch *et al.*[32]. Reproduced with permission from *Nature,* **269**, 29 (1977).)

DNA double helix. In principle, the distortion may be uniformly distributed along the molecule or it may take the form of a few sharp bends at regular intervals in the double-helical DNA. Early models preferred the latter structure, Crick and Klug[33] as well as Sobell *et al.*[34] presented models of kinked DNA, with bends at every 10 or every 20 bases. Recent results from several laboratories, based on physico-chemical investigations, model building and calculation of energy functions, revealed that DNA can be bent smoothly and uniformly into a superhelix of the dimensions required for nucleosome formation[35]. The model in *figure 1.18* also shows a smoothly bent superhelical DNA.

Histones are supposed to have an essentially structural role[36]. As formation of nucleosomes is not sequence specific, it seems probable that these particles serve the packaging of DNA rather than some special control function. However, attachment of histones does not only change the behaviour of DNA in a structural way. Histone-covered DNA is accessible to different enzymes which can still recognise specific sites and specific sequences within this nucleoprotein structure. However, the presence of histones does suppress transcription to an appreciable extent. Naked DNA, i.e. DNA devoid of all proteins, is a more efficient template for transcription than DNA associated with histones. A further reduction in template activity is observed in complete reconstituted chromatin which also contains NHCPs. These proteins seem to exert a double function, suppressing transcription of some genes and activating that of others. Histones are not species specific, whereas NHCPs are specific for the cell type. It can thus be expected that suppression of transcription by histones should be a nonspecific effect, whereas suppression by NHCPs should be cell specific. It has often been claimed that NHCPs are responsible for the selective transcription of specific genes. Many attempts have been made to substantiate this idea, but so far no unequivocal evidence has been obtained for a direct role of these proteins in directing transcription of specific genes. There is little doubt, however, that NHCPs are involved in the control of transcription. Apart from experiments with reconstituted chromatin, the interpretation of which is still controversial, a control function of NHCPs in transcription is also emphasised by the comparison of the structure of active, transcribed, and inactive, non-transcribed regions of chromatin. A greater susceptibility to nucleases and solubility in the presence of Mg^{2+} ions suggest that the DNA of active chromatin is present in a more open conformation than that of non-transcribed chromatin. The nucleosome structure of these two types of chromatin is very similar, but the active chromatin has been found associated with NHCPs and with RNA[37].

REFERENCES

1 Gilbert, W. *Nature,* **271**, 501 (1978).

2 Griffin, B. E., Fried, M. and Cowie, A. *PNAS,* **71**, 2077 (1974).

3 Nomura, M. in Proceedings of the 10th FEBS Meeting Chapeville, F., Grunberg-Manago, M. (eds) North Holland/American Elsevier, New York, 233 (1975).
Nomura, M. *Cell,* **9**, 633 (1976).

4 Sanger, F. *et al. Nature,* **265**, 687 (1977).
Sanger, F. *et al. J. Mol. Biol.,* **125**, 225 (1978).

5 Fiers, W. *et al. Nature,* **273**, 113 (1978).
Reddy, V. B. *et al. Science,* **200**, 494 (1978).

5a Soeda, E., Arrand, J. R., Smolar, N., Walsh, J.E. and Griffin, B. E. *Nature*, **283**, 445 (1980).

6 Schwarz, E., Scherer, G., Hobom, G. and Kössel, H. *Nature*, **272**, 410 (1978).

7 Valenzuela, P., Bell, G. I., Masiarz, F. R., DeGennaro, L. J. and Rutter, W. J. *Nature*, **267**, 642 (1977).
Maxam, A. M., Tizard, R., Skryabin, K. G. and Gilbert, W. *Nature*, **267**, 643 (1977).

8 Sures, J., Maxam, A., Cohn, R. H. and Kedes, L. H. *Cell*, **9**, 495 (1976).

9 Barrell, B. G., Air, G. M. and Hutchison, C. A. III *Nature*, **264**, 34 (1976).
Smith, M., Brown, N. L., Air, G. M., Barrell, B. G., Coulson, A. R., Hutchison, C. A. III and Sanger, F. *Nature*, **265**, 702 (1977).

10 Shaw, D. C., Walker, J. E., Northrop, F. D., Barrell, B. G., Godson, G. N. and Fiddes, J. C. *Nature*, **272**, 510 (1978).

11 Contreras, R., Rogiers, R., Van de Voorde, A. and Fiers, W. *Cell*, **12**, 529 (1977).

12 Celma, M. L., Dhar, R., Pan, J. and Weissman, S. M. *Nucl. Acids Res.*, **4**, 2549 (1977).

13 Jay, E., Roychoudhury, R. and Wu, R. *Biochem. Biophys. Res. Comm.*, **69**, 678 (1976).

14 Dickson, R. C., Abelson, J., Barnes, W. M. and Reznikoff, W. S. *Science*, **187**, 27 (1975).

15 Musso, R., DiLauro, R., Rosenberg, M. and deCrombrugghe, B. *PNAS*, **74**, 106 (1977).

16 Southern, E. M. *Nature*, **277**, 794 (1970).
Skinner, D. M., Beattie, W. G., Stark, B. P., Blattner, F. R. and Dahlberg, J. E. *Biochemistry*, **13**, 3930 (1974).
Retel, J. and Van Keulen, H. *Eur. J. Biochem.*, **58**, 51 (1975).

17 Cavalier-Smith, T. *Nature*, **250**, 467 (1974).
Wilson, D. A. and Thomas, C. A. Jr *J. Mol. Biol*, **84**, 115 (1973).

18 Watson, J. D. and Crick, F. H. C. *Nature*, **171**, 737 (1953).

19 Maxam, A. M. and Gilbert, W. *PNAS*, **74**, 560 (1977).

20 Schweizer, E., MacKechnie, C. and Halvorson, H. O. *J. Mol. Biol.*, **40**, 261 (1969).

21 Klein, W. H., Murphy, W., Attardi, G., Britten, R. J. and Davidson, E. H. *PNAS*, **71**, 1785 (1974).

22 Southern, E. M. *J. Mol. Biol.*, **98**, 503 (1975).

23 Cohn, R. H., Lowry, J. C. and Kedes, L. H. *Cell*, **9**, 147 (1976).
Wu, M., Holmes, D. S., Davidson, N., Cohn, R. H. and Kedes, L. H. *Cell*, **9**, 163 (1976).

24 Leder, P. *et al. Cold Spring Harbor Symp.*, **42**, 915 (1977).
Tilghman, S. M., Tiemeyer, D. C., Seichman, J. G., Peterlin, B. M., Sullivan, M., Maizel, J. V. and Leder, P. *PNAS*, **75**, 725 (1978).

25 Glover, D. M. and Hogness, D. S. *Cell*, **10**, 167 (1977).
Pellegrini, M., Manning, J. and Davidson, N. *Cell*, **10**, 213 (1977).
Bos, J. L., Heyting, C., Borst, P., Arnberg, A. C. and Van Bruggen, E. F. J. *Nature*, **275**, 336 (1978).
Wild, M. A. and Gall, J. G. *Cell*, **16**, 565 (1979).

26 Cohen, P. T., Yaniv, M. and Yanofsky, C. *J. Mol. Biol.*, **74**, 163 (1973).
Bronson, M. J., Squires, C. and Yanofsky, C. *PNAS*, **70**, 2335 (1973).
Platt, T. and Yanofsky, C. *PNAS*, **72**, 2399 (1975).

27 Lund, E., Dahlberg, J. E., Lindahl, L., Jaskunas, S. R., Dennis, P. P. and Nomura, M. *Cell*, **7**, 165 (1976).
Morgan, E. A., Ikemura, T., Lindahl, L., Fallon, A. M. and Nomura, M. *Cell*, **13**, 335 (1978).

28 Altman, S. *Cell,* **4**, 21 (1975).
Smith, J. D. *Progr. Nucl. Acid Res. Mol. Biol.,* **16**, 25 (1976).
Rossi, J. J. and Landy, A. *Cell,* **16**, 523 (1979).
29 Post, L. E., Arfsten, A. E., Nomura, M. and Jaskunas, S. R. *Cell,* **15**, 231 (1978).
Yamamoto, M. and Nomura, M. *J. Bacteriol.,* **137**, 584 (1979).
29a Levy, S., Childs, G. and Kedes, L. *Cell,* **15**, 151 (1978).
30 Gorenstein, C. and Warner, J. R. *PNAS,* **73**, 1547 (1976).
30aChow, L. T., Gelinas, R. E., Broker, T. R. and Roberts, R. J. *Cell,* **12**, 1 (1977).
Klessig, D. F. *Cell,* **12**, 9 (1977).
31 Kornberg, R. D. *Ann. Rev. Biochem.,* **46**, 931 (1977).
32 Finch, J. T., Lutter, L. C., Rhodes, D., Brown, R. S., Rushton, B., Levitt, M. and Klug, A. *Nature,* **269**, 29 (1977).
33 Crick, F. H. C. and Klug, A. *Nature,* **255**, 530 (1975).
34 Sobell, H. M., Tsai, C., Gilbert, S. G., Jain, S. C. and Sakore, T. D. *PNAS,* **73**, 3068 (1976).
35 Sussman, J. L. and Trifonov, E. N. *PNAS,* **75**, 103 (1978).
Levitt, M. *PNAS,* **75**, 640 (1978).
Kallenbach, N. R., Appleby, D. W. and Bradley, C. H. *Nature,* **272**, 134 (1978).
36 Bina-Stein, M. and Simpson, R. T. *Cell,* **11**, 609 (1977).
37 Gottesfeld, J. M. and Butler, P. J. G. *Nucl. Acids Res.,* **4**, 3155 (1977).
Gottesfeld, J. M. and Partington, G. A. *Cell,* **12**, 953 (1977).
38 Godson, G. N., Barrell, B. G., Staden, R. and Fiddes, J. C. *Nature,* **276**, 236 (1978).
39 Sauer, R. T. *Nature,* **276**, 301 (1978).
40 Valenzuela, P., Venegas, A., Weinberg, F., Bishop, R. and Rutter, W. J. *PNAS,* **75**, 190 (1978).
Goodman, H. M., Olson, M. V. and Hall, B. D. *PNAS,* **74**, 5453 (1977).
40aBrosius, J., Palmer, L. M., Kennedy, P. J. and Noller, H. F. *PNAS,* **75**, 4801 (1978).
41 Farabaugh, P. J. *Nature,* **274**, 765 (1978).
42 Schaffner, W., Kunz, G., Datwyler, H., Telford, J., Smith, H. O. and Birnstiel, M. L. *Cell,* **14**, 655 (1978).
Sures, J., Lowry, J. and Kedes, L. H. *Cell,* **15**, 1033 (1978).
43 Dugaiczyk, A., Woo, S. L. C., Lai, E. C., Mace, M. L. Jr, McReynolds, L. and O'Malley, B. W. *Nature,* **274**, 328 (1978).
Mandel, J. L., Breathnach, R., Gerlinger, P., Le Meur, M., Gannon, F. and Chambon, P. *Cell,* **14**, 641 (1978).

2 Replication of DNA

THE OVERALL PROCESS

When new DNA molecules are synthesised, the information contained in the parental molecule is passed on with very great accuracy to the daughter molecules. The basic principle on which such accurate copying occurs is suggested by the double-helical structure of DNA; each strand serves as a template for the synthesis of a complementary strand, thus producing two molecules identical to the original DNA. The diagram in *figure 2.1* shows how, by gradual separation of the two parental strands, a *replication fork* is formed, and how, on the two template strands thus exposed, complementary DNA strands are synthesised.

This replication mechanism is semi-conservative as each daughter molecule contains one strand of the parental molecule and one new strand. In a conservative replication mechanism the parental molecule would be retained in its original form and the daughter molecule would be formed of two new strands. The proof that replication operates by the semi-conservative mechanism was provided by Meselson and Stahl in 1958 (ref. 1). Bacterial cells were grown in a medium containing the heavy nitrogen isotope ^{15}N. Under these conditions the bacteria synthesised 'heavy', ^{15}N-containing DNA. These cells, which contained only heavy DNA, were then transferred to a normal, ^{14}N-containing medium. After one generation time,

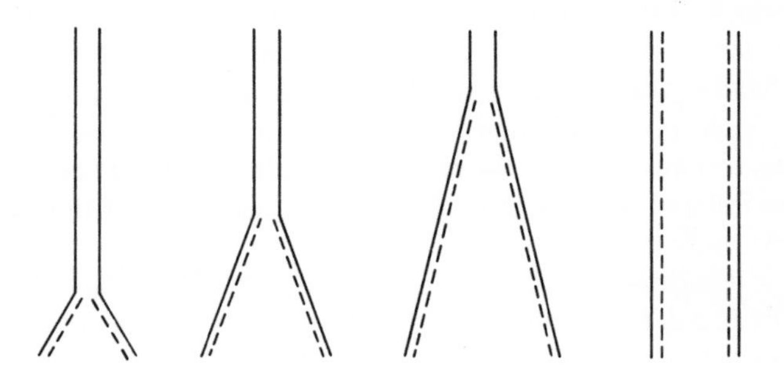

Figure 2.1 Different stages in the replication of a DNA molecule. Formation and propagation of the replication fork, leading to the synthesis of two complete daughter molecules.

Template strands ——— ; Newly synthesised strands — — — —

the DNA of the progeny cells was isolated and fractionated according to density on a CsCl gradient. 'Heavy' DNA, containing ^{15}N, and 'light' DNA, containing ^{14}N, can be separated and identified by this technique. It was found that at this stage all the progeny DNA was of intermediate density, indicating that each DNA molecule contained one heavy and one light chain. After a further generation time there were two kinds of DNA molecules present in equal amounts: DNA of intermediate density, as above, and light DNA, containing only the normal ^{14}N isotope. This

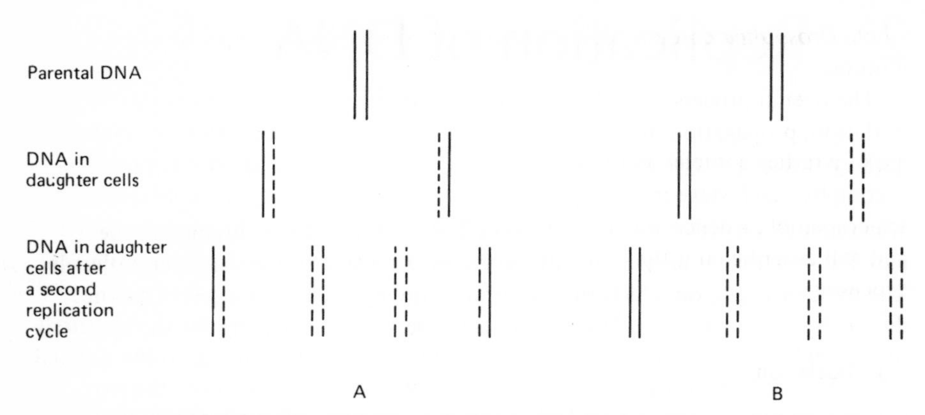

Figure 2.2 Results to be expected in the Meselson–Stahl experiment, in case of (A) semi-conservative replication; (B) conservative replication.

'Heavy' DNA strand ——— ; 'Light' DNA strand - - - -

agrees well with the products expected from a semi-conservative replication mechanism. As illustrated in *figure 2.2*, after one generation time each daughter molecule is composed of a parental (heavy) and a newly synthesised (light) chain. The second replication cycle produces a similar hybrid molecule from the heavy parental strand and a light molecule from the light parental strand. Conservative replication would produce only light molecules beside the original heavy variety.

The principle of the replication mechanism is thus inherent in the double-helical structure of DNA. The overall process seems simple, but the actual enzymic mechanism, the many individual steps involved in the synthesis of a new DNA molecule, is of great complexity. A great number of specific enzymes and protein factors are required, the carefully coordinated action of which brings about the fast and accurate transfer of genetic information from one DNA molecule to the other.

DNA is replicated at a very high speed and with amazing accuracy: the total genome of *E. coli* (consisting of 4×10^6 bp) is replicated in 40 min; this corresponds to a rate of 1700 base pairs per second. Mistakes leading to detectable mutations occur at a frequency of 1 in 10^6 genes. Such high speed and accuracy is achieved by highly active and specific enzymes and a sophisticated mechanism which makes it also possible that any mismatching should be repaired in the course of DNA synthesis. The replication of giant eukaryotic DNA molecules would be too slow, however, even with the most active polymerising enzymes: at a rate of 1700 base pairs per second, synthesis of a billion base pairs would still take 160 hours. Moreover, the eukaryotic enzymes are less active, the rate of fork movement is considerably slower; in animal chromosomes it is about 50 bp per second. In these organisms replication starts simultaneously at a very great number of different sites on the DNA molecule; many replicating units, *replicons*, are formed. The simultaneous synthesis of DNA in 3000 replicons makes it possible that the whole *Drosophila* chromosome (65×10^6 bp) should be replicated in a few minutes.

The overall process of replication can be divided into three main events: initiation, propagation of the replication fork and termination. All three events are of a rather complex nature, and a series of enzyme reactions are required to accomplish each stage in the replication process. We will first discuss here the experimental evidence which shed light on the mechanism of the overall process and will describe later the individual enzymic reactions underlying these complex processes.

(1) Initiation

Genetic studies on replicating cells as well as electron microscopic investigation of replicating chromosomes provided ample evidence that, at least in the relatively small genomes of prokaryotes and of eukaryotic viruses, replication starts at a specific site, a *fixed origin.* Modern techniques made it possible to locate, isolate and characterise this site in several chromosomes. In the *E. coli* chromosome the origin of replication has been mapped at position 82 or 83. A restriction fragment containing the replication origin was isolated by Marsh and Worcel[2]. *E. coli* was grown under conditions which ensured that initiation of replication was synchronous in the culture. When the cells were pulse-labelled with ^{3}H-thymidine for a very short period at the time that replication started, the label was incorporated only at the site around the replication origin. By restriction mapping, the labelled region could be located and a fragment comprising this region was isolated. By genetic and physical mapping, the origin of replication has also been located in several viral DNAs.

Recent development in electron microscopic techniques allows direct observation of the replicating chromosome. Initiation, formation and propagation of the replication fork can be followed on a series of electron micrographs taken of DNA molecules at different stages of replication. In *figure 1.7* it can be seen how first a little 'bubble' is formed at the origin. In the previous chapter (p. 22) it has been described how the origin of replication was mapped in polyoma DNA by this method. The same method has also been applied to SV40 DNA.

Isolation of restriction fragments containing the origin of replication makes nucleotide sequence determination of these regions also possible. The primary structure of the replication origin has been established in the *E. coli* chromosome[52], in human mitochondrial DNA[53] and in a number of DNA viruses: SV40 (*see* p. 27), polyoma[54] as well as in the genomes of coliphages ΦX174 (ref. 3), λ (ref. 4), G4 (ref. 49), fd (ref. 50). It is interesting that at least in three viral DNAs the replication origin is localised within a structural gene coding for a viral protein which itself initiates replication[3,4,51]. The nucleotide sequences around the origin in the different DNAs show no common characteristic features except for a varying degree of symmetry. In *E. coli* chromosomal DNA, the 422-bp-long stretch which contains the origin shows many repeated sequences[52]. At the replication origin of viral DNAs an A–T-rich region is present between two G–C-rich regions. The data are still insufficient to allow further conclusions as to the structural characteristics which determine the function of this site.

Replicating molecules of linear DNA have also been observed under the electron microscope. In the case of the DNA of T_7 phage, the startpoint of replication was located at a distance of 17 per cent from one end of the molecule. The replication forks proceed from here in both directions, reaching one end sooner than the other.

In the giant eukaryotic chromosomes of higher organisms, replication starts simultaneously at a great number of initiation sites, and there is a specific origin for each replicon. The number of these sites and their distance in the chromosome have been determined in a few cases (in the chromosome of *Drosophila* there are two to three thousand origins of replication about 30,000 bp apart) but these DNA molecules are still not accessible to more exact structural analysis.

(2) Propagation of the replication fork

If DNA molecules at different stages of replication are observed in the electron microscope, the growth of the replication bubble can be directly followed. In several viral DNAs the electron micrographs showed clearly that growth of the bubble occurred in both directions, suggesting bidirectional propagation of the replication forks. The replication bubble is, in fact, made up of two replication forks moving in opposite directions along the DNA molecule, as shown in *figure 2.3*. This is in good agreement with earlier experiments which, by less direct

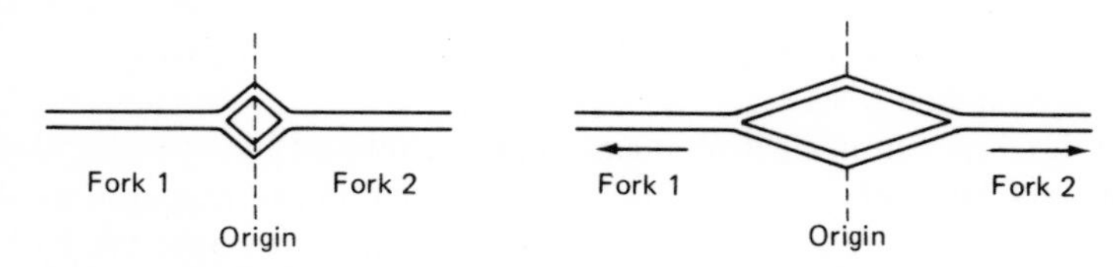

Figure 2.3 Replication bubble formed by movement of the replication forks.

methods, provided evidence for bidirectional replication in different organisms. Prescott and Kuempel[5] followed by autoradiography the uptake of labelled thymine into the replicating *E. coli* chromosome. Applying first pulse-labelling with low activity, followed by a short period of labelling with high activity, they obtained autoradiographs on which the low-activity region showed up in the middle, and the high-activity regions at both ends. This suggested that in the *E. coli* chromosome replication progressed in both directions. This bidirectional fork movement is characteristic of the majority of replicating organisms; few exceptions have been found so far.

Propagation of the replication fork is accompanied by continuous unwinding of the double-helical structure of the template molecule. The model of replication of double-helical DNA predicts that DNA synthesis must be preceded by strand separation (*figure 2.1*). Investigation of the properties of the enzymes responsible for DNA synthesis also confirmed that copying occurs on single template strands.

Unwinding of a double helix requires rotation of one strand around the other, a process which can take place in linear molecules only. In a closed circular structure, free rotation of the strands is not possible, and therefore if at the site of synthesis one turn of the helix is unwound, a twist will occur in another part of the molecule. This would eventually produce a strongly twisted structure (*positively* supertwisted, because it results from *overwinding*) which might prevent further propagation of the replication fork. This effect is counteracted by enzymes which can remove superhelical twists and can thus restore or preserve the original helical structure of the DNA. A number of such enzymes have been detected in different organisms; they act by nicking the DNA, thus allowing free rotation and resealing the nicked DNA strand when the supertwists have been abolished. Champoux[6] has shown that the untwisting enzyme of rat liver binds covalently to the broken DNA strand, thus preserving the energy of the phosphodiester bond which has to be re-formed when the single-strand break is resealed. In *E. coli*, the enzyme gyrase fulfils the function of preserving the original helical structure of the replicating DNA. *E. coli* DNA is negatively supercoiled; gyrase can introduce negative supercoils into the molecule. *Figure 2.4* shows a replicative intermediate of SV40 DNA. This DNA is underwound; it is obtained in negatively supercoiled structure when the histones are removed. During replication the original helical structure is retained, due to the action of a nicking–closing enzyme.

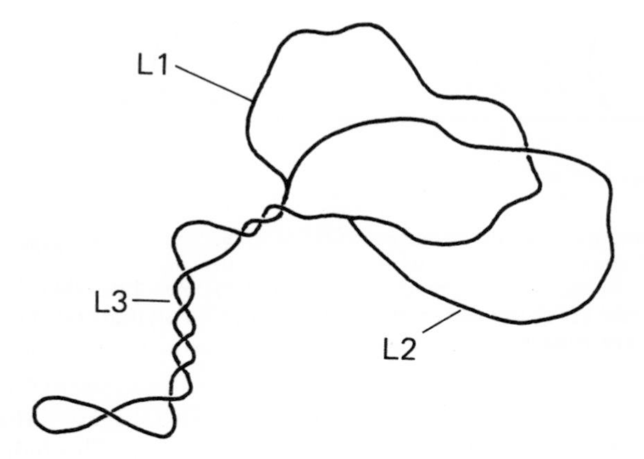

Figure 2.4 Replicating SV40 DNA. Diagram of the electron micrograph of Sebring *et al.*[58]. L1 and L2: parts of DNA already replicated; L3: unreplicated part of DNA in supercoiled conformation. (Reproduced with permission from *J. Virol.*, 8, 478 (1971).)

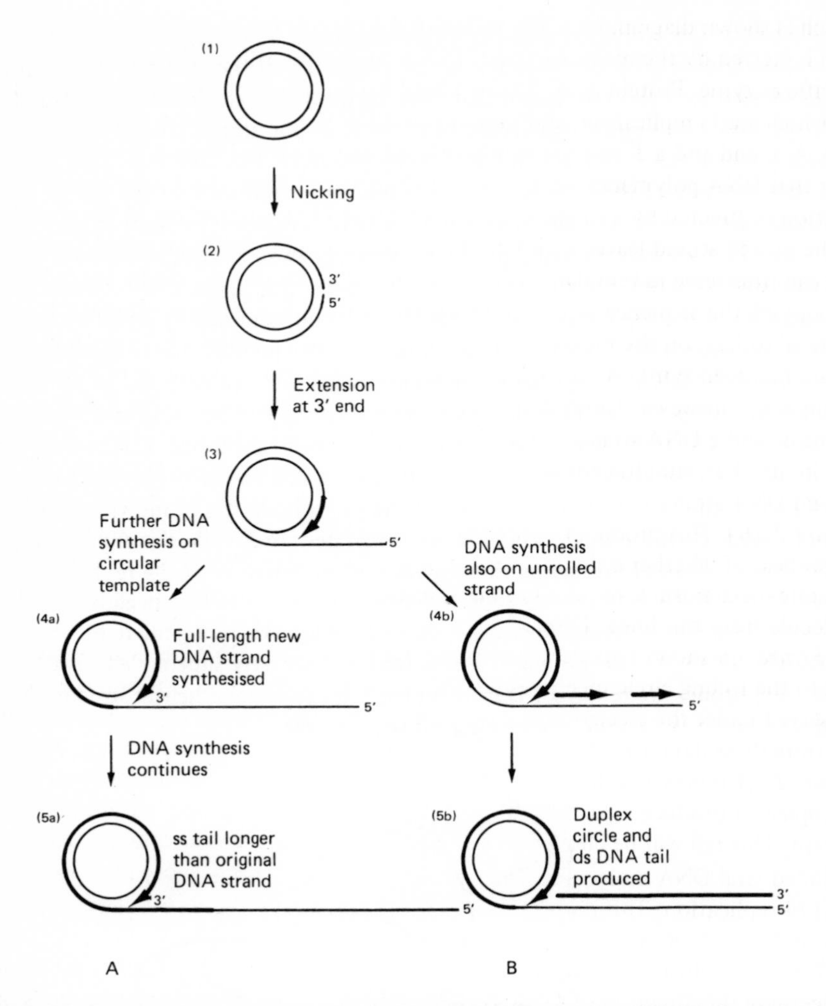

Figure 2.5 Diagrammatic representation of DNA replication by the rolling circle mechanism. Thin lines: old strands; thick lines: newly synthesised strands. A (4a and 5a): synthesis of (+)strand on a circular (−)strand template. B (4b and 5b): new DNA synthesis both on the circular ((−)strand) template and on the single-stranded ((+)strand) tail. (For explanation of discontinuous synthesis in 4b *see* p.53–56.)

An alternative mechanism of replication was proposed by Gilbert for circular DNA molecules. This mechanism is probably less widespread, but in some cases (e.g. in the replication of ΦX174 virus) it has been clearly demonstrated in electron micrographs of replicating viral DNA. This is the 'rolling circle' mechanism[7,8], which is shown diagrammatically in *figure 2.5*. According to this model, replication is started by the nicking of one DNA strand at the site of the origin by a specific enzyme. Protein A of ΦX174 phage has been shown to carry out the nicking which starts replication, and proteins of other viruses may have a similar function. A 5′ end and a 3′ end are thus produced on the nicked strand. We will see later that DNA polymerases can extend a DNA strand from the 3′ terminus if the reaction is directed by a single-stranded template. Here, unravelling of the 5′ end of the nicked strand leaves a part of the circular strand in single-stranded form; this can thus serve as template. The 3′ end of the nicked strand can be extended by copying the sequence of the intact circle. As this unravelling continues (the circle is 'rolling' on the nicked strand), a stage is reached when a full-length new strand has been synthesised (*figure 2.5A*). The replication process need not stop at this stage, however. Unravelling of one strand and extension at the 3′ end can continue and a DNA strand of indefinite length can be synthesised. There is also a possibility that, simultaneously with the process described above, synthesis of a second DNA strand may also take place, using the unravelled strand as template (*figure 2.5B*). This produces a partially duplex circle and a double-stranded tail. Regardless of whether a single-stranded or a double-stranded tail is produced, a separate mechanism is required to cut it to the proper size and to form a circular molecule from this linear DNA.

Figure 2.6 shows two electron micrographs of ΦX174 DNA replicating according to the rolling circle mechanism. Different intermediates of replication were observed under the electron microscope by Koths and Dressler[55], who reconstructed from these data the different steps in the replication of this phage (*see* p. 70–71). *Figure 2.6A* represents the usual intermediate form: the circular template strand is copied to produce a partially duplex ring and a single-stranded tail of unit length. This tail will be cleaved to size and circularised and will form new single-stranded viral DNA molecules. The structure in *figure 2.6B* is rarely observed in ΦX174 replication. This corresponds to the mechanism represented diagrammatically in *figure 2.5B*; here the tail is also double stranded.

The mechanism of propagation of the replication fork poses a further problem concerning the direction of DNA synthesis. As the two complementary strands of DNA are antiparallel, the movement of the replication fork coincides with the 3′ → 5′ direction of one template strand and with the 5′ → 3′ direction of the other. All enzymes capable of synthesising DNA are specific for the direction of the template and can copy it only from the 3′ towards the 5′ end. DNA synthesis and fork movement will thus occur in the same direction on one template strand but in opposite directions on the other. This apparent controversy presented a very serious difficulty in elucidating the mechanism of replication. The intriguing problem, how the enzymic synthesis of DNA strands and the overall growth of the DNA molecule can proceed in opposite directions, was eventually solved by Okazaki[9], who suggested that the synthesis of both DNA strands occurs in a dis-

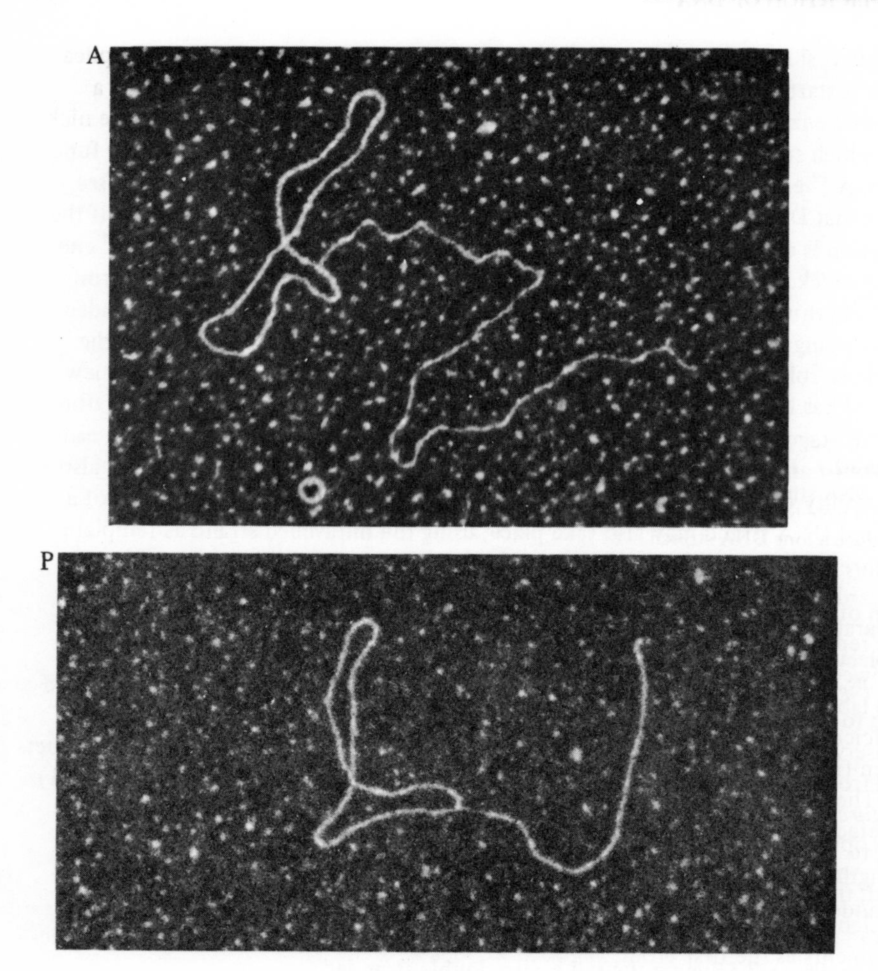

Figure 2.6 Electron micrographs of two structures detected in replicating ΦX174 DNA by Koths and Dressler[55]. A: Structure of main intermediate in ΦX viral DNA replication; partially duplex circle with long single-stranded tail. B: Rare structure; partially duplex circle with double-stranded tail. Corresponds to structure 5b in Figure 2.5. (Photographs by courtesy of Drs Koths and Dressler.)

continuous way, through intermediates of thousands of small DNA fragments which are all synthesised in the same direction, in agreement with the specificity of the enzymes carrying out this reaction. The small fragments are joined together at a later stage to form gradually bigger and bigger molecules until a full copy of the template strand is obtained (*figure 2.7*).

Direct proof for such discontinuous synthesis was soon found, although for some time the data were accepted with some reservations. At early stages of replication the production of small-size nucleic acid fragments with sedimentation coefficients around 20S could be detected by sedimentation analysis. Incorpora-

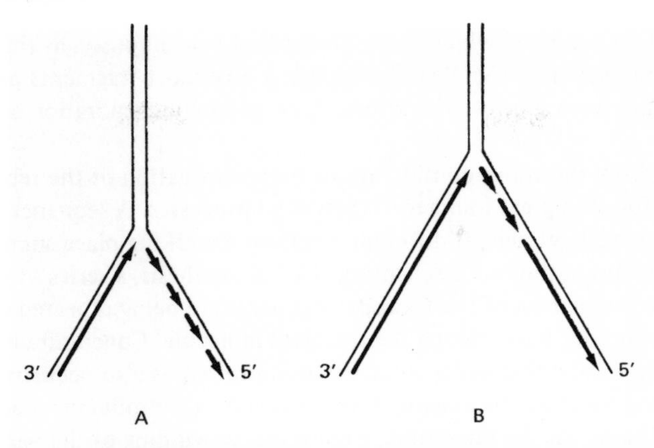

Figure 2.7 Discontinuous synthesis of DNA. Arrows show the direction of synthesis of the new DNA strands (5′ → 3′). A: Initial stage; on one of the template strands only short fragments have been synthesised. B: After maturation; some short fragments have been joined to produce a long DNA stretch.

tion of labelled nucleotides into small fragments could be shown after labelling the replicating cells for about 1 min. If, after a 1-min pulse-label, replication was allowed to continue for a longer period in the absence of radioactive substrates, the labelled nucleotides were found in larger DNA pieces with sedimentation coefficients of 70S to 120S, suggesting that the originally detected short pieces had been transformed into these intermediate, longer molecules.

The objection had been raised, however, that the small fragments may be artefacts, but in the following years, improved techniques seemed to rule this out[10]. The short fragments, called Okazaki pieces, could be isolated and characterised; they were estimated to consist of about 1000 to 2000 residue-long polynucleotide chains. In some cases, evidence has been presented for even shorter fragments as the first products of DNA chain elongation. In replicating polyoma virus, very short, 100–140 nucleotide-long Okazaki fragments have been detected.

Although the existence of Okazaki fragments has been confirmed, some doubts have arisen recently as to whether these DNA pieces are indeed produced in the way described above. It was discovered that dUTP can be incorporated into DNA in place of dTTP and the faulty nucleotide can subsequently be excised and replaced by the correct one, by way of a repair mechanism. The process of excision and repair would lead to the production of short fragments which in turn are transformed into larger molecules, i.e. it would give exactly the same experimental results as those expected from a discontinuous synthesis mechanism[11]. This ambiguity has been resolved by Olivera and by Tye *et al.*[56], who showed that both effects together were responsible for the detection of small DNA pieces in replicating DNA. They arrived at the conclusion that DNA synthesis is asymmetrical: it occurs discontinuously on one template strand and continuously on the other (*figure 2.7*). The necessity for discontinuous synthesis arises only on the strand where fork movement is opposite to the direction of polymerisation. (This is also the case in the rolling circle mechanism shown in *figure 2.5B*: the unravelled strand provides a template which cannot be copied continuously, as this would require polymerisation in the $3' \rightarrow 5'$ direction.) Any short fragments produced on the other strand were shown to be artefacts, caused by incorporation and excision of dUTP.

It follows from the above considerations that propagation of the replication fork is the result of a great number of individual processes. A sequence of events cannot be very well defined, as different reactions are taking place simultaneously. Because of the discontinuous mechanism of DNA synthesis, a series of reactions leading to the production of the Okazaki fragments are being repeated over and over again as the fork moves along the template molecule. Concomitantly with the synthesis of new short fragments, a 'maturation process' is also occurring, processing and joining together the existing fragments and thus producing gradually longer DNA chains. At the same time, a continual unwinding of the parental DNA is taking place at the growing point of the replication fork. This is also an enzymic process, brought about by the binding of unwinding protein to the DNA strands[12]. The action of the protein is stoichiometric: a large number of protein molecules bind in a cooperative way to the site where the double helix is unwound, ahead of the actual site of DNA synthesis. As with the propagation of the fork this site moves along the template, some molecules of the unwinding protein constantly become detached, to clear the way for the DNA synthesising enzyme, and attach again to a more distant site on the template. The diagram in *figure 2.8* gives a somewhat simplified picture of how different events occur at the same time at different sites of the replicating DNA molecule.

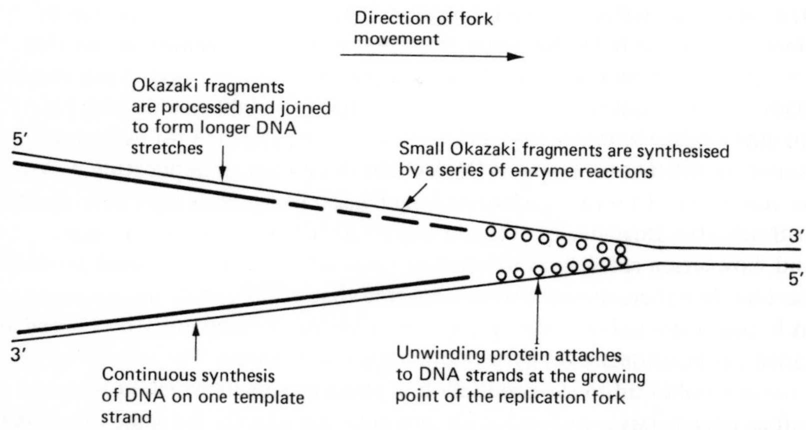

Figure 2.8 Different events occurring simultaneously in the course of the propagation of the replication fork.

(3) Termination of replication

Termination of replication is different depending on whether linear or circular molecules are being copied. In linear DNAs replication stops when both replication forks reach the end of the molecule. This does not necessarily occur at the same

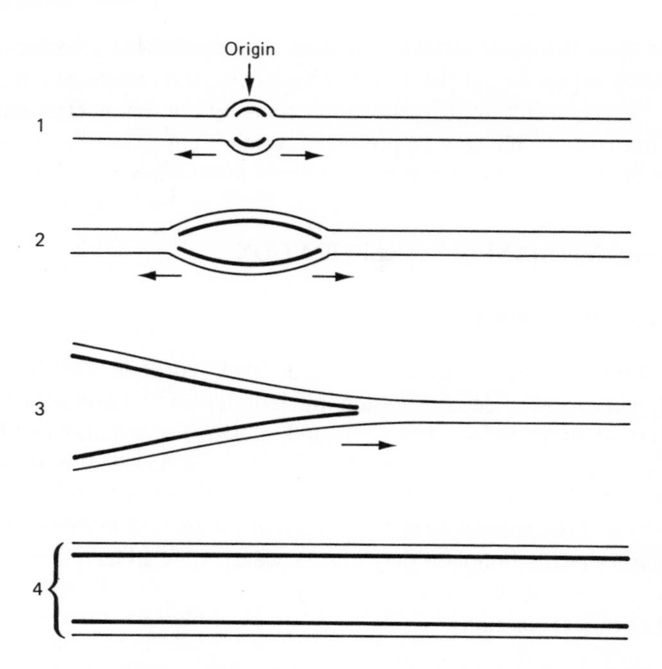

Figure 2.9 Different stages in the replication of a linear DNA molecule. Arrows show direction of fork movement.

time, as the origin is not in the middle of the DNA molecule. When one fork has reached the end of the DNA, thus transforming the bubble-shape into an open fork-shape, fork movement may still be in progress in the other direction, as shown in *figure 2.9*. In circular molecules the two forks may meet either at 180° from the origin (e.g. in SV40 DNA), a site reached at the same time by both forks, or at a specific site where one fork stops and 'waits' for the other which is slower or which has to travel a longer distance. In the first case, there is no need to postulate any specific structural signal on the DNA to specify the termination point. In the second, however, a specific termination signal must be present in the primary or in the secondary structure of the DNA which causes the faster fork to stop.

In *E. coli*, replication is normally terminated at a site diametrically opposed to the origin, at position 27. Louarn *et al.*[13] have shown, however, that the termination site is fixed in this region. If the origin of replication is displaced by integration of an episome, the replication forks still meet at the same site, although this means that the two forks have to travel unequal distances.

In linear molecules, when replication is terminated, the two daughter molecules are automatically separated. We will see later that some further enzyme reactions are needed to complete the exact replication of the parental DNA, but these are already being carried out on the separated daughter molecules. The situation is different in the case of circular DNA molecules: here the two template strands are intertwined until the end and cannot be separated without splitting one of them.

The enzymic mechanisms of termination must thus include the splitting of one template strand, separation of the two new molecules and resealing of the DNA to re-form the circular structure. If replication occurs by the rolling circle mechanism, a very long linear molecule may be produced which is cut to unit length by a nuclease and is circularised by an as yet unknown mechanism.

THE ENZYMIC MECHANISM OF REPLICATION

(1) Polymerisation of nucleotides

The chemical reaction which forms the basis of DNA synthesis is the sequence-specific polymerisation of deoxyribomononucleotides into a polynucleotide chain. A wide variety of enzymes have been detected in different organisms which can carry out the polymerisation of deoxyribonucleotides as directed by the nucleotide sequence of a template molecule. These enzymes, the DNA polymerases, may differ in many of their specific properties, even in the biological role they fulfil and in the nature of the template they use. However, they all have one basic property in common: they carry out the polymerisation of nucleotides according to the same chemical reaction:

$$\cdots \mathrm{P}N_1\,\mathrm{P}N_2\,\mathrm{P}N_3\text{-OH} + \mathrm{PPP}N_4\text{-OH} \longrightarrow \cdots \mathrm{P}N_1\,\mathrm{P}N_2\,\mathrm{P}N_3\,\mathrm{P}N_4\text{-OH} + \mathrm{PP_i}$$

The substrates are deoxyribonucleoside triphosphates (dNTPs). The α-phosphate group of the incoming nucleoside triphosphate is joined in $3'$, $5'$-phosphodiester linkage to the terminal $3'$-hydroxyl of the existing nucleotide chain, while the β and γ phosphates are split off in the form of inorganic pyrophosphate. It follows from this reaction mechanism that polymerisation can proceed only in the $5' \rightarrow 3'$ direction. The presence of all four dNTPs is required; the selection of the correct nucleotide to be joined, as directed by the nucleotide sequence of the template, is controlled by base pairing and probably also by specific selectivity of the enzyme itself (*see* p. 65).

This reaction mechanism is of universal importance in the synthesis of nucleic acids. It is valid not only for the enzymes synthesising DNA on different templates (double-stranded DNA, single-stranded DNA, RNA), but also for the sequence-specific polymerisation of ribonucleotides into RNA molecules by the different RNA polymerases. The only major difference between these two groups of polymerases is that while RNA polymerases can both initiate and elongate ribonucleotide chains, DNA polymerases can only extend existing oligo- or polynucleotides. DNA polymerases therefore require, in addition to the substrates and the template, a primer, an oligonucleotide which provides a 3′-terminal hydroxyl group to which further nucleotides can be attached. We will see later that this difference in the activities of the two kinds of polymerases is of great importance for the replication process.

(2) Enzymes participating in DNA replication

The first enzyme known to polymerise deoxyribonucleotides in a sequence-specific way was discovered in *E. coli* by Kornberg[14] in 1958. This enzyme, DNA polymerase I, has been purified to homogeneity and was assayed in an *in vitro* system in the presence of different templates. The fact that it synthesised a DNA strand complementary to the template used was at first accepted as direct evidence for its central role in DNA replication. Quite soon, however, a discrepancy between the results of *in vivo* and *in vitro* experiments became apparent. Genetic studies on mutants of *E. coli* revealed that DNA polymerase I (pol I) was not indispensable for the replication process. Mutants lacking this activity could still replicate even though slight alterations may have occurred in some processes connected with their DNA synthesis.

Two further DNA polymerase enzymes, pol II and pol III, showing very similar *in vitro* activities, were subsequently detected in *E. coli*[15]. While pol I is the main component, present in 400 molecules per cell, the respective amounts of pol II and pol III are about 10 and 40 times less. The specific activity of pol III, however, makes up for this difference, for it can polymerise about 40 times more nucleotides per second than can pol I. Because of the similar properties of the three enzymes, only genetic studies could shed light on the different biological roles played by each of them *in vivo*[16]. Genetic evidence, obtained mainly from work with temperature-sensitive mutants, showed clearly that mutant strains lacking pol III activity were unable to replicate, while mutants deficient in pol I activity produced shorter DNA molecules, intermediates of DNA synthesis. A need for pol II activity could only be detected when the other two enzymes were not present. This suggests that pol III is the enzyme responsible for the main polymerisation process leading to the synthesis of DNA, while pol I is somehow involved in the maturation of the molecule. (pol I also plays a role in DNA repair; this function will be discussed on p. 67.) The role of pol II in replication has not been established unambiguously; it has been proposed that it may provide an alternative pathway for the synthesis of DNA.

Studies on the enzyme activities of the three polymerases revealed some unexpected properties which seemed difficult to reconcile with the preconceived idea of their function in the synthesis of DNA. Only years later, since it has been realised that every step in DNA synthesis is brought about by the coordinated action of a complex system of enzymes and protein factors, has it become possible to correlate the *in vitro* observed characteristics of the enzymes with their role in different steps of the replication process.

One of the characteristics of DNA polymerases, mentioned earlier, is their inability to initiate DNA chains and their consequent requirement for primers. This suggests that in the initiation of DNA synthesis other enzymes must also be involved which produce suitably primed templates.

Another discrepancy between the mechanism of the *in vivo* process and the enzyme activities of the three purified polymerases concerns the nature of the template they require. None of these enzymes can attach to or work with intact natural double-stranded DNA. In *in vitro* assays they can utilise single-stranded DNA (with primer attached) or nuclease-treated DNA ('gapped' DNA) which contains single-stranded stretches. This is an indication for the involvement in DNA replication of other enzymes (or protein factors) which unwind the double-stranded structure and expose single template strands[12]. DNA polymerases detected in other organisms seem to share these properties with the *E. coli* enzymes.

It may be a less universal characteristic of DNA polymerases that they are multi-functional enzymes. pol I, pol II and pol III of *E. coli* possess exonuclease activity in addition to their polymerase activity: all three can degrade DNA in the $3' \rightarrow 5'$ direction, pol I and pol III are also active in the $5' \rightarrow 3'$ direction. Exonuclease activity has been detected in the three DNA polymerases of *Bacillus subtilis* which are similar in every respect to the *E. coli* enzymes and also to polymerases from other prokaryotes. Some eukaryotic enzymes, however, show no nuclease activity.

As we will see later, exonucleolytic breakdown plays a very important role in DNA replication: the $3' \rightarrow 5'$ nuclease activity is essential for repairing 'faulty' DNA sequences and thus represents a safeguard for the conservation of the correct genetic information over the generations; the $5' \rightarrow 3'$ exonuclease activity in conjunction with the polymerase activity is involved in the maturation of DNA molecules, a process made very efficient by the fact that one multi-functional enzyme can carry out both reactions. In higher organisms where DNA polymerases devoid of nuclease activity have been detected, this process may require the joint action of more than one enzyme.

The necessity of the involvement of further enzymes in DNA synthesis is clearly indicated by these properties of the DNA polymerases. A purely biochemical approach to the elucidation of the mechanism of DNA replication was made possible when a number of proteins contributing to the process had been isolated, and reconstituted *in vitro* systems had been constructed in which some stages of the replication process could be separately studied. Because of their smaller size and simpler structure, viral DNA templates have often been used in such studies. Useful information has been obtained in this way on the enzymic mechanism of initiation, elongation and maturation of DNA, although in some cases the viral process differed from chromosomal replication, and involved some virus-specific proteins. Small bacteriophages which contain single-stranded DNA genomes (e.g. ΦX, M13, G4) and utilise host enzymes for the synthesis of a complementary DNA strand, also helped to elucidate some steps in DNA replication. However, the complexity of the enzyme system and the great variety of individual steps involved make this approach rather difficult. Isolation and identification of all components required for replication have not yet been accomplished.

Table 2.1 Enzymes and protein factors required for DNA replication in *E. coli*

Gene product	Enzyme or protein factor identified	Required for
dnaA		
dnaB		Initiation
dnaC		Initiation
dnaD		
dnaE	DNA polymerase III	Elongation
dnaF		
dnaG	RNA polymerase	Initiation
dnaH		
dnaZ		Combines with pol III
polA	DNA polymerase I	Maturation, repair
polB	DNA polymerase II	
lig	Ligase	Maturation
	EF I, II, III	Combines with pol III
	Unwinding enzyme	Initiation
	Untwisting enzyme	Relaxes supercoiled structure

In the meantime, genetic studies on a great number of mutants defective in different steps of replication revealed that more than a dozen gene products are directly involved in the replication process. Column 1 of *table 2.1* lists the gene products required for DNA replication in *E. coli*. Some of these proteins have been identified with enzymes and factors isolated from this organism, e.g. DNA polymerases I, II and III are the products of genes polA, polB and dnaE, respectively; the dnaG product is an RNA polymerase. The nature and exact function of others are not yet known. On the other hand, the genetic loci of some of the well studied protein factors have not yet been established. The two sets of data shown in *table 2.1*, one obtained from genetic analyses, the other from enzymological studies on purified proteins, may together represent an almost complete list of all components of the replication system, although the overlap between the two is still incomplete.

While the overall process of replication has been studied in a wide variety of organisms, analysis of the exact molecular mechanisms involved has been restricted to a number of cells and tissues from which individual enzymes could be purified and their properties investigated. Most of our present information comes from work with *E. coli* enzymes or reconstituted systems containing some components of *E. coli*. The following outline of the enzymic mechanism of replication is based mainly on these data. It is an open question as to how far observations on *E. coli* (and mainly on phage-infected *E. coli*) systems apply to other organisms. With respect to DNA polymerase it is possible to make a comparison with other bacteria, with different eukaryotic cells, even with some mammalian systems. Other enzymes and factors have been identified only in very few organisms. The universal characteristics of DNA polymerases described on p. 58 are, of course, valid for every kind of cell. In other respects, however, their properties may vary to a great extent. In *B. subtilis*, three DNA polymerases have been detected which are very similar to the *E. coli* enzymes[17]. In eukaryotes there may be greater variations in different organisms[18,19]. Discontinuous synthesis and a requirement for priming also seem to be characteristic for replication in higher organisms, but less is known about the individual enzymes and their exact reaction mechanisms. In higher organisms, two DNA polymerases are usually present, one in the nucleus, the other in the cytoplasm. The latter is the major component, representing 80-90% of the total polymerase activity of the cell, while the nuclear enzyme amounts only to 10-15%. This allows no conclusions as to their biological role; we should recall that in *E. coli* the enzyme present in the lowest amount is responsible for the main polymerisation process. The two mammalian DNA polymerases differ in their molecular weight and other chemical properties. They definitely differ from the *E. coli* enzymes by showing no nuclease activity. Some other enzymes of the replication system, e.g. unwinding enzymes and untwisting enzymes, show some similar properties when obtained from different sources. However, generalisations are not always justified and we have to keep in mind that it is not known today exactly how far analogies can be taken concerning the individual steps in replication of different organisms.

(3) Individual steps in DNA replication in *E. coli*

Initiation

This is probably the most complex step in replication, requiring the coordinated action of 6 to 10 proteins in different systems.

First, the double-helical structure is unwound. As neither of the DNA polymerases can attach to or work on a double-stranded DNA template, their action must be preceded by that of another enzyme which separates the DNA strands and thus provides a suitable template for the polymerase. A protein fulfilling this function was first detected in T_4 phage-infected *E. coli*; similar proteins were later also isolated from uninfected bacterial cells and eukaryotic organisms[12]. They are called unwinding enzymes or DNA binding proteins. They bind strongly and preferentially to single DNA strands. It was assumed therefore that these proteins can bring about unwinding of the double stranded DNA and can stabilise the separated strands. In the light of recent results on the mechanism of replication of different coliphages, it seems, however, that in *E. coli* the DNA binding protein acts in conjunction with another protein, the product of the rep gene (rep protein) in unwinding the DNA strands at the replication fork. The rep protein utilises the energy of ATP for this process; it exhibits ATPase activity, coupled to its catalytic activity of strand separation. DNA binding protein acts stoichiometrically; a large number of protein molecules attach to the separated strands, covering a long stretch of the template, ahead of the newly synthesised DNA chain.

Unwinding of the double helix is still not sufficient to provide a suitable template for DNA polymerases. The inability of these enzymes to initiate DNA chains makes their action dependent on the presence of a primer, an oligonucleotide with a free 3′-hydroxyl group to which further nucleotides can be attached, according to the reaction mechanism described on p. 58. *In vitro*, pol I and pol III can both extend quite short oligonucleotides; octanucleotides have proved effective as primers. *In vivo*, much longer primer molecules seem to participate in the replication of *E. coli.*

The nature and synthesis of these primer molecules were among the intriguing problems in the initiation of DNA synthesis. As neither of the known DNA polymerases is capable of initiating the DNA chain, the possibility of a different type of oligonucleotide primer synthesised by a different type of enzyme had to be considered. The obvious idea was that oligomers of ribonucleotides may serve this purpose, as RNA polymerases, which are present in every organism, are able to start the synthesis of an oligo- or polynucleotide chain. Furthermore, in *in vitro* experiments, oligomers of ribonucleotides proved as effective in priming the synthesis of DNA as did oligomers of deoxyribonucleotides.

The first direct evidence that RNA polymerase is involved in the replication process came from studies on the replication of some coliphages which showed that rifampicin, a potent inhibitor of *E. coli* RNA polymerase, also prevents phage DNA synthesis. It has been discovered since that this is not a universal phenomenon, and replication in quite a few organisms is insensitive to rifampicin. Even in *E. coli*, replication of the bacterial chromosome can take place in the presence of this antibiotic. In the meantime, however, further evidence accumulated supporting the idea that ribonucleotide primers are the initiators of DNA chains. Newly synthesised DNA has been isolated from *E. coli* and the presence of RNA stretches covalently bound to the DNA chain could be detected in the nascent DNA fragments[10,29].

The controversial data of rifampicin sensitivity of the priming reaction in some and insensitivity in other organisms found an explanation when it was discovered that several different RNA polymerases may be present in the same cell and that the enzyme synthesising the primers may not be identical with the RNA polymerase responsible for the normal RNA synthesis of the cell. When studying the replication of bacteriophages in infected *E. coli* cells, Kornberg's group found that different enzymes were involved in the priming of different templates[20]: M13 phage DNA required a rifampicin-sensitive RNA polymerase resembling that which carries out transcription in the uninfected *E. coli* cell but containing one more subunit, while the RNA polymerase used by G4 and ΦX174 phages is the product of dnaG gene, a single polypeptide, resistant to rifampicin. This may be the enzyme responsible also for the priming of the replication of the bacterial chromosome. In higher organisms there is no unequivocal proof for the participation of RNA polymerase in the priming process, although there are indications for the involvement of ribonucleotides in the initiation of replication and the known DNA polymerases show the same primer requirement as the *E. coli* enzymes[21].

With the production of primed single-stranded template, the initiation process should be completed, and elongation, i.e. the actual polymerisation of deoxyribonucleotides, should follow. According to genetic evidence, this reaction is carried out by DNA pol III. *In vitro*, in purified, reconstructed systems, pol III does not bind to natural long template strands. In crude extracts, however, the enzyme exists in an 'active form' which can efficiently synthesise DNA on long template molecules[22]. It was found that this active form (often described, not quite correctly, as holoenzyme) contains three further protein factors needed to bind pol III to the primed template. The exact nature of these protein factors has been the subject of some controversy. Recently, Wickner has identified the components of this active protein complex: apart from pol III itself, it contains two elongation factors, EFI and EFIII, and the product of the dnaZ gene[23]. The role of the different proteins has also been established: first, the dnaZ gene product forms a complex with EFIII. This complex catalyses the binding of EFI to the DNA at the site where the primer is present. This reaction requires ATP or dATP.pol III can

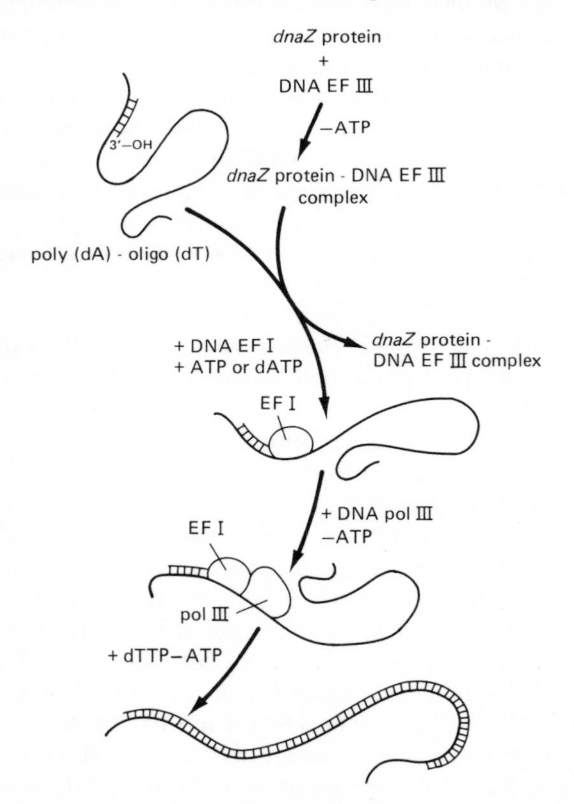

Figure 2.10 The mechanism of binding of DNA pol III to a primed single-stranded DNA template (poly(dA) primed with oligo(dT)) to start elongation of the DNA chain. dnaZ gene product and elongation factor EFIII first form a complex which, in the presence of ATP, interacts with elongation factor EFI and binds the latter to the DNA template, at the site where the primer ends. DNA pol III can then attach to this site and extend the nucleotide chain at the 3' hydroxyl. (By courtesy of S. Wickner[24].)

bind to this DNA–primer–EFI complex, and start the elongation process[24] (*figure 2.10*).

The discontinuous mechanism of DNA synthesis implies that initiation of single DNA chains occurs at many sites along one of the template strands, not only at the origin of replication. The requirement for a primed single-stranded template and for an active complex of pol III to bind to the primed template thus exists at each of these sites. The whole process of initiation, comprising the action of the unwinding protein, the synthesis of primer oligonucleotides by RNA polymerase and the binding of pol III to DNA by the combined action of EFI, EFIII and dnaZ product may thus be repeated many times along a long DNA molecule.

Unlike the origin of chromosomal replication, the initiation sites of the individual Okazaki fragments need not be specific, as they have only a transient function: in the maturation process the small fragments will be joined and transformed into large DNA molecules and the initiation sites will have no further significance. In good agreement with this idea, nucleotide sequence studies carried out on the short primer molecules on polyoma DNA indicate that Okazaki fragments are initiated at random sites along the DNA template[25].

Elongation

This is the stage at which mononucleotides are actually polymerised into a polynucleotide chain, and is easier to describe in terms of biochemical reaction mechanisms. Joining of the incoming nucleotide to the 3′ end of the growing DNA chain occurs according to the reaction shown on p. 58. The fidelity of the whole process depends on the specific selection of nucleotides complementary to those in the template. Base pairing to the template is a prerequisite for a nucleotide to be joined to the growing DNA chain, but the accuracy of this interaction would not be sufficient in itself to secure the precision of copying. Hydrogen bonds can also be formed between base pairs other than A–T and G–C, even though the stability of such base pairs may be lower. To exclude the possibility of such mis-matching, the polymerising enzyme itself must have a built-in specificity, which enables it to recognise the properly matched bases and to reject any mis-matched nucleotides.

A probable explanation of how DNA polymerases fulfil this selective function, is based on the requirement for exact stereochemical conditions under which the 3′ hydroxyl of the last nucleotide in the growing DNA chain, the α-β pyrophosphate bond on the incoming deoxyribonucleoside triphosphate, and the active site of the enzyme, are brought into close contact[26]. With non-complementary pairs, the spatial arrangement of the above sites will be different and will not allow a reaction between the 3′ hydroxyl of the growing chain and the α-phosphate of the NTP.

The frequency of *in vivo* mistakes is usually calculated from the frequency of certain mutations, also allowing for the fact that not all mutations have detectable biological effects. An estimate for the frequency of actual mistakes in copying, taking into account these undetectable 'silent mutations', is about 10^{-8} to 10^{-9}.

The actual accuracy of replication is thus much higher than could be predicted by stereochemical considerations. This very high accuracy, which is of vital importance for the conservation of the genetic information over many generations, is due to a second control mechanism, the 'proof-reading' capacity of DNA polymerases. All three polymerases of *E. coli*, as well as some similar enzymes found in other bacteria, possess the ability to repair mistakes made in the course of copying the template. This is the step where the $3' \rightarrow 5'$ exonuclease activity of pol III plays an important role[27]. If a wrong nucleotide is incorporated, it causes a slight distortion in the structure of the DNA duplex formed and also in the structure of the polymerase. This stimulates the action of the exonuclease, which removes the last, mis-matched nucleotide(s) from the 3′ end. Functioning again as a polymerase, the enzyme then replaces the eliminated nucleotide(s) with the correct complementary nucleotide(s). This repair activity is called proofreading.

As the result of this double safeguarding mechanism, the frequency of mutations as a consequence of mis-matching during replication is very low. However mutations can also occur as a consequence of outside effects, such as radiation and mutagenic chemicals. The cell also has a similar mechanism to correct these faulty sequences. An endonuclease attacks the double-stranded DNA molecule, causing a single-strand-nick in the faulty strand, and the mis-matched region is repaired by combined exonuclease and polymerase action, as above (*figure 2.11*); pol I is involved in this repair function: mutants deficient in pol I show a decreased

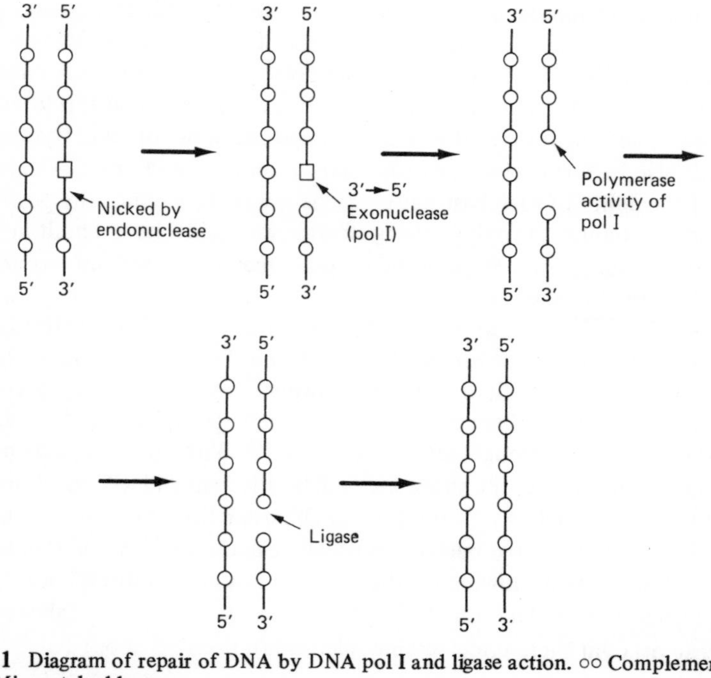

Figure 2.11 Diagram of repair of DNA by DNA pol I and ligase action. ○○ Complementary bases. ○□ Mis-matched bases.

resistance to mutating agents. When pol I has finished the repair-polymerisation, two adjacent nucleotides will still be left in a nicked structure, with no phosphodiester bond between them. This bond can be formed by DNA ligase, which seals the nick and completes the repair process[28].

Maturation

The short fragments produced by the discontinuous synthesis of DNA must be joined together in order to form a mature DNA molecule. Furthermore, each Okazaki fragment contains an RNA stretch which will have to be removed and replaced by deoxyribonucleotides in order to produce a perfect copy of the template DNA. The process transforming the small, primer-attached Okazaki fragments into large-size, continuous DNA molecules is called maturation.

The multi-functional character of pol I is of great importance in this process. In addition to the polymerase and 3′ → 5′ exonuclease activity, this enzyme also shows 5′ → 3′ exonuclease activity. Information as to how the primers are removed and the DNA fragments joined in the course of maturation, came from studies on mutants which were deficient either in the polymerase or in the 5′ → 3′ exonuclease activity of pol I. In such mutant strains short, RNA-linked Okazaki fragments accumulated and their transformation into longer DNA molecules was appreciably slowed down[29]. This suggests that *in vivo*, both the 5′ → 3′ exonuclease and the polymerase activity of DNA pol I are involved in the maturation of DNA. The active site for the 5′ → 3′ exonuclease is in a different part of the protein molecule from the active site for polymerase activity[30]. This enables the

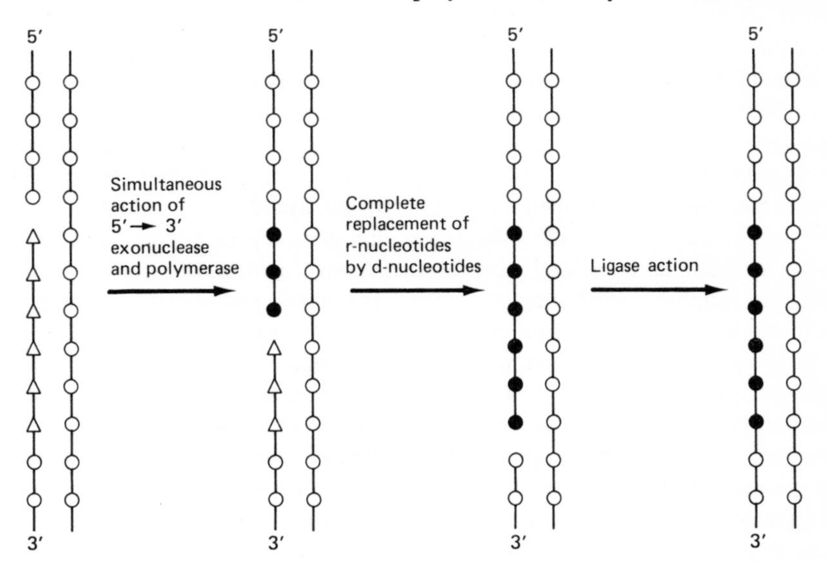

Figure 2.12 Diagram of maturation of DNA; removal of primers and replacement by DNA stretches by DNA pol I. ○ Deoxyribonucleotides. △ Ribonucleotides. ● Newly polymerised nucleotides.

enzyme to carry out two different reactions simultaneously on the DNA: acting as exonuclease, it digests off the primer, starting from the 5′ end; and acting as polymerase, it extends the neighbouring DNA chain at its 3′ end. As can be seen in *figure 2.12*, the result is that the ribonucleotides of the primer are replaced by a DNA stretch. The enzyme stops when it reaches the next DNA stretch.

It can also be seen from the diagram that this double reaction can take place at the primed end of each Okazaki fragment except for the first primer molecule at the 5′ terminus of a linear DNA. Here the exonuclease can remove the primer, but the polymerase, having no access to a pre-existing 3′-hydroxyl group, cannot synthesise a DNA stretch. As the result of pol I action, a long stretch of the template strand is covered by DNA fragments, except for this short gap at the 5′ terminus.

DNA pol I can fill in the gaps left by removing the primers; it cannot, however, join the extended fragments. It leaves 'nicks' in the structure, where two adjacent nucleotides are not linked to each other in phosphodiester linkage. DNA ligase[28] is the enzyme specific for the formation of this bond; it seals the nicks and thus completes the process of forming long continuous polynucleotide chains *(figure 2.12)*.

Termination

As the Okazaki pieces are joined together, gradually longer and longer DNA fragments are produced in the maturation process. Eventually, full-length DNA molecules are synthesised on both template strands. However, these are still not perfect copies of the parent molecule. For the completion and separation of the two daughter molecules some further enzyme reactions are required. The situation is different depending on whether circular or linear DNA molecules are being replicated. In the case of circular molecules replicated by the supercoiling and untwisting mechanism, the two template strands are present in the form of interconnected circles until the very end of the copying process. They can be separated only by nicking and thus splitting open one of the circular strands.

This, however, might lead to the production of an unnatural, inactive linear molecule which has completely lost its native conformation. Different mechanisms have been proposed which may overcome this problem and ensure that the proper conformation is preserved throughout the process of replication. They are based on the presence of complementary, possibly redundant, stretches at the site where nicking occurs. This would allow the formation of an annealed structure which could preserve the circular conformation in spite of the nick in the chain. This annealing should be only temporary, until the same enzyme which introduced the nick, or another ligase, reseals the strand. If replication occurs by the rolling circle mechanism, a full-length linear DNA molecule is cut off the tail and is circularised by an as yet unknown mechanism (*see figure 2.5*). Completion of the duplex circle does not require any special additional steps.

In linear DNA molecules, separation of the daughter molecules does not present difficulties, but it is the completion of the new DNA strands which causes some problems. Complete, end-to-end copying cannot take place with the aid of the enzymes so far considered. As mentioned above, a gap remains at the 5′ termini of both new strands where the primer molecules have been removed by pol I but could not be replaced by deoxyribonucleotides. This leaves short single-stranded tails on both template strands. At present, we have no clear evidence to show how these gaps are eventually filled in. In some viral DNAs, electron microscopic evidence suggests that dimers of the DNA molecules are present at some stage of replication. Watson[31] proposed that such dimers may be formed temporarily as intermediates at the stage of termination. Palindromic sequences are often present at the termini of DNA molecules, and if the above described tails are at such a region, they may anneal, forming a double-stranded DNA of double length, with no gaps (*figure 2.13*). An endonuclease may cut this double molecule into two, probably in the

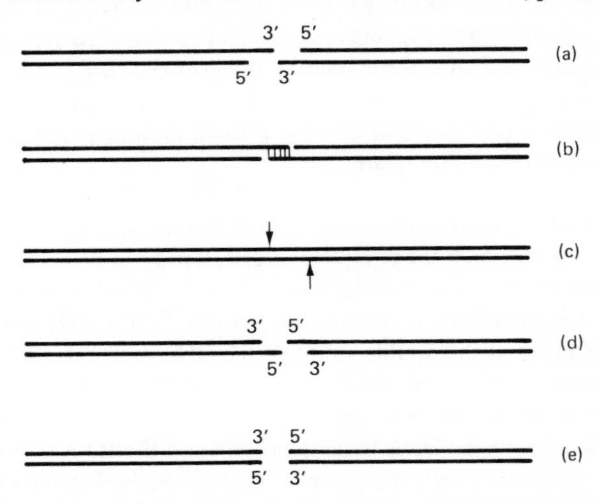

Figure 2.13 Termination of the replication of linear DNA molecules. Mechanism proposed by Watson[31], through temporary formation of dimers. Two DNA molecules with gaps at their 5′ termini and complementary sequences at their 3′ tails (a), anneal by base pairing of the 3′ termini (b). Filling of the gaps produces a double-stranded DNA molecule of double length, which, in turn, is cleaved by endonuclease at the sites marked by arrows (c). The cleavage products are DNA monomers with gaps at their 3′ termini (d). These gaps can be filled by DNA polymerase, leading to the production of two complete double-stranded DNA molecules (e).

way some restriction enzymes work, splitting the two strands at somewhat distant sites. This will again produce single-stranded tails, but this time with 3′-terminal gaps in each molecule. Such gaps can be filled in easily by pol I, as this process requires extension of the 3′ hydroxyl of the almost complete chain. Cavalier-Smith[32] suggested a somewhat different mechanism, also based on the presence of palindromic sequences. In his model, these sequences are folding back upon themselves, thus creating a hairpin loop at the terminus which can be joined to the other strand. The participation of redundant palindromic sequences at this stage is also assumed in other models[33].

REPLICATION OF VIRUSES WITH GENETIC MATERIAL OTHER THAN DOUBLE-STRANDED DNA

The small bacteriophages which contain single-stranded DNA as genetic material have been mentioned previously (p. 60). Their replication cycle starts with the synthesis of a complementary DNA strand, producing a double-stranded 'replicative form' (abbreviated to RF). It is interesting that in many cases (DNAs of ΦX174, G4, M13) this first stage, the transformation of a single-stranded genome into a double-stranded replicative form, shows more similarity with the bacterial replication process than the following stage, when the daughter molecules are produced from the double-stranded RF DNA. In fact, many of the enzyme reactions involved in bacterial replication have been studied and clarified by using as model systems the formation of RF DNA from one of the above single-stranded DNA viruses. The enzymes participating in this process of bacteriophage replication are often identical with those used by the host cell. Differences in the various viral replication systems are usually observed in the priming process, either in the RNA polymerase[20] (*see* p. 63) or in other protein factors which are involved in the production of the primed template[34].

In the replication of bacteriophage G4, only DNA binding protein and dnaG product, i.e. a rifampicin insensitive RNA polymerase, are required to produce the primed template on which the complementary strand can be synthesised. Initiation of this process occurs at a well defined site of the single-stranded G4 DNA. This site has been sequenced and located in the genome[49]. It is different from the origin of replication of the double-stranded RF DNA molecule of G4 phage.

The synthesis of the complementary strand of ΦX174 DNA is a more complex process. On this, DNA initiation can occur at different sites. Priming requires four proteins in addition to the DNA binding protein and the RNA polymerase (which is the product of dnaG gene), viz. protein i, protein n, dnaB product and dnaC product[34,35,57]. For the complete synthesis of the double-stranded RF DNA, more than 10 proteins are required which bring about priming, binding of DNA pol III, maturation, and closure of the circle.

In the conversion of single-stranded DNA to the double-stranded RF molecule, initiation often occurs at sites with strong secondary structure. This has been found in the DNAs of the phages G4 (refs 49, 51) and fd (ref. 50). It has been proposed that DNA binding protein may cover the whole DNA molecule except for the hairpin loops at the initiation site, and that only these free sites may be accessible for the priming reaction.

The second stage in the replication of these viruses is the synthesis of new DNA molecules on the double-stranded RF DNA. An excess of the viral (+)strand is produced, using the complementary (–)strand as template. The replication of ΦX174 DNA has been most thoroughly analysed. It has been mentioned before that this process occurs according to the rolling circle mechanism[8,55] (*see* also p. 53 and *figure 2.6*). The completely double-stranded circular RF DNA is nicked by protein A at a specific site in the (+)strand. The site of nicking, which is the origin of replication of the double-stranded DNA, is localised within the gene coding for protein A[3,35a]. It has been suggested that protein A is multi-functional and also takes part in subsequent steps of DNA replication. According to Eisenberg *et al.*[8] and to Koths and Dressler[55], the protein probably remains attached to the 5′ terminus of the nicked strand, which is unravelled, while the 3′ end of the same strand is extended on the circular template, as described on p. 53. The protein may be released at the last step, when full-length DNA molecules ((+)strands) are cleaved off the tail and circularised. Protein A may have a role in the closure of the circle. The partially duplex ring seen in *figure 2.5* and *figure 2.6* can either continue the synthesis of a long tail, producing further viral (+)strands, or the circle may be completed, and a fully double-stranded circular RF DNA may be formed.

The replication of RNA viruses usually involves the synthesis of RNA molecules only and is therefore essentially similar to the transcription process, described in Chapter 3. A complementary strand is first produced by a specific RNA-dependent RNA polymerase. In the second step this (–)strand is used as template to produce the viral RNA strand. The enzyme and factor requirements may not be identical for the two steps. In the replication of Qβ phage RNA, synthesis of the (–)strand requires two host factors which are not needed for the synthesis of the (+)strand. (The Qβ RNA polymerase is of special interest. It comprises one virus-induced subunit and three host factors. The latter are all proteins which normally take part in translation in the bacterial cell: two elongation factors of protein synthesis and ribosomal protein S1 (ref. 36).)

A special group of RNA viruses, the RNA tumour viruses, have an important, indirect way of reproducing themselves. In 1970, an unexpected discovery was made simultaneously in two laboratories: Themin[37] and Baltimore[38] found that the replication of these viruses is achieved in a two-step process: first, a DNA molecule is synthesised on the viral RNA template and this DNA serves in turn as template for the synthesis of viral RNA. The second step, synthesis of RNA on a DNA template, does not differ significantly from the process of transcription in other cells. The first step, however, is of very great importance and at the time of its discovery caused great excitement, as this is the only physiological process known in which genetic information is passed from RNA to DNA. The enzyme responsible for this process is called *reverse transcriptase*, and, according to its function, it can be described as an RNA-dependent DNA polymerase. It has been detected in a great number of different tumour viruses and has been purified from some of these sources[39]. The best studied species is the enzyme from Rous sarcoma virus (RSV).

The properties of reverse transcriptases have been extensively studied by Spiegelmann's group[39]. In *in vitro* studies the enzyme was shown to synthesise a complementary DNA (cDNA) on a primed single-stranded RNA template. Its action shows a close analogy with other DNA polymerases: it uses deoxyribonucleoside triphosphates as substrates and carries out their polymerisation according to the general reaction mechanism described on p. 58. It cannot initiate polymerisation but requires a primer which it extends by adding nucleotides to its

3′ hydroxyl. The significance of the rather surprising discovery that *in vitro* tRNA acts as primer in the transcription of RSV DNA is not yet fully understood[40]. tRNA is associated with the RNA genome of RSV which would emphasise its role in priming replication *in vivo*. However, the site of attachment is not far from the 5′ terminus of the viral RNA and thus leads to the synthesis of relatively short cDNA stretches only, complementary to the 5′-terminal part of the genome[41]. Still, the synthesis of larger molecules has been reported; possibly, the enzyme can carry on polymerisation by switching to the 3′ end of the template[42].

In vivo experiments showed that the newly synthesised DNA strand remains annealed to the template, and that the first reaction product is a DNA:RNA hybrid. In the second step of the reaction, circular double-stranded DNA is produced, the new DNA strand serving as template for the synthesis of a complementary strand. The mechanism of circularisation is not yet known. The RNA strand is eliminated, probably by an RNase H activity; this is a nuclease which degrades the RNA moiety of DNA:RNA hybrids.

In vitro, so far only the first step, the synthesis of single-stranded cDNA, has been achieved with the aid of reverse transcriptase. Artificial double-stranded DNA molecules could be synthesised by the combined action of reverse transcriptase and DNA polymerase I (*see* p. 73–74).

The discovery of reverse transcriptase opened great practical possibilities because it provided a simple way to synthesise DNA molecules with nucleotide sequences complementary to specific RNAs. Using radioactive nucleotides for the synthesis, these complementary DNAs (usually called cDNAs) can be obtained in highly radioactive form. Radioactive cDNAs have been synthesised on a wide variety of specific mRNA molecules as templates. These cDNAs have a very wide application: they are used as probes in hybridisation assays to determine the number and location of genes in the chromosome, in constructing physical maps of different genomes, and in the study of the organisation of eukaryotic genes, to locate coding sequences in a given DNA region (*see* p. 34 and p. 38). Reverse transcription is also applied in nucleotide sequence determination (*see* Chapter 4).

In conjunction with other enzymes, reverse transcriptase can be used for *in vitro* synthesis of double-stranded DNAs which have the same nucleotide sequence as a natural mRNA. In some respects, these double-stranded DNA molecules could be considered as artificial genes.

IN VITRO SYNTHESIS OF SPECIFIC DOUBLE-STRANDED DNAs

In the last few years different methods have been worked out with the aim of synthesising *in vitro* artificial genes. Test tube synthesis of a number of double-stranded DNA molecules carrying specific messages has indeed been achieved, but some of these products do not resemble the natural gene as it is present in the chromosome.

Maniatis' group was the first to carry out *in vitro* a fully enzymic synthesis of double-stranded DNA on rabbit β-globin mRNA as template[43]. Two enzymes seemed suited for this task. The obvious candidate was reverse transcriptase, which had been used in earlier attempts to synthesise cDNA on a globin mRNA template. The first attempts failed because the optimal conditions for producing full-length or almost full-length copies were not known. Often only cDNA fragments, much shorter than the mRNA molecule, were obtained. The technique of Efstratiadis and coworkers[43] has overcome this difficulty. A 580-nucleotide-long cDNA, only 80 nucleotides shorter than globin mRNA, was synthesised. The next task was to convert this single-stranded cDNA into a double-stranded molecule. This required a DNA polymerase, and pol I seemed a suitable enzyme to produce a complementary strand to the cDNA. The success of this approach depended on the property of cDNA that it can fold back on itself, forming a hairpin loop at the 3′ terminus. This hairpin loop served as a primer on the DNA strand (*see figure 2.14*), providing the free 3′ hydroxyl which could be extended by pol I. As copying occurs here in the 5′ → 3′ direction, the complementary strand could be synthesised continuously requiring pol I enzyme only. The product of this reaction was a double-helical molecule formed of one long DNA strand folding back on itself; it had only one free end. A very specific nuclease, nuclease S1, which can degrade single-stranded nucleic acids or split short, single-stranded stretches in DNA but does not attack any double-stranded structure, was used to split specifically the hairpin loop where a very short single-stranded stretch was present. The final product was thus a double-stranded DNA molecule the sequence of

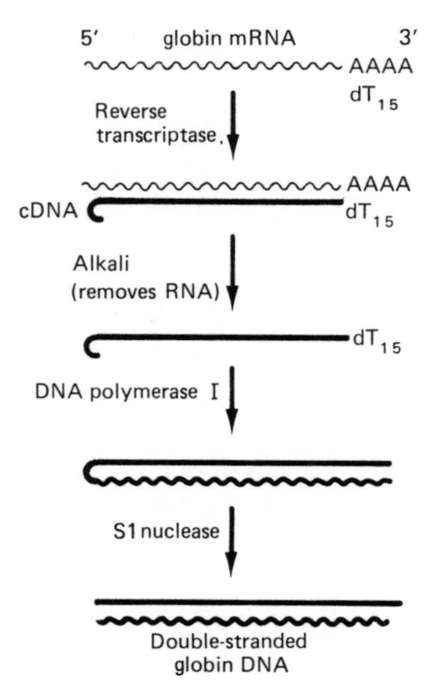

Figure 2.14 The different steps in the *in vitro* synthesis of a duplex complementary DNA to rabbit globin mRNA. Method of Efstratiadis *et al.*[43]. For explanation, see text. (Reproduced, with permission, from Maniatis *et al. Cell,* 8, 163 (1976).) Thin lines: RNA; thick lines: DNA; wavy lines: nucleotide sequences identical to RNA; straight lines: nucleotide sequences complementary to RNA.

which corresponded exactly to that of the globin mRNA, although a small part of the original nucleotide sequence was missing[43]. This synthetic DNA could be inserted into a plasmid and cloned in *E. coli*. Sequence determination proved that it was a faithful copy of rabbit β-globin mRNA[44]. Other duplex cDNAs have since been produced by the same or similar techniques. A practically complete duplex cDNA of ovalbumin mRNA was prepared and cloned by O'Malley's group[45]. Synthetic ovalbumin gene could be inserted into *E. coli* DNA. This resulted in the synthesis of an ovalbumin-like protein in the bacterial cell[46].

These artificial DNAs reflect the structures of the mRNAs only, not those of the genes coding for the mRNAs. First, it is not possible to include in the synthetic DNA any non-transcribed regions of the gene. This means that the *in vitro* products may not contain the different control signals which in the chromosome regulate the transcription and expression of the gene. Second, as mature mRNA is used for synthesising cDNA, the synthetic molecule cannot contain the intervening sequences which are present in the natural gene, but which have been eliminated during splicing of the mRNA. In the case of spliced mRNAs, as in the globin and ovalbumin mRNAs, the synthetic duplex DNA comprises only the expressed, structural sequences. This is not necessarily a drawback: these DNAs are very useful tools in detecting structural sequences within a natural gene and consequently, can be successfully used in preparing exact maps of the split genes in eukaryotic chromosomes[47].

A complete gene also containing regulatory sequences has been synthesised by Khorana's group using a more chemical approach and omitting any preformed template. Chemical synthesis of oligonucleotides, combined with the use of purified enzymes, has been applied to synthesise DNAs of known nucleotide sequence. Khorana used this method first for the synthesis of $tRNA^{ala}$ of yeast, the complete nucleotide sequence of which had been determined in 1965 by Holley. The second gene synthesised in Khorana's laboratory was that of tyrosine suppressor tRNA[48]. Nucleotide sequence determinations were extended this time to include the stretches of DNA preceding and following the tRNA sequence in the *E. coli* chromosome, and these sequences were also incorporated into the synthetic gene. When introducing this DNA into *E. coli*, Khorana's group obtained proof that the artificial gene was normally transcribed into functional suppressor $tRNA^{tyr}$ molecules. DNA molecules coding for A and B chains of human insulin were synthesised by Crea *et al.*[59] using chemical methods. The synthetic molecules could be fused to *E. coli* DNA and synthesis of insulin chains could thus be achieved in these cells[60].

REFERENCES

1 Meselson, M. and Stahl, F. W. *PNAS*, **44**, 671 (1958).
2 Marsh, R. C. and Worcel, A. *PNAS*, **74**, 2720 (1977).
3 Sanger, F. *et al. Nature*, **265**, 687 (1977).

4 Denniston-Thompson, K., Moore, D. D., Kruger, K. E., Furth, M. E. and Blattner, F. R. *Science,* **198**, 1051 (1977).
5 Prescott, D. M. and Kuempel, P. L. *PNAS,* **69**, 2842 (1972).
6 Champoux, J. J. *PNAS,* **74**, 3800 (1977).
Champoux, J. J. *J. Mol. Biol.,* **188**, 441 (1978).
7 Gilbert, W. and Dressler, D. *Cold Spring Harbor Symp.,* **33**, 473 (1969).
8 Eisenberg, S., Griffith, J. and Kornberg, A. *PNAS,* **74**, 3198 (1977).
9 Okazaki, R. *et al. Cold Spring Harbor Symp.,* **33**, 129 (1968).
Olivers, B. M. and Bonhoeffer, F. *Nature NB,* **240**, 233 (1972).
10 Okazaki, R., Hirose, S., Okazaki, T., Ogawa, T. and Kurosawa, Y. *Biochem. Biophys. Res. Comm.,* **62**, 1018 (1975).
11 Tye, B. K., Nyman, P. O., Lehman, I. R., Hochhauser, S. and Weiss, B. *PNAS,* **74**, 154 (1977).
12 Sigal, N., Delius, H., Kornberg, T., Gefter, M. I. and Alberts, B. *PNAS,* **69**, 3537 (1972).
Weiner, J. H., Bertsch, L. L. and Kornberg, A. *J. Biol. Chem.,* **250**, 1972 (1975).
Herrick, G. and Alberts, B. *J. Biol. Chem.,* **251**, 2124 (1976).
13 Louarn, J., Patte, J. and Louarn, J. M. *J. Mol. Biol.,* **115**, 295 (1977).
14 Lehman, I. R., Bessman, M. J., Simms, E. S. and Kornberg, A. *J. Biol. Chem.,* **233**, 163, 171 (1958).
15 Kornberg, T. and Gefter, M. L. *PNAS,* **68**, 761 (1971).
Kornberg, T. and Gefter, M. L. *J. Biol. Chem.,* **247**, 5369 (1972).
16 Gefter, M. L., Hirota, Y., Kornberg, T., Wechsler, J. A. and Barnoux, C. *PNAS,* **68**, 3150 (1971).
Tait, R. C. and Smith, D. W. *Nature,* **249**, 116 (1974).
17 Andersen, J. J. and Ganesan, A. T. *J. Mol. Biol.,* **106**, 285 (1976).
18 Weissbach, A. *Cell,* **5**, 101 (1975).
19 Kornberg, A. *DNA Synthesis,* W. H. Freeman, San Francisco, 151 (1974).
20 Schekman, R., Weiner, A. and Kornberg, A. *Science,* **186**, 987 (1974).
21 Reichard, P., Eliasson, R. and Söderman, G. *PNAS,* **71**, 4901 (1974).
22 Wickner, W. and Kornberg, A. *J. Biol. Chem.,* **249**, 6244 (1974).
23 Wickner, S. and Hurwitz, J. *PNAS,* **73**, 1053 (1976).
24 Wickner, S. *PNAS,* **73**, 3511 (1976).
25 Reichard, P., Eliasson, R. and Söderman, G. *PNAS,* **71**, 4901 (1974).
26 Topal, M. D. and Fresco, J. R. *Nature,* **263**, 289 (1976).
27 Brutlag, D. and Kornberg, A. *J. Biol. Chem.,* **247**, 241 (1972).
28 Lehman, I. R. *Science,* **186**, 790 (1974).
29 Kurosawa, Y., Ogawa, T., Hirose, S., Okazaki, T. and Okazaki, R. *J. Mol. Biol.,* **96**, 653 (1975).
30 Setlow, P., Brutlag, D. and Kornberg, A. *J. Biol. Chem.,* **247**, 224 (1972).
Setlow, P. and Kornberg, A. *J. Biol. Chem.,* **247**, 232 (1972).
Klenow, H. and Hemingsen, I. *PNAS,* **65**, 168 (1970).
31 Watson, J. D. *Nature NB,* **239**, 197 (1972).
32 Cavalier-Smith, T. *Nature,* **250**, 467 (1974).
33 Tattersal, P. and Ward, D. C. *Nature,* **263**, 106 (1976).
34 Wickner, S. and Hurwitz, J. *PNAS,* **71**, 4120 (1974).
Wickner, S. *PNAS,* **74**, 2815 (1977).
35 Schekman, R., Weiner, J. H., Weiner, A. and Kornberg, A. *J. Biol. Chem.,* **250**, 5859 (1975).
35a Baas, P. D., Jansz, H. S. and Sinsheimer, R. L. *J. Mol. Biol.,* **102**, 633 (1976).
36 Blumenthal, T., Landers, T. A. and Weber, K. *PNAS,* **69**, 1313 (1972).
Wahba, A. J. *et al. J. Biol. Chem.,* **249**, 3314 (1974).
37 Temin, H. M. and Mizutani, S. *Nature,* **226**, 1211 (1970).

38 Baltimore, D. *Nature,* **226**, 1209 (1970).
39 Spiegelman, S. *et al. Nature,* **227**, 563 (1970).
Cuatico, W., Cho, J. R. and Spiegelman, S. *PNAS,* **70**, 2789 (1973).
Gillespie, D., Saxinger, W. C. and Gallo, R. C. *Progr. Nucl. Acid Res. Mol. Biol.,* **15**, 1 (1975).
40 Dahlberg, J. E. *et al. J. Virol.,* **13**, 1126 (1974).
41 Cashion, L. M., Joho, R. H., Planitz, M. A., Billeter, M. A. and Weissmann, C. *Nature,* **262**, 186 (1976).
42 Haseltine, W. A., Kleid, D. G., Panet, A., Rothenberg, E. and Baltimore, D. *J. Mol. Biol.,* **106**, 109 (1976).
Shine, J., Czernilofsky, A. P., Friedrich, H., Bishop, J. M. and Goodman, H. M. *PNAS,* **74**, 1473 (1977).
43 Efstratiadis, A., Kafatos, F. C., Maxam, A. M. and Maniatis, T. *Cell,* **7**, 279 (1976).
Maniatis, T., Sim, G. K., Efstratiadis, A. and Kafatos, F. C. *Cell,* **8**, 163 (1976).
44 Efstratiadis, A., Kafatos, F. C. and Maniatis, T. *Cell,* **10**, 571 (1977).
45 Monahan, J. J., Woo, S. L. C., Liarakos, C. D. and O'Malley, B. W. *J. Biol. Chem.,* **252**, 4722 (1977).
McReynolds, L. A., Catterall, J. F. and O'Malley, B. W. *Gene,* **2**, 217 (1977).
46 Mercereau-Puijalon, O. *et al. Nature,* **275**, 505 (1978).
47 Dugaiczyk, A., Woo, S. L. C., Lai, E. C., Mace, M. L. Jr, McReynolds, L. and O'Malley, B. W. *Nature,* **274**, 328 (1978).
Garapin, A. C. *et al. Nature,* **273**, 349 (1978).
48 Khorana, H. G. *et al. J. Biol. Chem.,* **251**, 565 (1976).
49 Fiddes, J. C., Barrell, B. G. and Godson, G. N. *PNAS,* **75**, 1081 (1978).
50 Gray, C. P., Sommer, R., Polke, C., Beck, E. and Schaller, H. *PNAS*, **75**, 50 (1978).
51 Godson, G. N., Barrell, B. G., Staden, R. and Fiddes, J. C. *Nature,* **276**, 236 (1978).
52 Meijer, M., Beck, E., Hansen, F. G., Bergmans, H. E. N., Messer, W., von Meyenburg, K. and Schaller, H. *PNAS,* **76**, 580 (1979).
Sugimoto, K., Oka, A., Sugisaki, H., Takanami, M., Nishimura, A., Yasuda, Y. and Hirota, Y. *PNAS,* **76**, 575 (1979).
53 Crews, S., Ojala, D., Posakony, J., Nishiguchi, J. and Attardi, G. *Nature,* **277**, 192 (1979).
54 Soeda, E., Kimura, G. and Miura, K. *PNAS,* **75**, 162 (1978).
55 Koths, K. and Dressler, D. *PNAS,* **75**, 605 (1978).
56 Olivera, B. M. *PNAS,* **75**, 238 (1978).
Tye, B. K., Chien, J., Lehman, I. R., Duncan, B. K. and Warner, H. R. *PNAS,* **75**, 233 (1978).
57 Ueda, K., McMacken, R. and Kornberg, A. *J. Biol. Chem.,* **253**, 261 (1978).
58 Sebring, E. D., Kelly, T. J., Thoren, M. M. and Salzman, N. P. *J. Virol.,* **8**, 478 (1971).
59 Crea, R., Kraszewski, A., Hirose, T. and Itakura, K. *PNAS,* **75**, 5765 (1978).
60 Goeddel, D. V. *et al. PNAS,* **76**, 106 (1979).

PART 2

The Transcript

3 The Process of Transcription

SOME GENERAL CHARACTERISTICS OF THE TRANSCRIPTION PROCESS

All the genetic information for the whole organism is encoded in the DNA molecules of each cell. According to the Central Dogma, the first step in the expression of this information is the transfer of individual messages into RNA molecules which in turn can carry this information to the protein synthesising system. The messenger RNAs thus produced contain only one or a few cistrons, i.e. the copies of one or a few genes. They are synthesised according to the actual needs of the cell; different control mechanisms determine which genes will be expressed in a given cell at any one time. A flexible system is thus created; this is the basis of the high adaptability of microorganisms and of the differentiation in higher organisms.

In prokaryotes, messenger RNAs are very short-lived, and are newly synthesised whenever the cell requires the proteins they are coding for. The transcription of a great number of genes (but not all) can be switched on or off depending on changes in the environment which create or abolish the need for a specific enzyme. After some cycles of translation the messenger RNAs are degraded by nucleases of the cell and synthesis of new messenger RNAs is needed to keep up the enzyme level. In eukaryotes the mRNA profile of a given cell is more constant. In highly differentiated organisms specific proteins are synthesised in each type of cells, and the control mechanisms of transcription work here in a more permanent way, determining which gene will be transcribed in which cell. The mRNAs are more stable and need not be replaced at such a high rate as in prokaryotes.

The differences in the organisation of prokaryotic and eukaryotic genomes also affect the mechanism of mRNA synthesis in these organisms. In the prokaryotic chromosome, genes for several proteins which are functionally correlated are often clustered together and transcribed into one polycistronic mRNA. Genes for enzymes participating in the same metabolic pathway are often present in one structural-functional unit, the operon. Transcription of these genes is thus coordinated and expression of a group of genes is regulated by a single control mechanism. In eukaryotes the message for each protein is transcribed into a separate mRNA molecule: the mRNAs are monocistronic. The split gene structure detected in many eukaryotic genomes implies that special processes may be involved in the production of the mature spliced mRNAs.

Transcription does not only produce mRNAs, for the information for the other RNA species of the cell is also built into the chromosomal DNA. tRNA genes and rRNA genes are transcribed even more efficiently than protein genes. Some of these genes are also clustered together and are organised into transcription units both in eukaryotes and in prokaryotes. In eukaryotes the genes for three ribosomal RNAs (28S, 18S and 5.8S RNA) constitute a transcription unit. The product of transcription is a large RNA molecule which undergoes modifications and splitting in the course of post-transcriptional processing and thus yields the three mature rRNA species. In prokaryotes the ribosomal transcription unit comprises the 23S, 16S and 5S ribosomal RNA genes and several tRNA genes (*see* p.39). These are also transcribed together into an RNA precursor which is cleaved and processed to produce the mature RNAs.

Whether a single gene or a complex transcription unit is transcribed, the process leads to the transfer of information from a well defined part of DNA into an RNA molecule. This implies the need for exact copying of a definite stretch of DNA, from a specific start-site to a specific termination site. Detection of a number of split genes in eukaryotic genomes raised the question as to whether the same mechanism and the same requirements are also valid for the transcription of these interrupted DNA sequences. Although our present knowledge of the different steps which lead to the formation of spliced RNA molecules is still very limited, all the available evidence suggests that the primary process of transcription proceeds essentially in the same way whether the genes are interrupted or not. It seems that the DNA region comprising a split gene is also copied faithfully from an initiation site to a termination point, thus producing a giant primary transcript which contains both the structural sequences belonging to the message and the intervening sequences present in the DNA. The specific processes which eliminate the redundant sequences and bring about the splicing of the RNA molecule take place post-transcriptionally.

In the following, transcription will first be described as it occurs in prokaryotes and the more complex situation in eukaryotic organisms will be discussed separately, as far as it is known today.

TRANSCRIPTION IN PROKARYOTES

(1) The fidelity of transcription

Transcription produces a faithful copy of a continuous DNA stretch. The fidelity of the process depends on (a) a specific start-point of transcription; (b) precise copying of the DNA sequence on the basis of complementarity of nucleotides; (c) a specific termination site of transcription. The enzyme carrying out this process must thus recognise the control signals for initiation and termination and must exhibit a strict specificity to polymerise nucleotides exactly as defined by the nucleotide sequence of the template. These enzymes, which are present in every cell in which RNA synthesis occurs, are the DNA-dependent RNA polymerases.

The sites at which RNA polymerase interacts with DNA and initiates transcription, are the *promoter sites.* A promoter may control the transcription of one gene or of a whole transcription unit. Operons, which are subject to a more sophisticated control mechanism, contain a more elaborate control region. In addition to the promoter, they comprise further control signals which, by interaction with specific regulatory proteins, can switch transcription on or off (p. 16).

Promoter sites have been located in several DNA genomes and more recently some characteristic features of the primary structures of these sites have also been established[1,2,3,3a]. The nucleotide sequence of the complete control region of a few operons is also known[3,4,5]. These new data, although they are not sufficient to explain why these sites interact with the polymerase or with the specific regulatory proteins, have shed some light on the mechanism of how DNA-protein interactions can bring about regulation of the transcription process.

The promoters are also responsible for the efficiency of transcription. Promoter mutations may lead to the production of highly increased or decreased amounts of RNA[6]. Recent theories explain the high efficiency of some 'strong' promoters by binding of multiple molecules of RNA polymerase to these promoter sites[7].

The structure of termination signals and their exact role in the control of transcription is not quite clear yet. Nucleotide sequences have been described which probably cause termination of transcription[8], but it seems that several different structural features may play a role in causing the RNA polymerase to stop at a definite site. The question is complicated by the fact that more than one termination mechanism may operate at different sites, requiring more than one termination signal (*see* p. 93). It is certain, however, that in some cases transcription can be controlled at the stage of termination. Interaction of a termination signal with specific proteins (termination or antitermination factors) can determine whether transcription stops at this site or if readthrough occurs, thus allowing the expression of genes beyond the termination site.

'High-fidelity' copying of the DNA sequence is essential, the more so as, unlike in the replication process, in transcription there is no way to correct mistakes and to replace mis-matched nucleotides. The enzymes and protein factors involved in this process must therefore act with very great accuracy. In some organisms we know which enzymes and protein factors take part in this process and can describe the exact mechanism in detail. Still, a variety of enzymes carrying out transcription are present in the different cells and although many of these seem to have basic characteristics in common, the specificity and accuracy of the transcription process may be achieved by slightly different mechanisms in the different systems[9].

All the known cellular RNA polymerases carry out the same chemical reaction. Starting with nucleoside triphosphates as substrates, they build polymers by joining the α-phosphate group of the incoming nucleotide in phosphodiester linkage to the 3′ hydroxyl of the preceding one. Inorganic pyrophosphate is split off at each step of polymerisation.

N_1 N_2 N_1 N_2

-OH + -OH → -OH + PP_i

PPP PPP PPP P

N_3 N_1 N_2 N_3

+ -OH → -OH + PP_i

PPP PPP P P

According to this mechanism, the synthesis of RNA proceeds in the direction from the 5′ end to the 3′ end. DNA-dependent RNA polymerases require a DNA template for the synthesis of RNA. The template used by bacterial enzymes is double-stranded DNA, but only one strand of the molecule is copied.

The mechanism of transcription will be described below as it is carried out by *E. coli* RNA polymerase. In this system the different steps of transcription and the different DNA–protein interactions which ensure the fidelity of the process have been extensively studied. RNA polymerases from other bacteria show similar characteristics to the *E. coli* enzyme and seem to recognise the same control signals[9a].

(2) Initiation of transcription

(a) RNA polymerase: its structure and interaction with promoter sites

DNA-dependent RNA polymerase of *E. coli* is the best known enzyme of this group. It has been purified in several laboratories and its protein structure, physico-chemical characteristics and mode of action have been thoroughly investigated. (This enzyme may not be the only RNA polymerase in *E. coli*, but it is the one responsible for *in vivo* transcription. There are strong indications that another enzyme, carrying out a very similar polymerisation reaction, is active in synthesising the primers in the replication of DNA (p. 63).)

The RNA polymerase of *E. coli* has been isolated as a homogeneous protein of very high molecular weight[10]. It was found to consist of four different subunits which could be separated by gel electrophoresis (*table 3.1*). One enzyme

Table 3.1 Subunits of *E. coli* RNA polymerase

Subunit	Molecular weight
α	40 000
β	150 000
β'	160 000
σ	86 000

molecule is built of 2α, 1β, 1β' and 1σ subunit; its structure can be described by the formula $\alpha_2\beta\beta'\sigma$. The σ subunit (often called σ factor) dissociates easily from the rest of the molecule, leaving behind the 'core enzyme' of the structure $\alpha_2\beta\beta'$, which in itself also has RNA polymerase activity. The specificity of the core enzyme is, however, different from that of the holoenzyme, i.e. the enzyme molecule containing also the σ subunit.

It has been found in *in vitro* studies that the holoenzyme of RNA polymerase initiates transcription at the specific promoter sites, while the core enzyme does not possess such specificity. The core enzyme works better on a single-stranded DNA template than on a double-stranded DNA, and when it transcribes double-stranded DNA, can initiate at nicks in the molecule where a very short single-stranded part is available. The products of *in vitro* transcription with core enzyme are thus different from the RNAs synthesised *in vivo* in the *E. coli* cell. All this points to an important role of σ factor in controlling the specificity of the initiation process.

Chamberlin studied the kinetics of the binding of holoenzyme and core enzyme to the DNA template[11]. He determined the number of enzyme molecules bound to DNA and the stability of the complexes formed. The results, part of which are represented in *figure 3.1*, showed that the core enzyme forms a great number of very weak complexes, while in the interaction of DNA with

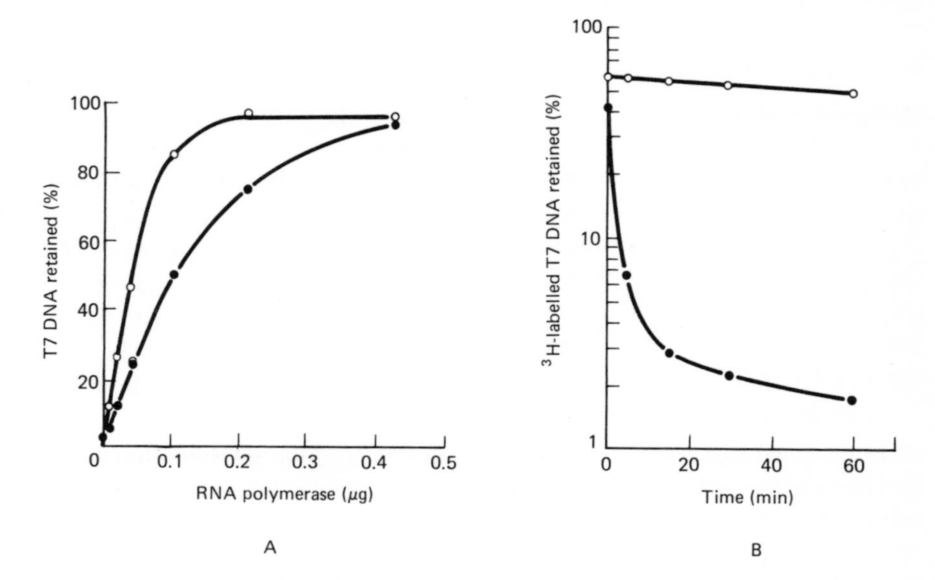

Figure 3.1 Kinetics of interaction of RNA polymerase with T7 phage DNA (Hinkle and Chamberlin[11]). A: Binding of RNA polymerase to ^{3}H-labelled T7 DNA. Enzyme-bound DNA plotted against the amount of enzyme present. Membrane filter assay. B: Dissociation of RNA polymerase-DNA complexes. Enzyme was bound to ^{3}H-labelled T7 DNA and dissociation of the complex followed in the presence of excess unlabelled DNA. ^{3}H-labelled DNA remaining in complex form is plotted against time. Assay as above. ○, Results with holoenzyme; ●, results with core enzyme, in both diagrams. (Reproduced with permission from *J. Mol. Biol.*, 70, 157 (1972). Copyright by Academic Press, London.)

holoenzyme two types of complexes are formed: a major part of weak complexes and a number of very stable complexes with association constants up to 10^{14} M^{-1}. The actual number of the latter corresponded well with the number of promoter sites in the DNA used in the binding reaction. This suggests that the core enzyme cannot recognise the promoter sites while the holoenzyme can do so, and that the latter can form stable complexes specifically at these sites. It has been confirmed since that the sites occupied by the holoenzyme in this tight complex are indeed the sites at which transcription is initiated[12]. Still, the holoenzyme molecules also initially attach randomly to DNA, but as they form only loose complexes at the unspecific sites, they can move along the DNA chain until they find a promoter site where they interact specifically and form a stable complex. Both the random and the specific tight complexes of RNA polymerase holoenzyme formed with T7 phage DNA have been visualised under the electron microscope[13]. The results confirmed that for the latter, the enzyme does specifically select the promoter sites.

All this evidence shows clearly that for both the recognition of the specific binding sites and the formation of the stable complex, the presence of σ factor is required. The function of σ factor is limited, however, to the recognition of the promoter site and the stabilisation of the complex; the actual binding of the enzyme to DNA is not brought about by σ factor. Subunit β' is involved in the binding of RNA polymerase to the DNA template[14].

(b) The structure and specificity of promoter sites

The question arises as to which structural features of the promoter site are recognised by the RNA polymerase holoenzyme. Modern techniques made it possible to isolate well defined DNA fragments containing the promoter site from different viral DNAs and also from a few operons of the *E. coli* chromosome. The nucleotide sequence of these DNA stretches could also be determined and the results of these studies are shown in *table 3.2*. It can be seen that there is a homologous region in the sequences of all promoter sites. Pribnow[1], when establishing the nucleotide sequence of promoters of the coliphage T7, called attention to a 7-nucleotide-long characteristic stretch

TATPuATPu (Pu: purine)

in the promoter sequence near the initiation site. Most of the promoter sites so far known in the different coliphages and in the *E. coli* chromosome contain a heptanucleotide sequence with a very high degree of homology to the above 'ideal' structure in the same position, i.e. preceding the initiation site by four or five nucleotides. This homologous stretch forms part of the RNA polymerase binding site. RNA polymerases from other bacteria bind to the same promoter sites, although with different efficiency[9a]. Some phage-induced enzymes, however, recognise different promoters.

Another characteristic of the RNA polymerase: DNA interaction is that the enzyme shows an increased affinity for negatively supercoiled DNA.

Table 3.2 Nucleotide sequences of promoter sites

DNA	
T7 phage, A3 promoter	AAGTAAACACGCTACGATGTACCACATGAAACGACAGTGAGTCA
fd phage	TGCTTCTGACTATAATAGACAGGGTAAAGACCTGATTTTTGA
phage λ, P_L promoter	CCACTGGCGGTGATACTGAGCACATCAGCAGGACGCACTGAC
phage λ, P_R promoter	TCTGGCGGTGATAATGGTTGCATGTACTAAGGAGGTTG
ΦX174 phage, promoter A	TTGTATGTTTTCATGCCTCCAAATCTTGGAGGCTTT
ΦX174 phage, promoter D	TGGATTACTATCTGAGTCCGATGCTGTTCAACCA
ΦX174 phage, promoter B	CTTATGGTTACAGTATGCCCATCGCAGTTCGCTAC
E. coli, $tRNA^{tyr}$ gene	TCATTTGATATGATGCGCCCCGCTTCCCGATAA
E. coli, lac operon	CTTCCGGCTCGTATGTTGTGTGGAATTGTGAGCG
SV40 virus†	TTGCAGCTTATAATGGTTACAAATAAAGCAATAGCA

The sequences underlined are required for binding RNA polymerase. The nucleotides marked by an asterisk are the initiation sites of transcription.

†SV40 is an animal virus; its DNA contains promoter sites which are recognised *in vivo* by the RNA polymerase of the host cell. There is, however, a different site on the DNA which is recognised *in vitro* by *E. coli* RNA polymerase. The sequence of this site is shown here.

Binding of the RNA polymerase is, however, not the only function of the promoter, and the binding site of the enzyme is not the only functional site in the promoter region. The first indication of the necessity for an additional functional site came from the above mentioned work of Pribnow. He isolated the RNA polymerase binding site at a promoter region of the DNA of T7 phage. The binding site of protein can be isolated and characterised by binding the protein to DNA and digesting off the free DNA with nucleases. The DNA stretch to which the protein is bound is not accessible to the nucleases and is therefore protected from digestion. The protected fragment can then be isolated and its nucleotide sequence determined. Pribnow found that if he isolated the binding site in the form of a 49 base pair-long fragment and brought this DNA fragment together with the enzyme, no binding occurred. As the same enzyme formed a stable complex with the same DNA stretch while the latter was still part of the complete DNA molecule, the only explanation for this negative result was that some further information was missing, a further site in the DNA which, although it did not take part in the actual binding of the enzyme, was still required to enable the enzyme to interact with the binding site. This is called the recognition site of RNA polymerase. The recognition site is also part of the promoter, but it does not overlap with the polymerase binding site. Studies on the structure of control regions in different operons suggest that it is located at a site which precedes the start-site of transcription by about 35 nucleotides. Recent work of Yanofsky's group on the structure of the operator–promoter region of the tryptophan operon[3] provided further evidence that the functionally complete promoter site extends over a region more than 39 and probably up to 59 base pairs long which precedes the start-site. Some characteristic sequence homologies have been detected in this region of promoter sites in ΦX174 and G4 phage DNAs[2].

The specificity of promoter recognition can also be studied in heterologous systems. *E. coli* RNA polymerase can also transcribe eukaryotic DNAs *in vitro*. Transcripts obtained from the DNA of SV40 virus under such conditions were compared with *in vivo* transcripts of the same DNA. It was found that transcription by *E. coli* RNA polymerase was initiated at a different site: the promoters recognised by the eukaryotic RNA polymerase were not recognised by the *E. coli* enzyme. The promoter site for *E. coli* RNA polymerase on SV40 DNA has a nucleotide sequence similar to those of the *E. coli* promoters[15] (*see* also *table 3.2*).

(c) Open and closed initiation complexes

Recognition of a promoter site and binding of the RNA polymerase in the form of a stable complex are the first events in the initiation of transcription. The structure of this first complex is, however, not suitable to accommodate the incoming ribonucleotide and to allow base pairing with the corresponding complementary nucleotide in the template. As in the double-helical DNA structure the bases are on the inside and are held together by hydrogen bonds, a local disruption of these base pairs is necessary to allow interaction between the nucleotides of the template and the nucleotides incorporated into the transcript. Local distortion of the double helical structure has been demonstrated in early studies[16]. The length of the DNA region which becomes unwound has been recently determined: it extends over a stretch of 11 base pairs[17].

Chamberlin[11] proposed a model of the different events in initiation (*figure 3.2*). This includes as the last step the transition from a non-functional primary complex which he called a 'closed complex' into an 'open complex' which can bind ribonucleotides and can thus initiate transcription. In the closed complex the enzyme attaches to the DNA double helix from the outside. Formation of the open complex is accompanied by a distortion of the double-helical structure which enables the enzyme to come into contact with the bases. This does not mean that the enzyme binds in between the two DNA strands, the distortion is not sufficient to allow that. In the open complex the RNA polymerase still binds

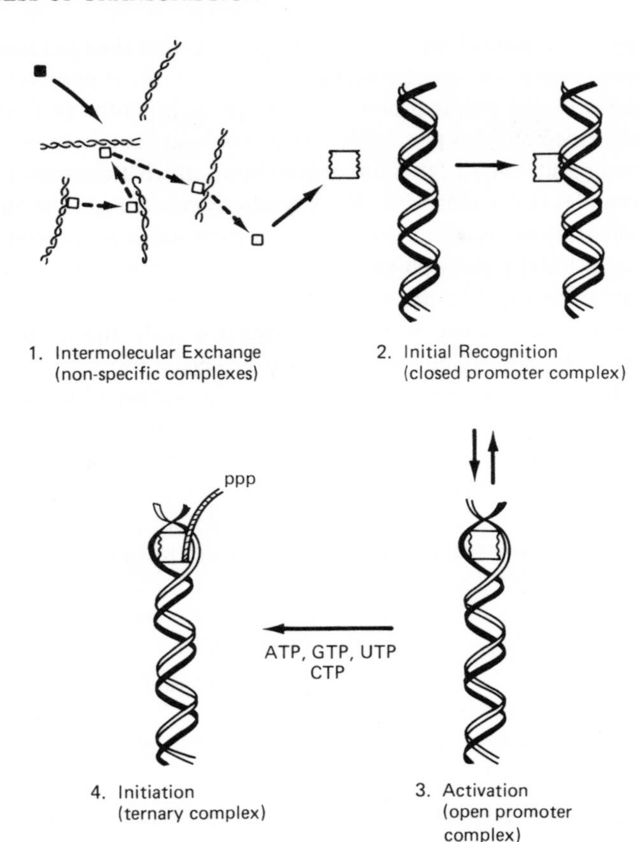

Figure 3.2 Initiation complex formation with *E. coli* RNA polymerase. (According to Chamberlin[64].) After an initial stage of nonspecific interactions, the enzyme recognises a promoter site and binds to it tightly to form a 'closed complex'. By opening of a few base pairs, the enzyme makes contact with the bases, an 'open complex' is formed and polymerisation of nucleotides can start. The presentation is highly schematic: the enzyme molecule is, in fact, not wedged in between the two DNA strands. (Reproduced, with permission, from *Ann. Rev. Biochem.*, **43**, 721. © 1972 by Annual Reviews, Inc.)

asymmetrically to one DNA strand. Such asymmetrical binding is in agreement with the function of the polymerase, viz. that it copies the sequence of one DNA strand.

(d) Start-site of transcription

With the formation of the open complex, the enzyme is ready to start transcribing the DNA template. The first question which arises is: at exactly which site will transcription start? The enzyme itself covers a DNA stretch at least 14 base pairs long and it is not even *a priori* obvious that it must start copying the DNA at the same site where it formed a stable complex with it. In fact, there have been suggestions that the enzyme may bind to a stretch of DNA at some distance from the start-site and then slide along the DNA template to find the specific start-point where the synthesis of RNA actually begins.

It was Pribnow's work on the promoter of the DNA of coliphage T7 (ref. 1) which definitely ruled out this possibility. He determined the structure of the polymerase binding site on T7 DNA under conditions in which initiation does not occur and also under conditions when RNA synthesis is initiated. The nucleotide sequences of the stretches of DNA isolated under these different conditions were identical, proving that the RNA polymerase does not move from its original binding site when it initiates transcription. Pribnow also compared the nucleotide sequence of this unique binding site with the sequence of the first 19 nucleotides of the RNA transcript and found that these two sequences overlap: the first nucleotide of the transcript is in the middle of the binding site of the enzyme (*figure 3.3*). Overlapping of the startsite of transcription with the binding site of RNA polymerase has since been demonstrated in the case of other viral promoters as well as of bacterial promoters.

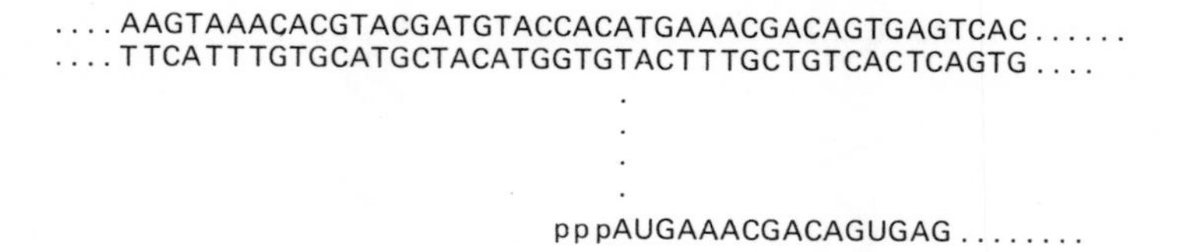

Figure 3.3 Nucleotide sequences of a promoter site of T7 DNA and of the 5′ end of the mRNA transcribed from this promoter[1].

Comparison of the 5′ termini of different transcripts also confirmed a regularity with respect to the nucleotides with which transcription can start: they are always purines, G or A (*table 3.2*). This result could be predicted from previous *in vitro* studies on *E. coli* RNA polymerase which showed that the enzyme initiates polynucleotide chains with purine nucleotides.

(3) Elongation of the polynucleotide chain

With the insertion of the first nucleotide the process of initiation is completed, and the actual polymerisation of nucleotides, i.e. the elongation of the RNA chain, begins. There are, however, important changes in the enzyme molecule connected with this transition from the stage of initiation to the stage of elongation.

The transition from initiation to elongation is accompanied by conformational changes in all participant molecules, the template, the enzyme and probably also the incorporated nucleoside triphosphate. The changes in the template involve the above described local distortion, which is reversible: when the enzyme moves away from the initiation site, the perfect double helix is re-formed and the distorted area moves along the DNA molecule, together with the enzyme. At the same time the conformation of the enzyme is also altered, as is indeed required by its changing function. RNA polymerase must fulfil very different requirements at the stage of initiation and at the stage of elongation of the RNA chain. At the former stage the enzyme forms a stable complex with the DNA, and shows a strong specificity for the primary structure of the binding site. In order to be able to move along the template at the stage of elongation, it must bind loosely to DNA. As it copies every nucleotide sequence in the DNA with equal efficiency, it cannot show any preference for a specific sequence at this stage.

These changes in the interaction of RNA polymerase with the DNA template are brought about by the release of σ factor from the enzyme after the process of initiation has been completed. It is assumed that release of σ factor occurs after the first internucleotide bond has been formed. The presence or absence of σ factor affects the conformation of the β and β' subunits. While associated with σ factor (initiation stage), β and β' take up a conformation which favours strong and specific binding to DNA. When σ factor dissociates, leaving behind the core enzyme (elongation stage), the conformation changes so as to exhibit the properties of the core enzyme, already discussed; the binding to DNA is loose and unspecific. Recognition or preference for any specific nucleotide sequence as well as the ability to form stable complexes with DNA are lost with the loss of σ factor.

The fact that σ factor operates in the initiation of transcription within an association–dissociation cycle was first revealed by Travers and Burgess[18], who studied the effect of addition of σ factor to an *in vitro* system in which the DNA of T4 phage was transcribed with core enzyme only. As could be expected, the poor transcription obtained by core enzyme could be strongly activated by adding σ factor to the system. It was, however, unexpected, that σ factor did not act in a stoichiometric but in a catalytic way. A small amount of σ factor could activate a large excess of core enzyme molecules. It seemed that σ factor must be used and re-used several times by several core enzyme molecules. Knowing that σ factor dissociates easily from the holoenzyme, the plausible explanation of the results was that σ factor is associated with the enzyme only temporarily, during the step of specific initiation, and after being released, it can associate with other core enzyme molecules.

The proof that this is indeed the case was obtained in an experiment in the same system, when transcription was carried out with minimal amount of holoenzyme, so that all enzyme molecules were engaged in transcription. A constant, slow rate of RNA synthesis was observed. As expected, addition of more RNA polymerase to this system resulted in a sudden increase in RNA synthesis, as new RNA molecules could be initiated. However, the important point was that addition of core enzyme or holoenzyme had the same effect: there was no need also to add σ factor, as free σ subunits, released from the polymerase molecules which were past the initiation stage, were available in the system. These σ subunits could combine with the added core enzyme molecules and enable them to initiate new RNA molecules.

The mechanism of elongation of the polynucleotide chain is relatively simple: the incoming nucleotides are joined in 3′, 5′-phosphodiester linkage, according to the reaction described on p. 82*. However, quite a few aspects of the process have yet to be clarified. We do not know which properties of the enzyme are responsible for the high fidelity exhibited in the copying of the DNA sequence. The incoming nucleotides are selected on the basis of complementarity to the template and mistakes occur only at a very low frequency. Contrary to the case of replication of DNA, there is no mechanism in the transcription process which provides for replacing mismatched bases (RNA polymerase is not capable of proofreading), and so the selection of the nucleotides themselves must occur in a strictly specific way. As discussed earlier, in connection with a similar problem in the specific selection of deoxyribonucleotides for the accurate replication of DNA (p. 65), the specificity of base pairing alone does not seem sufficient to ensure the required precision of the process. The enzyme must play an active part in the selection of the complementary and in the rejection of non-complementary nucleotides. As proposed in the case of DNA polymerase, here, too, the specificity may be based on the fulfilment of stereochemical conditions, i.e. a favourable spatial arrangement of the reacting nucleotides with respect to the active centre of the enzyme, which can be achieved only if exact base pairing occurs between the incoming nucleotide and its complementary partner in the template.

It follows from the mechanism of the polymerisation reaction that the first, 5′-terminal nucleotide of the RNA chain will retain its triphosphate group while all the other nucleotides will lose their β and γ phosphates, which will be split off in the form of inorganic pyrophosphate. This provides an easy way to distinguish the process of initiation from that of elongation in *in vitro* experiments: if NTPs labelled in the γ-phosphate are added to the system, the label will be incorporated only at the stage of initiation, while with $\alpha[^{32}P]$-labelled NTPs as substrates, radioactivity will also be incorporated during the whole elongation process.

During elongation of the RNA chain, the newly synthesised polynucleotide does not remain attached to the template strand. There is only one attachment point at any time and this moves along the DNA in the course of elongation: this

*This reaction mechanism is of general validity: the joining of the α-phosphate group of the incoming nucleoside triphosphate to the 3′ hydroxyl of the last nucleotide of the growing polynucleotide chain is the basis of the sequence-specific synthesis of all nucleic acids in every organism. Basically the same kind of reaction is also taking place when DNA molecules are synthesised on a DNA or an RNA template (*see* Chapter 2). There is, however, one great difference between the reactions carried out by RNA polymerases or DNA polymerases. RNA polymerase can also initiate the synthesis of RNA, it can start with one nucleoside triphosphate, add the second nucleotide, the third, etc., while DNA polymerases can only extend an existing polynucleotide chain. This difference is also expressed in the chemical equations describing the polymerisation reactions (p. 58 and p. 82).

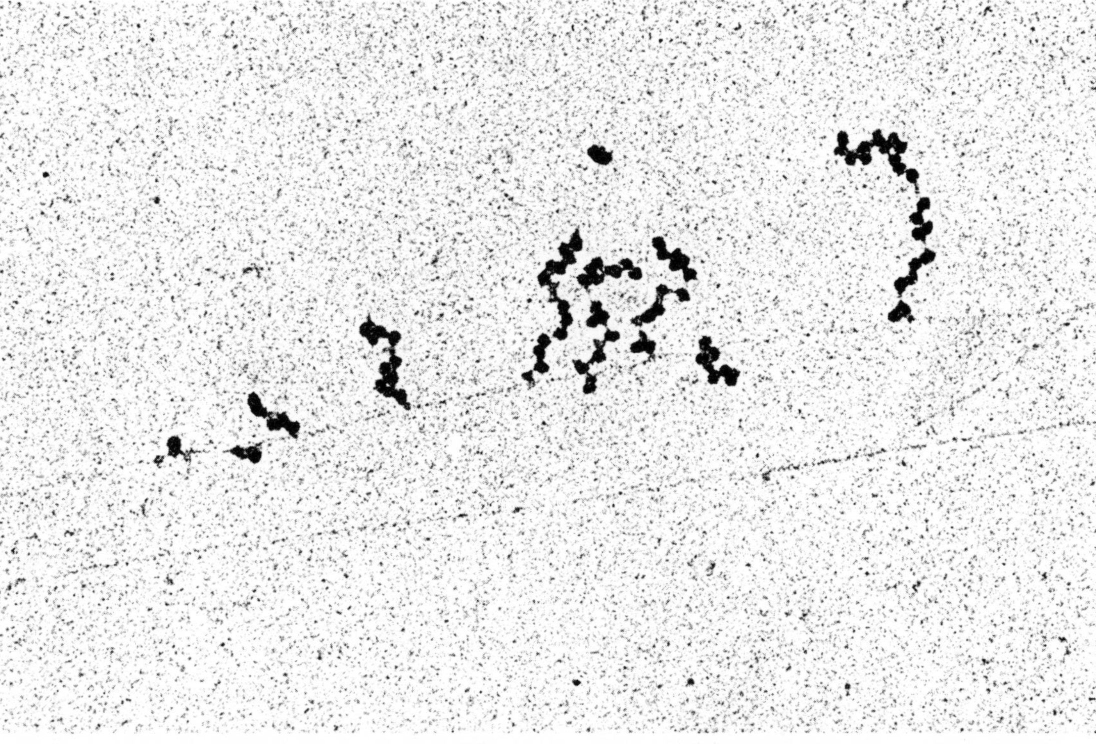

Figure 3.4 Electron micrograph of *E. coli* DNA in the process of transcription (Miller *et al.*[65]). Several RNA polymerase molecules are simultaneously transcribing the same DNA. It can be seen how the length of the transcripts increases as the RNA polymerase proceeds farther from the start-site. Ribosomes are associated with the transcripts: transcription and translation take place in a coupled process (*see* Chapter 8). (Copyright 1970 by the American Association for the Advancement of Science. Photograph by courtesy of Professor O. L. Miller, Jr.)

is the point where the enzyme is actually joining the incoming nucleoside triphosphates to the 3′ terminus of the growing chain, i.e. at the growing point of the polynucleotide chain. *Figure 3.4* shows an electron micrograph of *E. coli* DNA in the process of transcription. It can be seen that several RNA chains are being synthesised simultaneously by several RNA polymerase molecules on the same template. When one enzyme molecule has moved away from the promoter site, another enzyme molecule can attach to this site and start the synthesis of a new RNA chain. The figure also shows how the transcript is growing longer as the enzyme is moving farther away from the promoter site.

(4) Termination of transcription

Termination of transcription also occurs at specific sites, although the specificity is not as strict at this stage as it is at initiation. In both *in vitro* and *in vivo* experiments it was observed that under certain conditions transcription can go beyond a termination site; 'readthrough' may occur. When RNA polymerase alone was used in *in vitro* experiments on the transcription of T4 phage DNA, the transcripts were found to differ significantly from the *in vivo* products. They were usually longer, suggesting that termination did not occur at the specific sites. Roberts detected a protein factor in *E. coli* which had a specific effect on the termination of transcription[19]. He isolated the protein, called rho (ρ) factor, and studied its effect in the above *in vitro* system. In the presence of ρ factor, the products of transcription were shorter RNA chains, and their length corresponded well to the *in vivo* transcripts. It was therefore suggested that specific termination of transcription is dependent on ρ factor which, in an unknown way, recognises a specific stop signal and prevents the RNA polymerase from continuing copying the template beyond this point.

The situation, however, seems more complicated than was first assumed. It was found that *in vitro*, ρ factor is not always needed for specific termination and this led to the assumption that *in vivo* too, there exist ρ-dependent and ρ-independent termination sites. At the same time, specific regulatory proteins were discovered which promote the expression of genes encoded in phage DNAs by bringing about readthrough. One well studied example of such proteins is the N gene product of λ phage, which interferes with the termination of transcription by *E. coli* RNA polymerase and thus causes transcription to proceed beyond the termination site into the genes of the phage DNA. Proteins which have such an effect are called 'anti-termination factors'.

Roberts[20] recently re-investigated a great number of different control systems which involved ρ-dependent and ρ-independent termination, and came to the conclusion that *in vivo*, probably all processes are ρ-dependent, and that the *in vitro* differences may be artefacts reflecting only the strength of a termination site. The 'weak' termination sites which *in vitro* function only in the presence of ρ factor, may also be weaker *in vivo* and thus susceptible to regulation via anti-termination factors which can cause readthrough at these sites. The 'strong' termination sites which *in vitro* function also without adding ρ factor, may still require ρ factor *in vivo*, the presence of which makes them resistant to anti-termination factors. Readthrough cannot occur at these sites. According to this theory, termination of transcription would be regulated mainly by protein effectors which in some cases can antagonise the termination of transcription at the specific termination sites.

These specific termination sites can be identified on physical maps of genomes or in the nucleotide sequence of DNAs, but the exact nature of the termination signal which is recognised by ρ factor or by the RNA polymerase itself, is still uncertain. At the 3′ end of a number of transcripts a sequence containing several U residues has been detected[21]. This strong homology suggests that the enzyme or a protein factor may recognise a complementary sequence containing a number of A residues in the template as a termination signal. Several authors assume that a specific secondary structure is required to cause termination of transcription. A symmetrical sequence in the DNA which allows the formation of a stem and loop structure is supposed to inhibit further progression of the RNA polymerase[22,23]. Another common feature at several termination sites is the presence of alternating A–T-rich and G–C-rich regions[24,25].

A termination signal which stops RNA polymerase at a specific site on the DNA template might be present at the 3′-terminal part of the transcribed region or shortly beyond the 3′ terminus of the primary transcript. Subramanian considers both possibilities[8]; Lee and Yanofsky find that termination actually occurs at the fifth or sixth U residue in the termination signal[23]. Rosenberg *et al.*[21] studied the nucleotide sequences of λ phage DNA beyond the 3′ termini of two mRNAs. They found no homology between the DNA sequences in the two untranscribed regions while they detected a stretch of U residues at the 3′ end of both transcripts. This supports their view that the stop signals are encoded into the transcribed parts of the genes.

A very interesting regulatory mechanism based on a termination signal at a rather unusual site has been detected in the trp operon by Yanofsky's group[26]. When determining the sequence of the mRNA from this operon, they found that, as also occurs in many other mRNAs, a long untranslated sequence precedes the first cistron which codes for the trp E polypeptide, the first protein belonging to the trp operon. This 160 nucleotide-long untranslated sequence, called leader sequence (*see* Chapter 5), contains a termination signal . . .CUUUUUUUU, about 30 nucleotides before the start of the trp E cistron.

The presence of this termination signal, called *'attenuator'*, has been established in further studies on the nucleotide sequences of the trp operon control regions in *E. coli* and *Salmonella typhimurium* DNAs. In *in vitro* experiments transcription of the trp operon stops at this site, although a minor fraction of RNA polymerase molecules escape this termination signal and proceed beyond it into the first structural gene. The attenuator plays a regulatory role: it can prevent or allow expression of the genes of the trp operon. Its function is affected by the tryptophan level in the cell. The genes are expressed if the trp level is low. The attenuator represents an additional control mechanism in the trp operon; this makes the cell's response to progressive tryptophan starvation more sensitive and more flexible[23] There are indications that attenuators may also control mRNA production in other genomes. Similar structures have been detected in the phenylalanine and histidine operons[27].

Termination of transcription also involves the release of both the completed transcript and RNA polymerase from the template. It seems that ρ factor can accomplish only the first of these events and some other cell component may be required for the release of the enzyme.

(5) Transcription units

It has been mentioned in Chapter 1 that in the prokaryotic genome the genes for several related proteins or for several related RNAs are often organised into one transcription unit (p. 39). Transcription of these structural-functional units occurs under common control. There are many ways in which coordinated transcription can be controlled. In the relatively simple rRNA transcription units, the genes which will be expressed are separated by spacer regions and the whole length of this structural unit is transcribed under the control of one promoter. The redundant sequences are eliminated from the transcript in a post-transcriptional process (p. 101). Any regulatory effect which influences the efficiency of rRNA synthesis affects the promoter.

It is less well understood how seemingly independent transcription units can be subject to common regulation: the genes for ribosomal proteins are organised into several transcription units but they seem to respond to the same control signal. The transcription of ribosomal protein and ribosomal RNA genes is also coordinated in such a way that neither component of the ribosome is produced in excess (*see* Chapter 6).

Transcription units comprising genes for several proteins can be transcribed into one polycistronic mRNA. In addition to the actual coding sequences, the transcript also contains untranslated regions at the two ends of the molecule and between the cistrons. Untranslated sequences are characteristic of the structure of every mRNA (Chapter 5); these are not eliminated by post-transcriptional processing. Although they are not expressed in translation, they may be of great importance for the control of the translation process. Regulation of the transcription of such organised protein genes may be brought about in many different ways and may involve very complex regulatory mechanisms.

In the following, a few examples will be given of the ways in which the synthesis of inducible enzymes can be regulated according to the needs of the cell. This is the most important form of transcriptional control, and forms the basis of the adaptability of microorganisms. Several enzymes taking part in the same metabolic pathway are often organised into special transcription units, *operons*, in the prokaryotic genome. Although the genes are present, they are not expressed, or only at a very low level, if there is no need for this metabolic pathway to operate.

If some change occurs in the environment which induces a requirement for this metabolic pathway (e.g. substitution of lactose for glucose in the growth medium necessitates the metabolism of lactose), transcription of the operon will be switched on, mRNA will be produced and enzymes synthesised.

Control of transcription in the lac and gal operons

The basic concept of the operon as the genetic unit which regulates the expression of inducible enzymes, was created by Jacob and Monod[28] in 1961. They worked out an elegant theory which explained how the synthesis of three enzymes, required for lactose metabolism, is switched on or off in the lac operon depending on the need of the cell for using this carbon source. Their ingenious model is still essentially valid, although the large amount of information which has since accumulated in this field revealed that in addition to the relatively simple control mechanism they proposed, there are further ways in which the transcription of different operons can be regulated, and further factors which can affect, in a positive or negative way, the transcription of operons.

In the past few years, it has become possible to isolate genes and control regions from different operons and to characterise them by biochemical techniques. In a few cases, the exact chemical structure of the DNA regions exercising these controls has been established. This now gives us a possible way of exploring the exact molecular mechanisms which regulate the expression of genes in a number of different operons.

Two examples of such control mechanisms will be described below, in the case of two operons where the chemical structure of the control regions is known and where the interactions of regulatory proteins with the different sites in DNA can therefore be exactly defined.

(1) The lac operon

In addition to the structural genes coding for the three enzymes (gene z for β-galactosidase, gene y for galactoside permease, gene A for galactoside acetylase), the operon contains a control region 122 nucleotides long, the primary structure of which is shown in *figure 3.5*. The control region comprises the operator (nucleotides 78 to 112) which exercises control over the whole operon; the promoter (starts at nucleotide 1 and overlaps with the operator); and a structural gene, i, which codes for the repressor protein (leftwards from nucleotide 1). In the absence of lactose, when the operon is not transcribed, the repressor protein binds to the operator. This specific binding enabled Gilbert and Maxam[29] to isolate the lac operator and to determine its nucleotide sequence. A DNA stretch 24 base pairs long was protected by the repressor (*see* Chapter 4, *figure 4.6*) which does not cover the complete operator region (this is 35 base pairs long). It was shown later by Bahl *et al.*[30] that the DNA sequence which is actually recognised by the repressor is even shorter, only 17 base pairs (nucleotides 87 to 103) and shows a highly symmetrical palindrome structure. This 17 base pair-long oligonucleotide

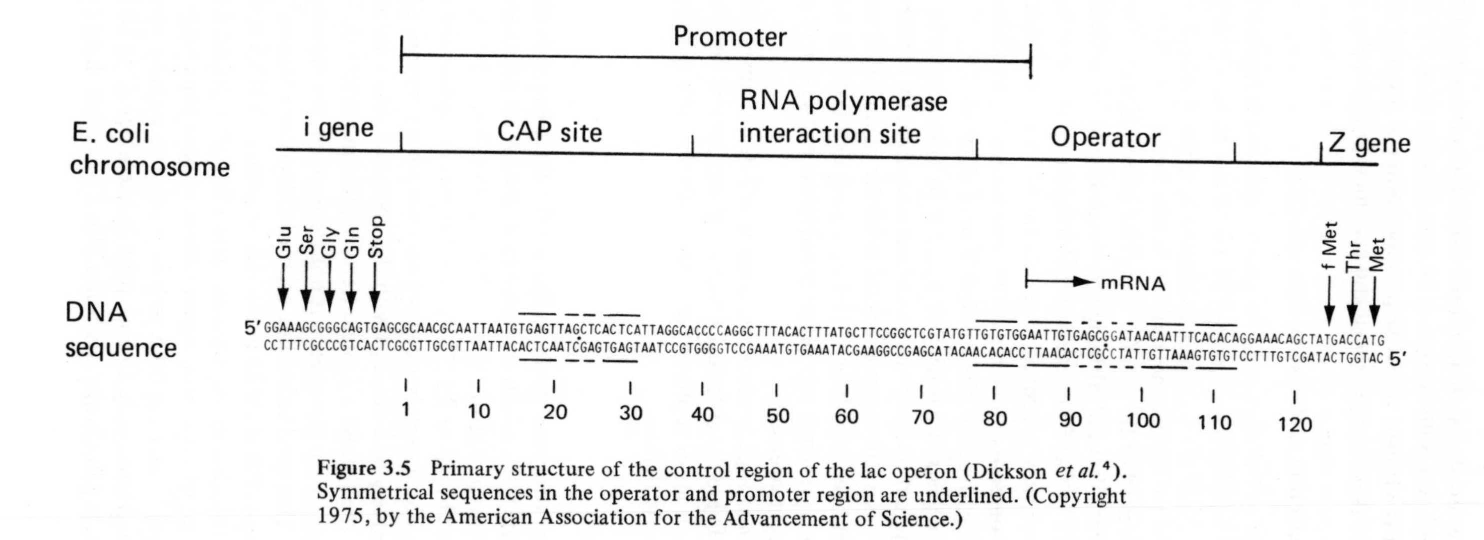

Figure 3.5 Primary structure of the control region of the lac operon (Dickson *et al.*[4]). Symmetrical sequences in the operator and promoter region are underlined. (Copyright 1975, by the American Association for the Advancement of Science.)

could also be inserted into a plasmid and could be cloned in *E. coli*. The cloned fragment was functioning like a natural lac operator.

The promoter is adjacent to, and partly overlapping with, the operator region and therefore this region is also blocked when the repressor binds to the operator. The RNA polymerase cannot attach to the promoter, and so transcription cannot take place (*figure 3.6A*): the operon is repressed. If an inducer is present, whether the natural inducer, lactose, or an artificial inducer of similar structure, this binds and inactivates the repressor with the result that the operator is unblocked and the promoter site becomes accessible to RNA polymerase (*figure 3.6B*). The enzyme attaches to the promoter site and transcribes the whole length of the operon, producing a polycistronic mRNA which contains the messages for all three enzymes. The start-point of transcription is at nucleotide 84, which means that part of the operator sequence is also transcribed[31].

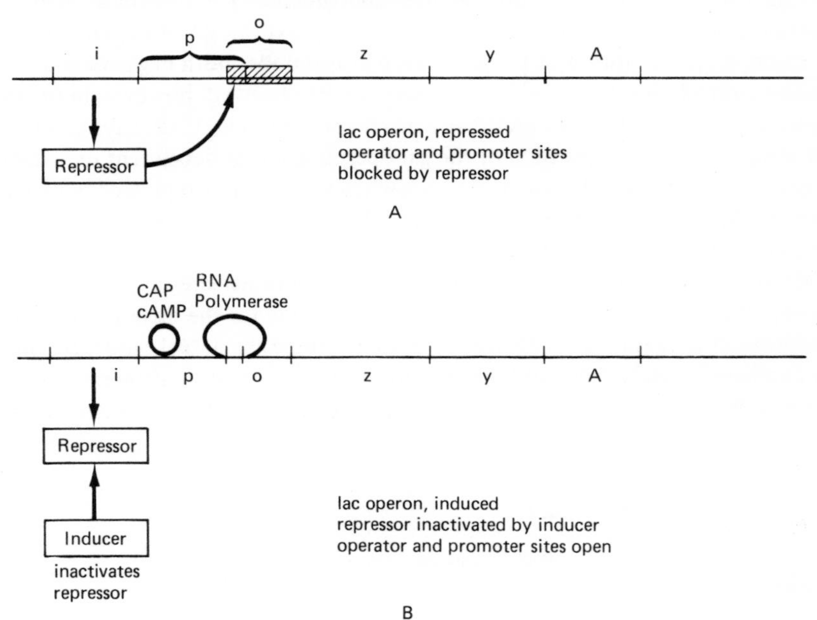

Figure 3.6 Diagram of the regulation of the lac operon by repressor and inducer. p, promoter site; o, operator site; z, y, A, structural genes; i, gene coding for repressor. Scale strongly distorted.

While the repressor exercises negative control, the transcription of the lac operon is also subject to positive control by another regulatory protein, the catabolite gene-activator protein (CAP). CAP, in conjunction with cyclic AMP (cAMP), activates transcription of the operon[32]. The site where this protein interacts with the DNA (nucleotides 1 to 37) is part of the promoter region and is adjacent to the site where RNA polymerase forms the open complex with the DNA. The interaction of CAP with the promoter facilitates the formation of the RNA polymerase-promoter complex. This might be brought about by affecting the recognition site of the RNA polymerase[4] which is at some distance from the actual binding site of the enzyme (p. 86).

If glucose, a more efficient carbon source, is present in the medium, there is no need for the metabolism of lactose. Under these conditions catabolite repression ensues: glucose acts by decreasing the level of cAMP which in turn inactivates CAP. The above mechanism of positive control is blocked and the efficiency of transcription is greatly reduced.

(2) The gal operon

The gal operon contains three structural genes, E, T and K, and a 93-nucleotide-long regulatory region the nucleotide sequence of which has been determined by Musso and coworkers[5]. The three genes code for enzymes which convert galactose into glucose-1-phosphate and uridine-diphosphoglucose. The control of their expression shows basic similarities with that of the expression of the lac operon; for example, transcription of the gal operon is also under both negative and positive control. Some apparent differences can be observed, however, in *in vivo* experiments. First, although cAMP and its receptor protein, CAP, have a definite activating effect, the expression of the gal operon is not as dependent on catabolite activation as is lac expression. Second, when an operon is repressed, a very low basal synthesis of the enzymes still occurs. This basal synthesis is much higher in the case of the gal enzymes than in other operons.

Musso *et al.*[5,33], by identifying the different functional sites in the regulatory region, revealed the molecular mechanisms responsible for the different ways of regulation of gal expression. *Figure 3.7* does not show the actual sequence, only the numbers of nucleotides and the regions of interaction with different regulatory proteins. When the enzymes are induced and CAP and cAMP are present, trans-

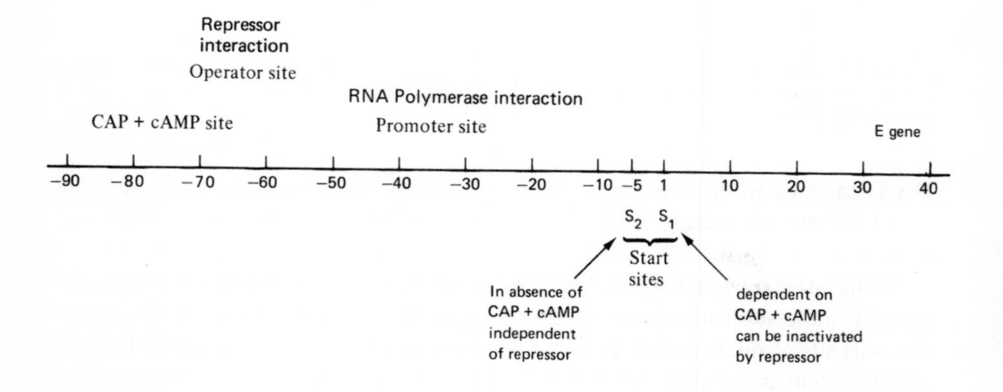

Figure 3.7 Diagram of the regulation of the gal operon. (According to Musso *et al.*[5,33,34].)

cription of the operon starts at nucleotide 1. The whole regulatory region is to the left of this nucleotide (marked by negative numbers). The site of interaction with RNA polymerase is somewhere within the region between nucleotides 1 and –60. Adjacent to this region, to the left, CAP interaction site has been located. Nucleotides –60 to –90 seem to play an essential role in the binding of this protein.

So far, the structure is similar to that of the lac operon. The position of the gal operator is, however, different from operators in many other operons. Musso and coworkers found that the gal repressor interacts with the DNA in an area preceding the transcription startpoint by 60 to 66 nucleotides. This means that, unlike the lac operator, which overlaps with the RNA polymerase binding site, the gal operator overlaps with the CAP binding site. It follows that repression interferes primarily with the catabolite-activating process and does not completely abolish transcription of the operon, as it still allows interaction between RNA polymerase and the promoter site. This effect may be responsible for the relatively high basal enzyme synthesis.

Transcription of the gal operon can also take place, with reduced efficiency, in the absence of CAP and cAMP. The transcripts obtained in the presence and absence of the catabolite activating system were compared[33,34], and it was found that the start-points were different in the two cases. *Figure 3.8* shows part of the

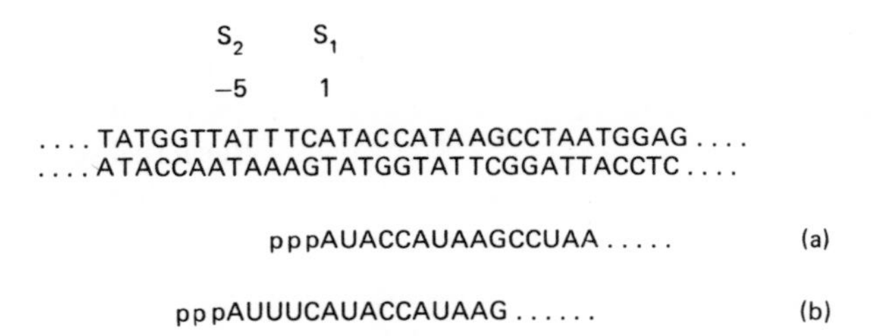

Figure 3.8 Nucleotide sequence of the DNA around the startpoints of transcription of the gal operon and 5′-terminal sequences of the mRNAs transcribed (a) in the presence of CAP and cAMP; (b) in the absence of CAP and cAMP. S_1 and S_2 : start-sites, see text.

DNA sequence around the start-points of transcription and the 5′-terminal sequences of the two transcripts. It can be seen that the two initiation sites are five base pairs apart; in the absence of CAP and cAMP, transcription starts at nucleotide –5. This latter start-site, S_2, is not affected by the repressor, but it is inactivated by CAP and cAMP, and so can function only in their absence. Transcription from the start-site at nucleotide 1, S_1, on the other hand, is activated by the CAP and cAMP system and is inhibited when the repressor binds to the operator. This dual control of transcription may serve the purpose of allowing a basal level of gal enzymes which are required for the synthesis of some components of the cell wall.

POST-TRANSCRIPTIONAL PROCESSING OF RNAs IN PROKARYOTES

The primary product of transcription is not always the mature functional RNA molecule. Transcription is often followed by a maturation process, whereby the primary transcript is modified to produce the final RNA molecule. The nature of these post-transcriptional modifications is different depending on the type of RNA molecule concerned.

(a) mRNA

In prokaryotes, mRNAs are usually not processed. The extreme lability of bacterial mRNAs means that some of these molecules hardly live long enough to exist even as complete primary transcripts; their degradation from the 5′ terminus may start before synthesis of the 3′ terminus has been completed. Bacterial mRNAs of a somewhat longer half-life (a few minutes) do not seem to undergo any post-transcriptional modification, either. Considering that transcription and translation of these RNAs occur simultaneously, i.e. ribosomes attach to and start translating the mRNA when only part of it has been synthesised, and move along the growing polynucleotide chain as more and more of the message is becoming available (*see* Chapter 8), it could not really be expected that any further modification should occur after transcription. The native mRNA is obviously fully functional. The mRNAs of some bacteriophages form an exception to this rule: T7 DNA, for example, is transcribed into larger precursor RNA which is eventually split into smaller-size mature mRNA molecules[35].

(b) Processing of tRNA

The main steps in the maturation of tRNA involve cleavage of the precursor, which is a larger molecule than the mature tRNA, and modification of a number of nucleotides.

All tRNA molecules contain a high percentage of so-called ‘minor bases’, i.e. nucleotides which differ from the four basic building blocks of nucleic acids by carrying different substituents (most often methyl groups) or by possessing a saturated bond or an isomeric form of the glycosidic linkage (*see* Chapter 7). All these modifications are introduced after transcription has been completed. tRNA methylases, enzymes which substitute methyl groups into different positions of the nucleotides in tRNA, have been detected in a variety of prokaryotic and eukaryotic organisms. The mechanism of the more complicated modifications and the enzymes responsible for these reactions are not yet known.

tRNA precursors are about 20% larger than the mature molecule, and contain redundant sequences at both the 5′ and the 3′ end. These are removed by specific nucleases. In prokaryotes specific enzymes, RNase P and RNase Q (or RNase PIII) have been identified which carry out cleavages at the 5′ and the 3′ end, respectively[36]. The complete nucleotide sequences of a $tRNA^{tyr}$ species from *E. coli* and of its precursor are shown in *figure 3.9A*, with the cleavage sites marked on the latter.

The CCA group at the 3′ end is missing from some tRNA precursors, and is replaced by a specific nucleotidyl transferase[37].

The primary transcript of tRNA genes is sometimes a very large precursor molecule which comprises two or more tRNA sequences, separated by spacers. These multimeric precursors are degraded into single copies and trimmed to size by the same enzymes, according to the same cleavage mechanism as for the monomeric precursors (*figure 3.9B*).

(c) Processing of ribosomal RNAs

The genes for three ribosomal RNAs are arranged within one transcription unit in the bacterial genome (*see* Chapter 1). In prokaryotes, processing occurs concomitantly with transcription; the precursor therefore never exists as a complete molecule in the cell. Large precursors could be isolated from mutants deficient in one of the processing enzymes, RNase III. Such studies revealed that the primary transcript, a 30S RNA molecule, contains 16S RNA sequences at the 5′-proximal part, separated by a rather long spacer region from the 23S RNA sequences which is again separated from the 3′-proximal 5S RNA sequences[37a]. Recent studies on the structure of ribosomal RNA genes and on the 30S precursor RNA revealed that tRNA sequences are also present in the spacer region and at the distal end[38,39] (*see* also p. 39). In addition to the three rRNA molecules, the tRNAs are also processed out of the 30S RNA precursor.

Processing of the primary transcript occurs via cleavage by RNase III in the spacer sequences, followed by trimming of the RNA precursors by several enzymes, not all of which have been identified[37a,40]. RNase III is an enzyme specific for splitting double-stranded RNA and this leads to the assumption that strongly base-paired sequences may be present in the redundant regions of the rRNA precursor. Young and Steitz[67] proposed that the 16S RNA sequence may be present in a looped-out structure within the primary transcript. They determined the nucleotide sequences flanking the 16S RNA gene and found that a strongly base-paired stem could be formed by these distant regions, providing a cleavage site for RNase III. By this mechanism a 17S RNA could be produced which had been known before as a precursor of 16S rRNA. This 17S precursor is trimmed at the 5′ and 3′ ends by two different enzymes[68], eventually leading to the production of the mature 16S rRNA. Gegenheimer and Apirion[69], studying the intermediates of rRNA processing in mutants of *E. coli* which were deficient in RNase III and contained a thermolabile RNase P, concluded that the latter enzyme plays an important role in the release of tRNA molecules from the spacer regions.

The final trimming to size of 16S RNA and 23S RNA precursors is preceded by the attachment of ribosomal proteins to the precursor RNAs. Pre-ribosomal particles are thus formed and the final steps of maturation are carried out in these particles[68].

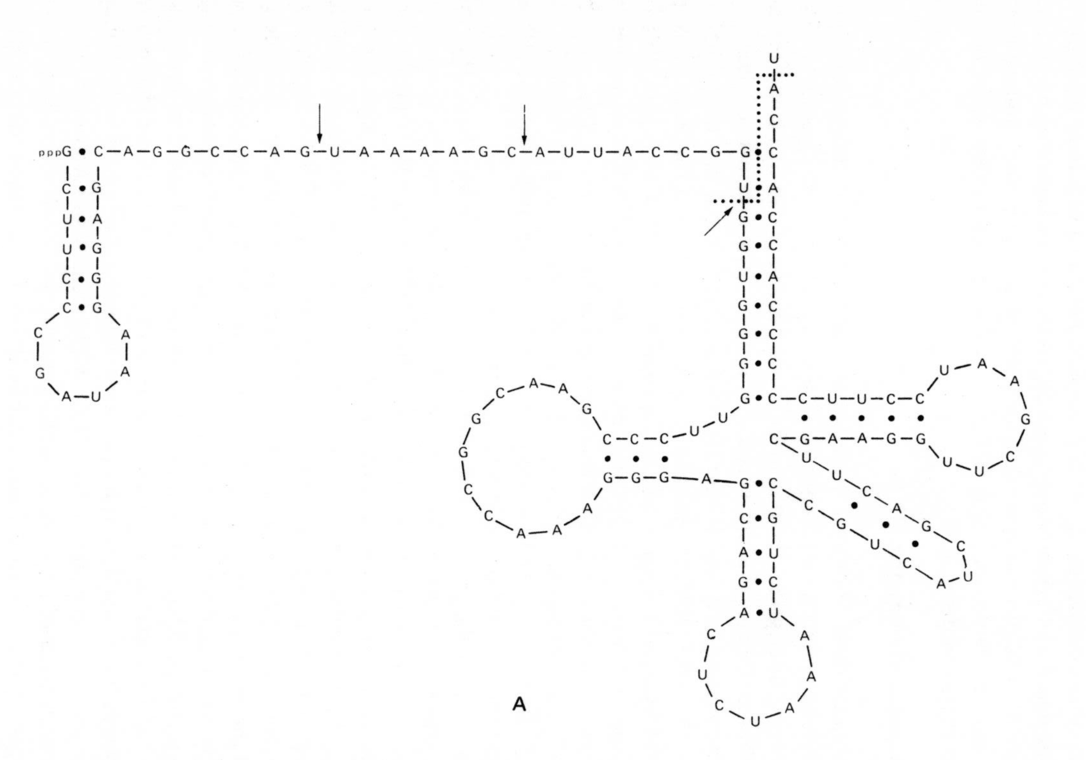

A

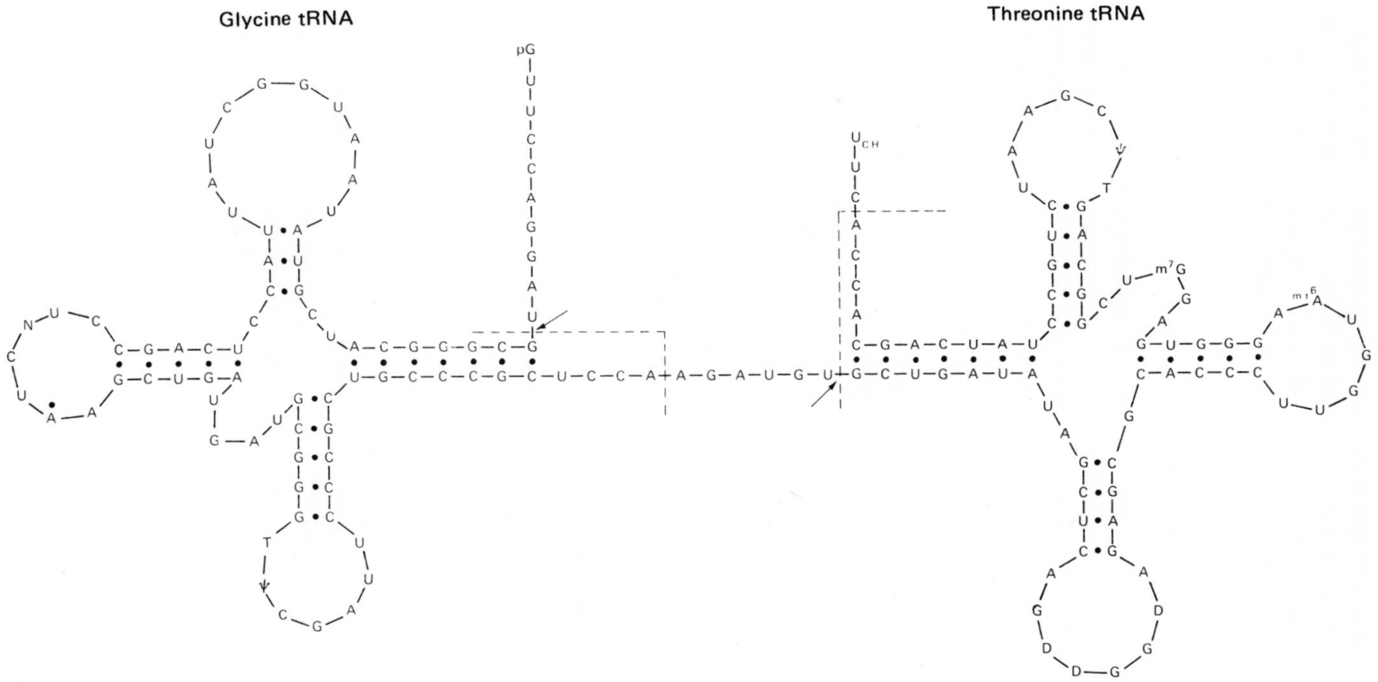

Glycine tRNA

Threonine tRNA

B

Figure 3.9 Nucleotide sequences of tRNA precursors. A: Precursor of a $tRNA^{tyr}$ species (Altman and Smith[66]). The dotted line delineates the mature tRNA sequence. Arrows show the cleavage sites of the RNases involved in maturation. The bases are still unmodified. B: tRNA dimer; common precursor of $tRNA^{gly}$ and $tRNA^{thr}$ (Carbon *et al.*[66]). Cleavage sites and sequence of the mature tRNAs marked as above.

Ribosomal RNA also contains some minor bases, although not in such high amounts as tRNAs. Modification of the bases probably occurs at early stages of maturation and involves rRNA methylases, detected in a variety of organisms. The mechanism of methylation is the same as in the case of tRNA; *S*-adenosylmethionine is used as methyl donor.

TRANSCRIPTION IN EUKARYOTES

Transcription in eukaryotes differs in several respects from transcription in prokaryotic organisms. The enzymes which carry out transcription are different with respect to both their protein structure and their specificity. The structures of promoter sites are also dissimilar in prokaryotes and eukaryotes. The importance of post-transcriptional processing and the complexity of this process are much greater in eukaryotes, especially as far as the synthesis of mRNA molecules is concerned.

The discovery of split genes and spliced mRNAs directed interest to the special problems and specific processes connected with the transcription of interrupted genes and the production of spliced RNAs. Very little is known today about the molecular mechanisms involved in the synthesis of such RNAs; there are many important questions which still await an answer. The importance of these phenomena, however, justifies that in the following we should concentrate on this aspect of eukaryotic transcription. Other differences between eukaryotic and prokaryotic systems, e.g. differences between the RNA polymerases or between the structures of promoters, will be mentioned only briefly.

(1) Eukaryotic RNA polymerases

In a wide variety of eukaryotic organisms, three different classes of DNA-dependent RNA polymerases have been detected in the nucleus. The protein structure of all these enzymes is very complex; they are built of a large number of subunits (10 to 12) of different sizes. The largest subunits, with molecular weights of over 100 000, seem to be unique and characteristic for each class, while some of the small ones may be interchangeable between the different enzymes[41,42]. The function of the three enzymes is different; they carry out transcription of different types of RNAs. DNA-dependent RNA polymerase I is located in the nucleolus, at the site of ribosomal RNA synthesis. It is supposed to transcribe rRNA genes, producing a large transcript which contains the 18S, 5.8S and 28S RNA cistrons. The other two enzymes are present in the nucleoplasm. The function of the different enzyme classes can be studied by making use of their different sensitivity to the inhibitor, α-amanitin. The enzymes have been purified and transcription with exogenous RNA polymerases has also been studied. The difficulty in such investigations originates mainly from the complex controls operating in chromatin: in artificial systems faulty initiations and uncontrolled interactions may occur. Our present knowledge of the action of these enzymes is therefore still limited. RNA polymerase II probably produces the primary transcripts of mRNAs. Although there is no direct proof for this function, there is much supporting evidence. The role of this enzyme in the transcription of histone genes is strongly suggested[70] and transcription of polyoma virus DNA has also been carried out by RNA polymerase II in *in vitro* studies[71]. RNA polymerase III is engaged in the synthesis of small RNA molecules, of tRNAs and of 5S RNA[41,43].

(2) Promoter sites and their recognition by RNA polymerases

Eukaryotic RNA polymerases recognise promoter sites which are different from the prokaryotic promoter sequences shown in *table 3.2* and are differently located in the DNA. Control sites of RNA polymerase II may include far stretches of the flanking sequence, and transcription by RNA polymerase III is controlled both by the flanking sequence and by a DNA region inside the gene. *E. coli* RNA polymerase does not recognise these control sites, but it can transcribe eukaryotic DNA *in vitro* or *in vivo*, if an *E. coli* promoter or a similar nucleotide sequence is present.

Eukaryotic genes can be inserted into phages or plasmids and can be cloned and expressed in prokaryotic cells, and expression of several eukaryotic genes in *E. coli* has been accomplished in this way[72]. Transcription of these genes usually occurs under the control of a prokaryotic promoter. Plasmids can be constructed which contain the control region of a prokaryotic operon and the eukaryotic gene can be fused to this sequence.

Transcription is specifically controlled in eukaryotes also by the proteins of chromatin. Non-histone chromatin (NHC) proteins are supposed to direct RNA polymerase to initiator regions of specific genes, by affecting the structure of DNA or the accessibility of different sites on the DNA. Several attempts have been made to establish the effect of NHC proteins on the *in vitro* transcription of specific genes using reconstituted chromatin with different NHC protein preparations. The results obtained so far are somewhat equivocal, the role of these proteins in the control of transcription is therefore not clear.

(3) The role of post-transcriptional processing

While in prokaryotes most mature mRNAs are produced directly by transcription, and post-transcriptional modifications affect only rRNA and tRNA molecules, in eukaryotes the significance of post-transcriptional processing is much greater. The structure of the primary transcript is usually not identical with that of the mature RNA. The giant precursors of mRNAs undergo a series of extensive modifications in the nucleus, until the RNA structure is produced which can be transported to the cytoplasm and can function in translation. Processing affects the 5′ and 3′ termini and involves a drastic reduction in size (p. 110-113) which includes 'splicing' of mRNAs, i.e. elimination of the intervening sequences and joining of the structural regions.

(4) Spliced mRNAs and the mosaic structure of genes in eukaryotes

Since the earliest studies on the molecular mechanisms of gene expression, our ideas and theories on the transfer of genetic information were based on the concept that genes and gene products are co-linear, i.e. that the nucleotide sequence of any mRNA is an exact complementary copy of the nucleotide sequence of the corresponding gene. Elucidation of the mechanism of transcription in prokaryotes, and comparison of the structures of DNA and RNA confirmed this belief. Earlier studies on eukaryotic systems also seemed to agree with this idea.

New improved methods for the mapping of eukaryotic genomes, for molecular hybridisation studies and for sequencing RNA and DNA allowed more exact comparison of the structures of mRNAs and of the DNA regions coding for them. As discussed in Chapter 1, such studies led to the discovery that in eukaryotes, mRNAs and their genes are in many cases not co-linear. The nucleotide sequences present in some mature mRNA molecules do not correspond to a linear stretch in the DNA but to two or more separate, distant regions in the genome. A major part of the DNA sequence which connects these stretches is absent from the mRNA. These structures have been described as 'split genes' or 'interrupted genes' and 'spliced mRNAs'. Within a very short time spliced structures have been established in a number of mRNAs: in SV40 mRNAs[44,45], adenovirus mRNAs[46], globin mRNA[47], ovalbumin mRNA[48] and immunoglobulin mRNA[49].

From the data available so far, it seems that there are two different types of spliced mRNAs. In adenovirus, SV40 and polyoma virus, most of the mRNAs contain coded messages which are faithfully copied from the corresponding regions in the DNA. However, mRNAs also contain untranslated sequences in addition to the coding sequence (*see* Chapter 5). In the above viruses, the 5'-terminal untranslated sequences of the mRNAs (called leader sequences, p. 162) were found to originate from distant parts of the viral genome. The arrangement of the nucleotide sequences in two late mRNAs of SV40, 19S RNA and 16S RNA, have been investigated by nucleotide sequence analysis[44] and molecular hybridisation techniques[45]. This revealed that both mRNAs have a common leader sequence[50] about 200 nucleotides long, copied from the same site in the viral DNA. This is at a distance of about 200 nucleotides from the beginning of one, and about 1000 nucleotides from the beginning of the other, coded message. Recently, a third late mRNA, 18S RNA, has been found to contain the same leader sequence[50]. Splicing of this RNA also occurs near the junction of the leader sequence and the coding sequence (*figure 3.10; see* also *figure 1.1*).

In adenovirus mRNAs, the arrangement of leader and coding sequences is even more complex. The leader sequence itself is composed of three or more stretches, transcribed from non-contiguous regions of the viral genome[51]. Here, too, the same leader sequence is attached to the coded messages of several mRNAs, as shown in *figure 3.11*. It is rather significant that the same leader sequence is present in different mRNA species. One may speculate whether some structural

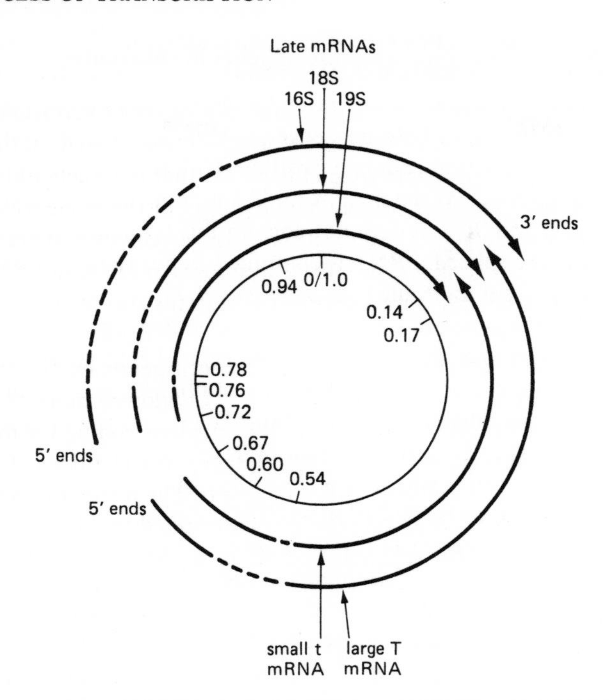

Figure 3.10 Mapping of early and late mRNAs in the SV40 genome. (Data from Hsu and Ford[45], Crawford *et al.*[52] and Fiers *et al.*[50].) Inner circle: DNA genome. Outer circles: Full lines represent sequences present in the mature mRNAs, broken lines are intervening sequences. Protein VP1 is translated from 16S RNA, VP3 from 18S RNA, VP2 from 19S RNA. The nucleotide sequence of 19S RNA comprises the messages for all three proteins. The structure and function of the early mRNAs, coding for small t and large T antigens, are described in the text.

feature of this untranslated sequence is essential for the translation of the viral message.

An interesting interruption occurs in the early mRNAs of SV40 virus[50,52]. Two early mRNA species code for two proteins, the small t antigen and the large T antigen. The molecular weights of the two proteins are very different (about 20 000 daltons for t and 90 000 daltons for T), but both mRNAs span the entire region of the genome from position 0.67 to position 0.14 (*figure 3.10*). The N-terminal parts of the two proteins are identical[73]. This reflects the fact that the 5′-terminal parts of the two mRNAs are identical, they are copied from the same

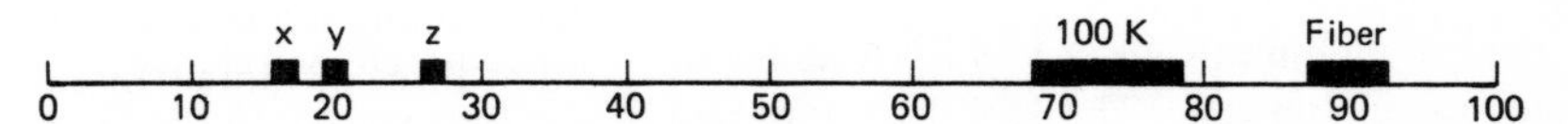

Figure 3.11 Mapping of the coding sequences of two mRNAs (100K and Fiber) and of the three segments of their common leader sequence (x, y, z) in the adenovirus genome. Each mRNA comprises the sequences x, y, z and either the '100K' or the 'Fiber' sequence, but not the sequences which separate these regions in the genome. (Data from Klessig[51].)

stretch of the genome. There is, however, a long interruption in the mRNA for the large T antigen, from map position 0.60 to position 0.54, while the mRNA for small t antigen has only a short interruption between positions 0.55 and 0.54. The start-point for translation of both messengers is identical. Translation of t is terminated at a UAA termination codon at position 0.55, thus producing a protein 174 amino acids long. As the mRNA for T does not contain the sequences between positions 0.60 and 0.54, it does not contain this termination codon. Translation of T from the spliced mRNA is continued to position 0.17, thus producing a much larger protein molecule (*see* also *figure 1.1*).

The interruption in the gene for T antigen is in the coding region and in this respect this structure resembles those of the second type of split genes which are usually found in non-viral chromosomes. In globin, ovalbumin and immunoglobin genes, insertions have been detected mostly within the protein coding regions. Jeffreys and Flavell[47] mapped a segment about 600 base pairs long in the globin gene region of rabbit liver DNA, which interrupts the β-globin coding sequence somewhere around the codons for the 101th–120th amino acids in the globin molecule. A similar insert is present in the same region of the globin gene from other tissues. Leder and coworkers[53] located an insert 550 base pairs long, in the gene for mouse β-globin, immediately after the codon for the 104th amino acid, and detected a smaller intervening sequence close to the 5′ end of the coding region (*see figure 1.14*). Dugaiczyk *et al.*[74] mapped seven interruptions in the chicken ovalbumin gene; six of these intervening sequences are within the coding region[75] (*see figure 1.15*).

The full implications of these findings are not yet clear. As far as synthesis of such spliced mRNAs is concerned, our present knowledge is still limited. Recent results with RNA precursors suggest that the primary transcript is a complete copy of the DNA, comprising also the 'silent' sequences, and that excision of redundant stretches occurs during the post-transcriptional processing of these RNAs[76,77,78,79]. As calculated from the size of the inserts, these primary transcripts should be giant molecules. The primary transcript of the globin gene should be twice the size of the mature mRNA, that of the ovalbumin gene more than three times as long as the ovalbumin mRNA. This agrees quite well with the sizes of the giant mRNA precursors found in the nucleus.

A 27S nuclear RNA, precursor of immunoglobulin light chain mRNA, was shown to contain both coding and intervening sequences[76]. There is also evidence that the nuclear RNA precursor of ovalbumin mRNA contains sequences which are deleted from the mature mRNA[77]. In the case of mouse β-globin mRNA, Tilghman and coworkers obtained direct proof that the mRNA precursor contains all the redundant sequences which are present in the DNA[78]. They hybridised a precursor of globin mRNA, a 15S RNA molecule, to a mouse DNA fragment carrying the globin gene. The 15S RNA annealed to one long stretch in the DNA (*figure 3.12*), thus proving that the RNA and DNA sequences were co-linear. (For comparison, see *figure 1.14*, the result of a similar experiment, carried out with

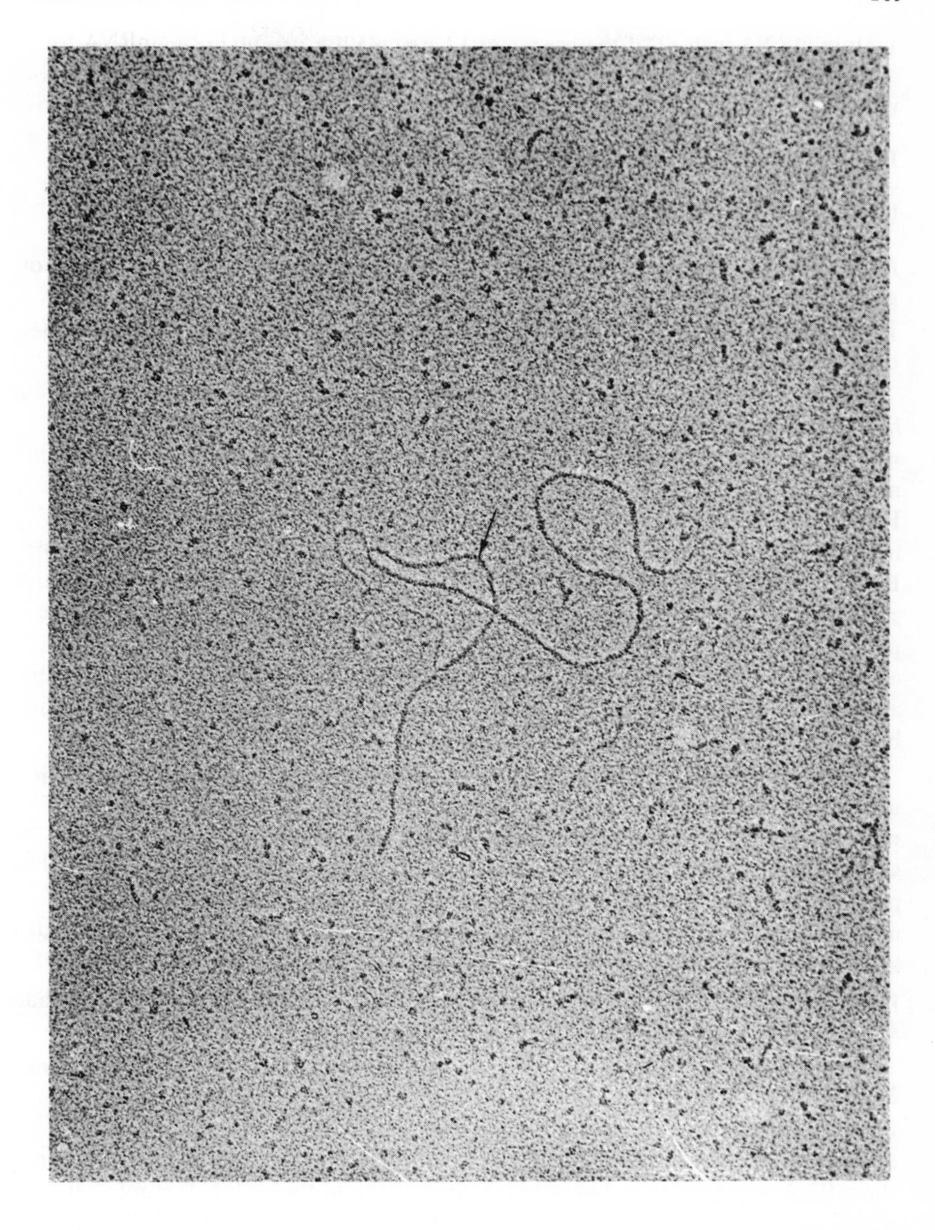

Figure 3.12 Hybridisation of a 15S RNA precursor of globin mRNA to a segment of DNA containing the globin gene. The same technique was used to visualise molecular hybrids as in *figure 1.14*. The precursor RNA forms only one long R-loop (marked by arrow); it anneals to a continuous stretch in the DNA. This proves that the complete sequence of the globin gene, structural sequences as well as interrupting sequences, has been transcribed into the precursor. (By courtesy of Dr P. Leder.)

mature mRNA instead of the precursor.) Large mRNA precursors containing redundant sequences have also been detected in the nuclear RNA fraction of adenovirus[79].

POST-TRANSCRIPTIONAL PROCESSING OF RNAs IN EUKARYOTES

(a) Processing of mRNA

mRNA synthesis takes place in the nucleoplasm where a heterogeneous mixture of giant RNA molecules accumulates; this fraction is called heterogeneous nuclear RNA (Hn RNA). It contains the precursors of different mRNAs at different stages of post-transcriptional processing[54].

The first evidence that Hn RNA contains precursors of cytoplasmic mRNAs has been obtained by identifying nucleotide sequences in the Hn RNA fraction which were present also in the mature mRNA. cDNA, prepared from β-globin mRNA with the aid of reverse transcriptase (*see* Chapter 2), was hybridised to Hn RNA from erythroid cells. The extent of hybridisation provided a strong indication for the presence of an RNA species which contained globin sequences[55]. By using more refined techniques, Curtis and Weissmann detected a 15S RNA species, which has been identified as a precursor of globin mRNA[56]. The precursor–product relationship has also been confirmed by Perry *et al.*[57], who calculated the extent of homology between Hn RNA and mRNA from mouse cells and found that a high proportion of the large molecular fraction of Hn RNA was built of sequences homologous to cytoplasmic RNA.

Precursors of several mRNAs have been recently isolated and found to contain intervening sequences (see p. 108). An important step in post-transcriptional processing is the elimination of these sequences by splicing. The intervening sequences are cut out and the structural sequences joined. This is a late event in processing, it occurs after modification of the 5′ and 3′ termini[80]. The exact mechanism of splicing and the enzymes which take part in this process are not yet known. A model for post-transcriptional processing of adenovirus mRNAs was proposed by Klessig[51]. *Figure 3.13* shows how different mRNAs with the same leader sequence can be produced according to this mechanism.

Processing of mRNAs also includes modifications at both the 5′ and 3′ termini. Modification at the 3′ terminus involves the addition of A residues by a polymerising enzyme, to form a long poly(A) tail[58]. Studies on the processing of adenovirus mRNAs revealed that RNA polymerase may not stop at the site where the mRNA sequence ends, but may transcribe beyond this point. Trimming at the 3′ terminus must therefore precede the synthesis of the poly(A) tail[80]. Polyadenylated molecules have been detected in both Hn RNA and mature mRNA. The length of the poly(A) tail is rather difficult to assess as it seems to decrease during the lifetime of the mRNA. About 150 to 200 nucleotides can probably be accepted as an average length.

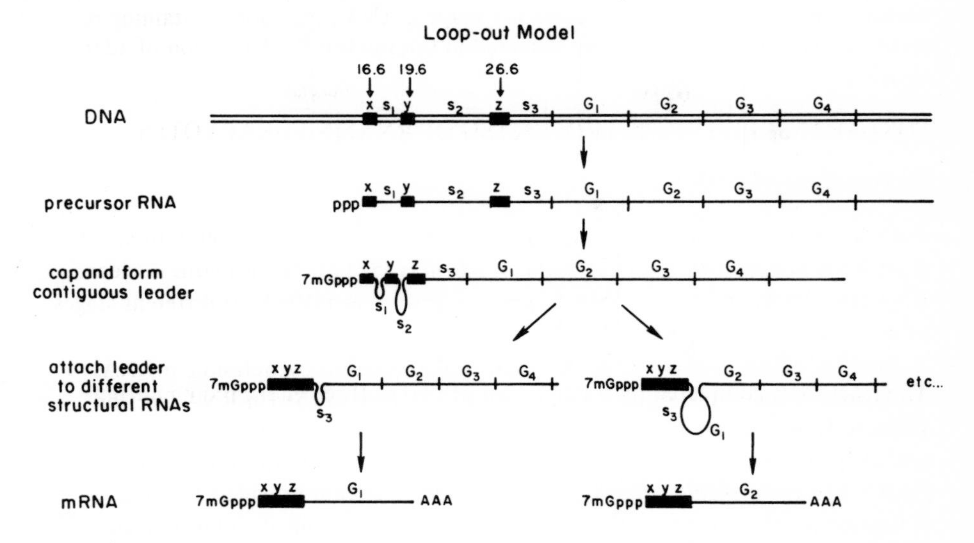

Figure 3.13 Model for the mechanism of transcription of split genes. Model of Klessig[51], constructed to explain the production of spliced mRNAs in adenovirus. x, y, z represent the three segments which comprise the common leader sequence; G1, G2, G3, G4 represent different coding regions. The complete DNA sequence is copied into the primary transcript. The redundant sequences are eliminated and the different segments joined during post-transcriptional processing of the mRNAs. (By courtesy of Dr D. F. Klessig.)

The modification at the 5′ end was discovered only a few years ago[59,60]. A very specific and rather unusual structure was first detected in some viral mRNAs and was later found to exist in most but not all eukaryotic mRNAs. Although there is some variation in the structure from one mRNA to the other, the common features of these 5′-terminal structures, called caps, are a 7-methylated G in the terminal position, attached in 5′, 5′-triphosphate linkage to the next nucleotide, which is methylated in the ribose moiety.

Two possible mechanisms have been proposed for the synthesis of these capped structures[60], based on evidence obtained with viral mRNAs, but it is not yet clear whether either or both of these pathways are operating in the capping of cellular mRNAs. According to the first mechanism, cap synthesis occurs soon after initiation of transcription, beginning with the removal of one phosphate group from the triphosphorylated 5′ end:

$$\text{pppG-P-}N_1\text{-P-}N_2\text{-P-}N_3\text{-P}\cdots \longrightarrow \text{ppG-P-}N_1\text{-P-}N_2\text{-P-}N_3\text{-P}\cdots + P_i$$

followed by reaction with GTP which leads to the formation of the 5′, 5′-triphosphate linkage, with release of inorganic pyrophosphate:

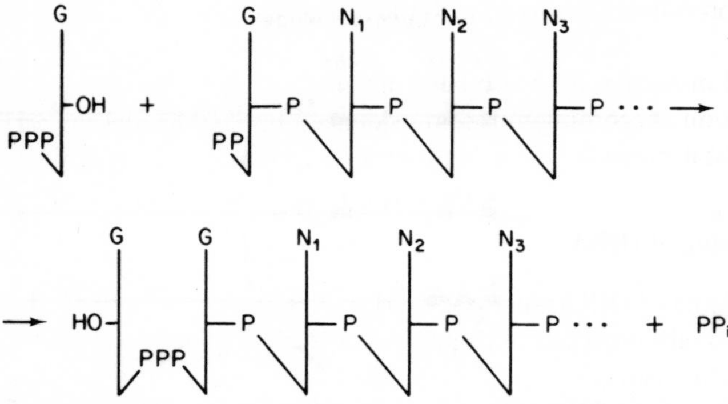

The reaction is completed by methylation of two or three end-standing nucleotides, to form

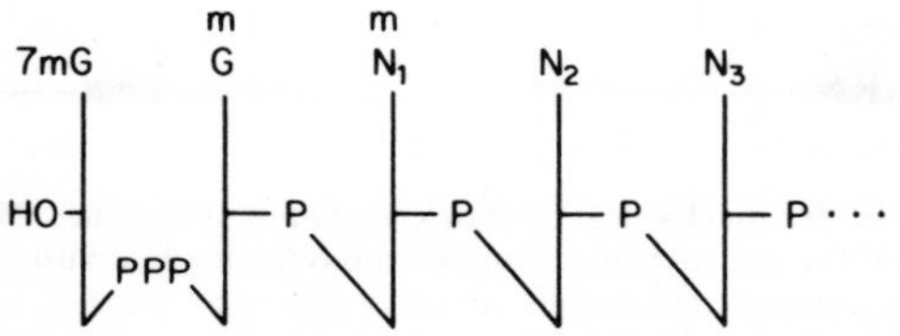

According to the second mechanism, capping might be a later event, preceded by cleavage of the precursor molecule to produce a new 5′ end which contains a 5′-monophosphate group

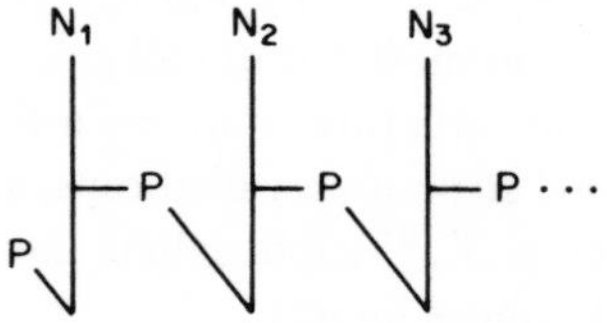

This molecule reacts with GTP to form the 5′, 5′-triphosphate bridge, with release of inorganic phosphate:

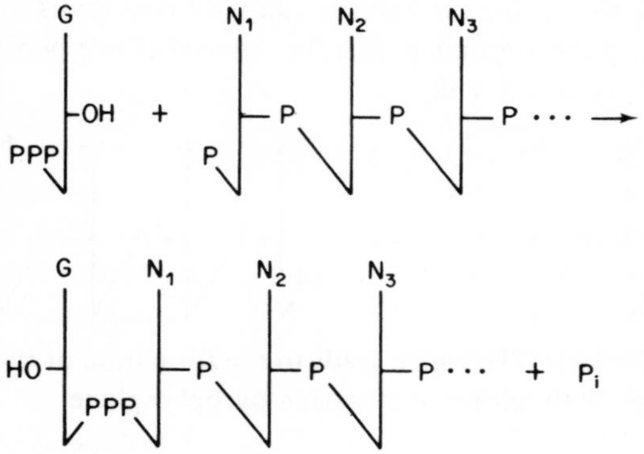

Methylation follows as above, with the aid of *S*-adenosylmethionine as methyl donor.

Capped molecules have also been detected in large precursor RNAs in the Hn RNA fraction. According to recent evidence, capping may be the earliest event in maturation; it may take place before transcription has been terminated.

(b) Processing of tRNA

The processing of tRNAs in eukaryotes occurs along similar lines to that in prokaryotes, although less is known about the individual steps and enzymes involved. The main reactions are cleavage of the precursor to a smaller size and modification of the bases. Dimeric and multimeric precursors are not found in eukaryotes, probably because the individual tRNA genes are separated by very long spacer sequences.

Redundant sequences may be present in the primary transcripts, originating from inserts in the tRNA genes. Goodman *et al.* and Valenzuela *et al.* detected short (14–19 bp) interruptions in tRNA genes in yeast[61]. The inserts were in the middle of the gene, near the anticodon site. Knapp *et al.* isolated precursors of $tRNA^{tyr}$ and $tRNA^{phe}$ from a temperature-sensitive mutant of yeast and found that the precursor molecules contained these intervening sequences[62]. Using these precursors as substrates, they could also show that wild-type yeast contains an enzyme activity which specifically removes the redundant sequences and produces mature tRNAs.

(c) Processing of rRNA

In eukaryotes, rRNA genes are highly redundant. In different organisms they are present in from 50 to several thousand copies. They are arranged in clusters, one repeat unit comprising 18S RNA, 5.8S RNA and 28S RNA sequences, separated from each other and from the next transcription unit by spacer regions. Each repeat unit is transcribed into one long RNA transcript.

Interruptions within rRNA genes have also been observed: very long spacer sequences are present in the middle of rRNA genes in the *Drosophila* chromosome[63]. We do not yet know how these split rRNA genes are transcribed and processed.

Transcription of non-interrupted rRNA transcription units produces large precursor molecules, with a sedimentation coefficient of 45S in mammals, and 36–38S in lower eukaryotes. Both transcription and processing of the 45S RNA precursor occur within the nucleolus. Processing of this molecule consists of four cleavages, as shown in *figure 3.14.* First, a 5′-terminal redundant stretch is removed, followed by a cleavage within the spacer region separating the two rRNAs. Further trimming produces the mature 18S RNA, while the part containing the 28S RNA undergoes more complex processing. A 140-nucleotide-long (5.8S) stretch, present in the spacer region, anneals to a complementary part of

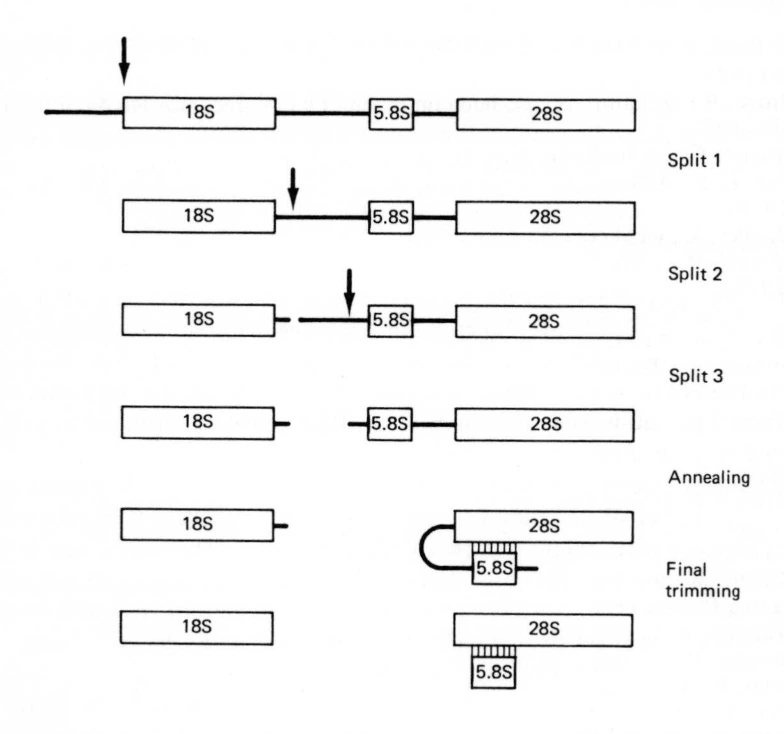

Figure 3.14 Processing of 45S ribosomal RNA precursor in eukaryotes. Diagrammatic presentation, scale distorted.

the rRNA molecule and the redundant sequences are removed so as to leave this small RNA attached to the mature 28S RNA molecule. Not much is known about the nucleases involved; RNase III-like activities have been detected in some eukaryotic organisms. In eukaryotes, as in prokaryotes, the precursor is not processed in the form of a free RNA molecule, but with ribosomal proteins already attached to it. Here, too, maturation takes place in pre-ribosomal particles.

REFERENCES

1 Pribnow, D. *PNAS*, **72**, 784 (1975).
2 Sanger, F. *et al. Nature*, **265**, 687 (1977); *J. Mol. Biol.*, **125**, 225 (1978).
Godson, G. N., Barrell, B. G., Staden, R. and Fiddes, J. C. *Nature*, **276**, 236 (1978).
3 Bennett, G. N., Schweingruber, M. E., Brown, K. D., Squires, C. and Yanofsky, C. *J. Mol. Biol.*, **121**, 113 (1978).
Brown, K. D., Bennett, G. N., Lee, F., Schweingruber, M. E. and Yanofsky, C. *J. Mol. Biol.*, **121**, 153 (1978).
Bennett, G. N. and Yanofsky, C. *J. Mol. Biol.*, **121**, 179 (1978).
3a Post, L. E., Arfsten, A. E., Reusser, F. and Nomura, M. *Cell*, **15**, 215 (1978).

4 Dickson, R. C., Abelson, J., Barnes, W. M. and Reznikoff, W. S. *Science,* **187**, 27 (1975).
5 Musso, R.; Di Lauro, R., Rosenberg, M. and de Crombrugghe, B. *PNAS,* **74**, 106 (1977).
6 Bruenn, J. and Hollingsworth, H. *PNAS,* **70**, 3693 (1973).
Post, L. E., Arfsten, A. E., Nomura, M. and Jaskunas, S. R. *Cell,* **15**, 231 (1978).
7 Mueller, K., Oebbecke, C. and Förster, G. *Cell,* **10**, 121 (1977).
Sümegi, M., Udvardy, A. and Venetianer, P. *Mol. Gen. Genet.,* **151**, 305 (1977).
8 Subramanian, K. N., Ghosh, P. K., Dhar, R., Thimmappaya, B., Zain, S. B., Pan, J. and Weissman, S. M. *Progr. Nucl. Acid Res. Mol. Biol.,* **19**, 157 (1976).
9 Fukuda, R., Ishihama, A., Saitoh, T. and Taketo, M. *Mol. Gen. Genet.,* **154**, 135 (1977).
9a Wiggs, J. L., Bush, J. W. and Chamberlin, M. J. *Cell,* **16**, 97 (1979).
10 Burgess, R. R. *J. Biol. Chem.,* **244**, 6168 (1969).
11 Hinkle, D. C. and Chamberlin, M. *J. Mol. Biol.,* **70**, 157 (1972).
Mangel, W. F. and Chamberlin, M. J. *J. Biol. Chem.,* **249**, 2995 (1974).
Chamberlin, M. in *RNA Polymerase* (eds Losick, R. and Chamberlin, M.) Cold Spring Harbor Press, New York, 159 (1976).
12 Gilbert, W. in *RNA Polymerase* (eds Losick, R. and Chamberlin, M.) Cold Spring Harbor Press, New York, 193 (1976).
13 Williams, R. C. and Chamberlin, M. J. *PNAS,* **74**, 3740 (1977).
14 Burgess, R. R. *Ann. Rev. Biochem.,* **40**, 711 (1971).
15 Dhar, R., Weissman, S. M., Zain, B. S., Pan, J. and Lewis, A. M. Jr *Nucl. Acids Res.,* **1**, 595 (1974).
16 Saucier, J. M. and Wang, J. *Nature NB.* **239**. 167 (1972).
17 Siebenlist, U., *Nature*, **279**, 651 (1979).
18 Travers, A. A. and Burgess, R. R. *Nature,* **222**, 537 (1969).
19 Roberts, J. W. *Nature,* **224**, 1168 (1969).
20 Roberts, J. W. *PNAS,* **72**, 3300 (1975).
21 Rosenberg, M., de Crombrugghe, B. and Musso, R. *PNAS,* **73**, 717 (1976).
Sugimoto, K., Sugisaki, T., Okamoto, T. and Takanami, M. *J. Mol. Biol.,* **111**, 487 (1977).
22 Schwarz, E., Scherer, G., Hobom, G. and Kössel, H. *Nature,* **272**, 410 (1978).
Rosenberg, M., Court, D., Shimatake, H., Brady, C. and Wulff, D. L. *Nature,* **272**, 414 (1978).
23 Lee, F. and Yanofsky, C. *PNAS,* **74**, 4365 (1977).
24 McMahon, J. E. and Tinoco, J. Jr *Nature,* **271**, 275 (1978).
25 Küpper, H., Sekiya, T., Rosenberg, M., Egan, J. and Landy, A. *Nature,* **272**, 423 (1978).
26 Bertrand, K., Korn, L., Lee, F., Platt, T., Squires, C. L., Squires, C. and Yanofsky, C. *Science,* **189**, 22 (1975).
Squires, C., Lee, F., Bertrand, K., Squires, C. L., Bronson, M. J. and Yanofsky, C. *J. Mol. Biol.,* **103**, 351 (1976).
Lee, F., Bertrand, K., Bennett, G. and Yanofsky, C. *J. Mol. Biol.,* **121**, 193 (1978).
27 Zurawski, G., Brown, K., Killingly, D. and Yanofsky, C. *PNAS,* **75**, 4271 (1978).
Di Nocera, P. P., Blasi, F., Di Lauro, R., Frunzio, R. and Bruni, C. B. *PNAS,* **75**, 4276 (1978).
Barnes, W. M. *PNAS,* **75**, 4281 (1978).

28 Jacob, F. and Monod, J. *J. Mol. Biol.,* **3**, 318 (1961).
29 Gilbert, W. and Maxam, A. *PNAS,* **70**, 3581 (1973).
30 Bahl, C. P., Wu, R., Stawinsky, J. and Narang, S. A. *PNAS,* **74**, 966 (1977).
Bahl, C. P. and Wu, R. *Gene,* **3**, 123 (1978).
31 Maizels, N. M. *PNAS,* **70**, 3585 (1973).
32 de Crombrugghe, B., Chen, B., Anderson, W., Nissley, P., Gottesman, M., Pastan, I. and Perlman, R. *Nature NB,* **231**, 139 (1971).
33 Musso, R. E., Di Lauro, R., Adhya, S. and de Crombrugghe, B. *Cell,* **12**, 847 (1977).
34 Musso, R. E., de Crombrugghe, B., Pastan, I., Sklar, J., Yot, P. and Weissman, S. *PNAS,* **71**, 4940 (1974).
35 Dunn, J. J. and Studier, F. W. *Brookhaven Symp. Biol.,* **26**, 267 (1975).
Rosenberg, M., Kramer, R. A. and Steitz, J. A. *Brookhaven Symp. Biol.,* **26**, 277 (1975).
36 Perry, R. P. *Ann. Rev. Biochem.,* **45**, 605 (1976).
37 Deutscher, M. P. *Progr. Nucl. Acid Res. Mol. Biol.,* **13**, 51 (1973).
37aDunn, J. J. and Studier, F. W. *PNAS,* **70**, 3296 (1973).
38 Nomura, M. *Cell,* **9**, 633 (1977).
Morgan, E. A., Ikemura, T., Lindahl, L., Fallon, A. M. and Nomura, M. *Cell,* **13**, 335 (1978).
39 Lund, E. and Dahlberg, J. E. *Cell,* **11**, 247 (1977).
40 Ginsburg, D. and Steitz, J. A. *J. Biol. Chem.,* **250**, 5647 (1975).
Westphal, H. and Crouch, R. J. *PNAS,* **72**, 3077 (1975).
41 Roeder, R. G. in *RNA Polymerase* (eds Losick, R. and Chamberlin, M.) Cold Spring Harbor Press, New York, 285 (1976).
42 Sklar, V. E. F., Schwartz, L. B. and Roeder, R. G. *PNAS,* **72**, 348 (1975).
Valenzuela, P., Hager, G. L., Weinberg, F. and Rutter, W. J. *PNAS,* **73**, 1024 (1976).
43 Weinmann, R. and Roeder, R. G. *PNAS,* **71**, 1790 (1974).
Sklar, V. E. F. and Roeder, R. G. *Cell,* **10**, 405 (1977).
44 Celma, M. L., Dhar, R., Pan, J. and Weissman, S. M. *Nucl. Acids Res.,* **4**, 2549 (1977).
Ghosh, P. K., Reddy, V. B., Swinscoe, J., Choudary, P. V., Lebowitz, P. and Weissman, S. M. *J. Biol. Chem.,* **253**, 3643 (1978).
Bina-Stein, M., Thoren, M., Salzman, N. and Thompson, J. A. *PNAS,* **76**, 731 (1979).
45 Hsu, M. T. and Ford, J. *PNAS,* **74**, 4982 (1977).
Lavi, S. and Groner, Y. *PNAS,* **74**, 5323 (1977).
46 Berget, S. M., Moore, C. and Sharp, P. A. *Cell,* **12**, 37 (1977).
Dunn, A. R. and Hassell, J. A. *Cell,* **12**, 23 (1977).
Lewis, J. B., Anderson, C. W. and Atkins, J. F. *PNAS,* **74**, 37 (1977).
47 Jeffreys, A. J. and Flavell, R. A. *Cell,* **12**, 1097 (1977).
48 Breathnach, R., Mandel, J. L. and Chambon, P. *Nature,* **270**, 314 (1977).
Doel, M. T., Houghton, M., Cook, E. A. and Carey, N. H. *Nucl. Acids Res.,* **4**, 3701 (1977).
Weinstock, R., Sweet, R., Weiss, M., Cedar, H. and Axel, R. *PNAS,* **75**, 1299 (1978).
49 Brack, C. and Tonegawa, S. *PNAS,* **74**, 5652 (1977).
50 Fiers, W. *et al. Nature,* **273**, 113 (1978).
51 Chow, L. T., Gelinas, R. E., Broker, T. R. and Roberts, R. J. *Cell,* **12**, 1 (1977).
Klessig, D. F. *Cell,* **12**, 9 (1977).
Chow, L. T. and Broker, T. R. *Cell,* **15**, 497 (1978).

52 Crawford, L. V., Cole, C. N., Smith, A. E., Paucha, E., Tegtmeyer, P., Rundell, K. and Berg, P. *PNAS,* **75**, 117 (1978).
Berk, A. J. and Sharp, P. A. *PNAS,* **75**, 1274 (1978).
53 Leder, P. *et al. Cold Spring Harbor Symp.,* **42**, 915 (1978).
Tilghman, S. M. *et al. PNAS,* **75**, 725 (1978).
54 Lewin, B. *Cell,* **4**, 11 (1975).
55 Imaizumi, T., Diggelmann, H. and Scherrer, K. *PNAS,* **70**, 1122 (1973).
Spohr, G., Imaizumi, T. and Scherrer, K. *PNAS,* **71**, 5009 (1974).
Macnaughton, M., Freeman, K. B. and Bishop, J. O. *Cell,* **1**, 117 (1974).
56 Curtiss, P. J. and Weissmann, C. *J. Mol. Biol.,* **106**, 1061 (1976).
57 Perry, R. P., Bard, E., Hames, B. D., Kelley, D. E. and Schibler, U. *Progr. Nucl. Acid Res. Mol. Biol.,* **19**, 275 (1976).
58 Spohr, G., Dettori, G. and Manzari, V. *Cell,* **8**, 506 (1976).
59 Rottman, F., Shatkin, A. J. and Perry, R. P. *Cell,* **3**, 197 (1974).
60 Shatkin, A. J. *Cell,* **9**, 645 (1976).
61 Valenzuela, P., Venegas, A., Weinberg, F., Bishop, R. and Rutter, W. J. *PNAS,* **75**, 190 (1978).
Goodman, H. M., Olson, M. V. and Hall, B. D. *PNAS,* **74**, 5453 (1977).
62 Knapp, G., Beckmann, J. S., Johnson, P. F., Fuhrman, S. A. and Abelson, J. *Cell,* **14**, 221 (1978).
63 Glover, D. M. and Hogness, D. S. *Cell,* **10**, 167 (1977).
64 Chamberlin, M. J. *Ann. Rev. Biochem.,* **43**, 721 (1974).
65 Miller, O. L. Jr, Hamkalo, B. A. and Thomas, C. A. Jr *Science,* **169**, 392 (1970).
66 Altman, S. and Smith, J. D. *Nature NB,* **233**, 35 (1971).
Carbon, J., Chang, S. and Kirk, L. L. *Brookhaven Symp. Biol.,* **26**, 6 (1974).
67 Young, R. A. and Steitz, J. A. *PNAS,* **75**, 3593 (1978).
68 Dahlberg, A. E. *et al. PNAS,* **75**, 3598 (1978).
69 Gegenheimer, P. and Apirion, D. *Cell,* **15**, 527 (1978).
70 Levy, S., Childs, G. and Kedes, L. *Cell,* **15**, 151 (1978).
71 Lescure, B., Chestier, A. and Yaniv, M. *J. Mol. Biol.,* **124**, 73 (1978).
72 Struhl, K. and David, R. W. *PNAS,* **74**, 5255 (1977).
Dickson, R. C. and Markin, J. S. *Cell,* **15**, 123 (1978).
Mercereau-Puijalon, O., Royal, A., Cami, B., Garapin, A., Krust, A., Gannon, F. and Kourilsky, P. *Nature,* **275**, 505 (1978).
Goeddel, D. V. *et al. PNAS,* **76**, 106 (1979).
73 Paucha, E., Mellor, A., Harvey, R., Smith, A. E., Hewick, R. M. and Waterfield, M. D. *PNAS,* **75**, 2165 (1978).
74 Dugaiczyk, A., Woo, S. L. C., Lai, E. C., Mace, M. L. Jr, McReynolds, L. and O'Malley, B. W. *Nature,* **274**, 328 (1978).
75 Mandel, J. L., Breathnach, R., Gerlinger, P., LeMeur, M., Gannon, F. and Chambon, P. *Cell,* **14**, 641 (1978).
76 Rabbits, T. H. *Nature,* **275**, 291 (1978).
77 Roop, D. R., Nordstrom, J. N., Tsai, S. Y., Tsai, M. J. and O'Malley, B. W. *Cell,* **15**, 671 (1978).
78 Tilghman, S. M., Curtis, P. J., Tiemeyer, D. C., Leder, P. and Weissmann, C. *PNAS,* **75**, 1309 (1978).
79 Goldenberg, C. J. and Raskas, H. J. *Cell,* **16**, 131 (1979).
80 Ziff, E. B. and Evans, R. M. *Cell,* **15**, 1463 (1978).
Nevins, J. R. and Darnell, J. E. *Cell,* **15**, 1477 (1978).

4 Nucleic Acid Sequence Determination

RELATIONSHIP BETWEEN STRUCTURE AND FUNCTION OF NUCLEIC ACIDS

It is often emphasised how closely the structure and function of biological materials are related to each other. This is certainly true for all biological macromolecules, but to some extent it is also true for any chemical substance: whether it takes part in a simple chemical reaction or in a complex biological process, the function of the molecule is determined by its structure. In the case of nucleic acids, however, the relationship of structure and function is unique: they are two aspects of the same basic property of the molecule; structure and function are, in fact, identical.

The function of nucleic acids is the conservation and transfer of genetic information. The genetic information itself is encoded in their primary structure, it is represented by the nucleotide sequence of the molecule. Should we thus consider the nucleotide sequence as part of the structure or as part of the function? The same applies to at least some aspects of the secondary structure. The double-helical model gives as much information on the structure of DNA as on the mechanism by which genetically determined characteristics are passed on from one generation to the next. It follows that structure and function of nucleic acids cannot be discussed separately. The methods applied in structural and functional studies are different and will be described in separate chapters. But the interpretation of the results cannot be divided, it will cover both aspects.

We are primarily interested here in the structural features of messenger RNA which reveal how the genetic code operates and how control signals are recognised. But the methods applied for the investigation of the structure of mRNA are also used for studying other RNAs and are closely related, in some cases even identical, to the methods used for studying DNA. In this chapter we will therefore discuss the techniques for nucleic acid sequence determination, covering the whole field of all the different nucleic acids.

METHODS OF NUCLEOTIDE SEQUENCE DETERMINATION IN RNA AND DNA

The number of nucleic acid sequences known today is well over a hundred and is increasing fast all the time. Quite a few RNA molecules have been completely sequenced, mainly the smaller ones like tRNAs, 5S RNAs and 5.8S RNAs, but the primary structures of a number of large messenger RNAs have also been established. The nucleotide sequences of rabbit and human β-globin mRNAs, rabbit α-globin mRNA, chicken ovalbumin mRNA as well as the mRNA of immunoglobulin light chain and the RNA of MS2 virus are all known. The nucleotide sequence of 16S ribosomal RNA of *E. coli* has been determined and the sequences of long stretches of 23S ribosomal RNA are also known. In other RNAs some regions of special functional interest have been sequenced.

In the DNA field, nucleotide sequence studies started later, but results caught up rapidly with RNA sequencing. The use of restriction enzymes to produce well defined DNA fragments and the introduction of the modern direct readout sequencing techniques opened new ways for DNA sequence determination. In the last three years complete nucleotide sequences of some viral genomes have been established: the primary structures of the complete DNAs of coliphages ΦX174 and G4 and of the eukaryotic viruses, SV40 and polyoma, have been determined.

Progress has been exceedingly fast in this field. It is remarkable to remember that this line of research started only 15 years ago, when Holley and coworkers published a short paper presenting the complete nucleotide sequence of yeast $tRNA^{ala}$ (ref. 1). This was the first nucleic acid for which the exact structure had been determined. It was the result of several years' work and it was neither the only nor the first attempt to determine nucleotide sequences. Protein sequencing was a widely used technique at that time and the rapidly growing interest in nucleic acids made the search for a similar approach to nucleic acid structure an obvious aim of modern research. The practical difficulties were, however, much greater than those in studies on proteins. Because of the great variability of amino acid sequences, even a short stretch in a polypeptide chain may be characteristic of the molecule; the sequence of three or four amino acids may be unique. In long polynucleotide chains, built of only four nucleotides, it is not easy to pick out characteristic stretches. The same oligonucleotide sequences may occur many times and this makes it difficult to locate them in the molecule.

Considering these difficulties, tRNA was the obvious candidate for the first sequence studies. Transfer RNAs are among the smallest RNA molecules, they consist of only 75 to 90 bases and contain also a number of substituted nucleotides, 'minor bases'. These can act as structural markers, yielding fragments of a unique sequence which can be easily recognised and located in the molecule. In Holley's work the presence of inosine and thymine in $tRNA^{ala}$ helped to sort out which fragment belonged to which part of the molecule. The exact technical details of Holley's method are of less importance today, as new approaches, more widely applicable, have been worked out since and these are being used now for studying the structure of large RNA molecules and DNA.

All modern sequencing techniques use ^{32}P-labelled nucleic acids. This is a great advantage because it allows a complete analysis to be carried out on microgramme quantities of the preparation. It is often impossible to isolate some RNA or DNA species in larger quantities; the minute amounts of the nucleic acid fragments which have to be analysed could not be detected in any chemical or physical assay. If, however, the specific radioactivity is high enough, even trace amounts of nucleotides can be detected by autoradiography.

Fingerprinting technique

The first breakthrough in RNA sequence determination came with the introduction of Sanger's 'fingerprinting' technique[2]. This is best applicable to small RNA molecules. The nucleotide sequence is determined in four steps:

(1) Degradation of RNA by specific nucleases.
(2) Separation of the degradation products.
(3) Analysis of the structure of each degradation product.
(4) Location of each oligonucleotide in the complete RNA molecule.

In step (1), two endonucleases are used (these are enzymes which split the polynucleotide chain at internal sites): either RNase T_1, which is specific for splitting after G residues, or pancreatic RNase, which splits specifically after pyrimidine residues. The specificity of the digesting enzyme already provides some information on the structure of the digestion products: in an RNase T_1 digest each oligonucleotide has a G residue at the 3′ end, while in a pancreatic RNase digest each oligonucleotide has a 3′-terminal U or C.

In step (2), Sanger introduced new techniques for the separation of all digestion products in one operation: two-dimensional paper electrophoresis or electrophoresis combined with chromatography[3]. Recent modifications of the technique also use gel electrophoresis in both dimensions[4]. Conditions for the separation in the two dimensions are different, thus yielding a 'nucleotide map' in which practically each oligonucleotide of the digest takes up a different position, according to its size and composition. This nucleotide map is called the 'fingerprint' and the techniques involving two-dimensional fractionation of digestion products are called fingerprinting. The fingerprints of RNase T_1 digests of three RNA preparations are shown in *figure 4.1*.

Regarding step (3), it follows from the above considerations that the position of each spot on the fingerprint gives some information on the composition of this oligonucleotide. Mononucleotides and very simple oligonucleotides can be identified by their position on the map. For the determination of the structure of more complex oligonucleotides, further digestion is required, this time using a different nuclease with a different specificity, as can be seen in the examples below.

Example 1

An oligonucleotide present in an RNase T_1 digest has the base composition (A_5, U, G). (The base composition can be determined by alkaline hydrolysis. Incubation in dilute NaOH leads to disruption of the phosphodiester bond and the release

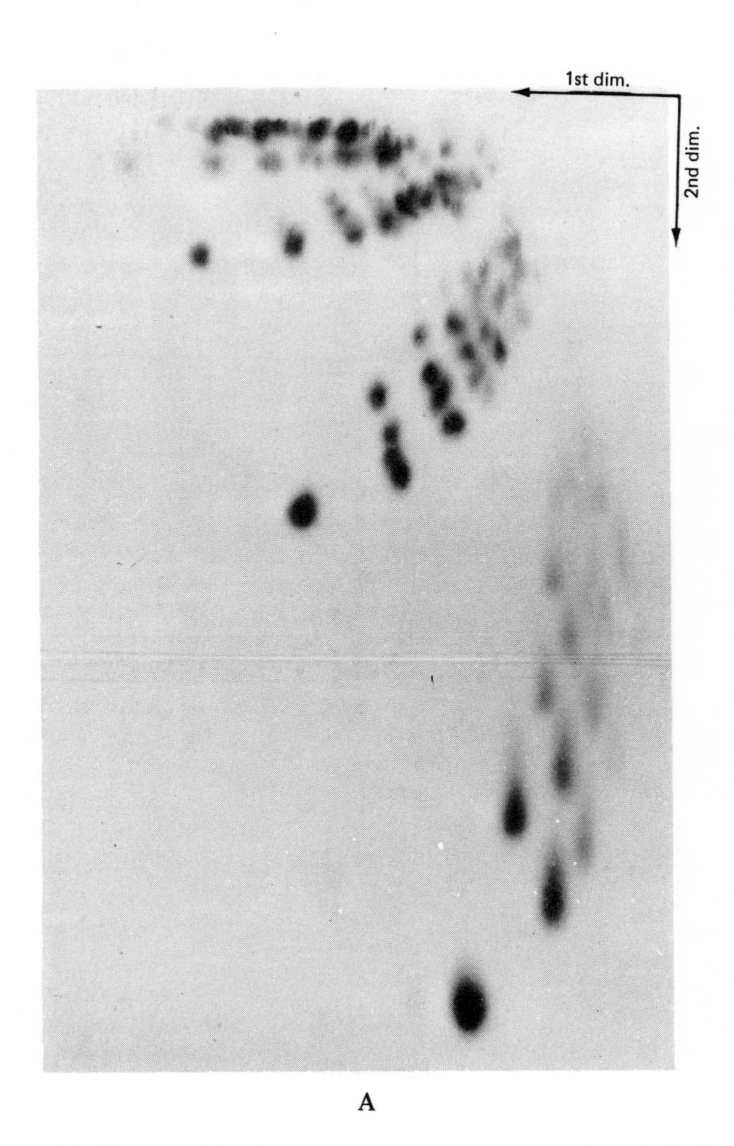

A

Figure 4.1 Fingerprints of MS2 RNA. (A) Fingerprint of complete MS2 RNA molecule, obtained by two-dimensional electrophoresis; (B) fingerprint of a fragment of MS2 RNA (the binding site of IF3, *see* Chapter 8), obtained by the same technique; (C) fingerprint of complete MS2 RNA molecule obtained by electrophoresis in the first dimension, thin layer homochromatography in the second. ((A) by courtesy of Mr M. J. F. Fowler; (B) and (C) by courtesy of Dr B. Johnson.)

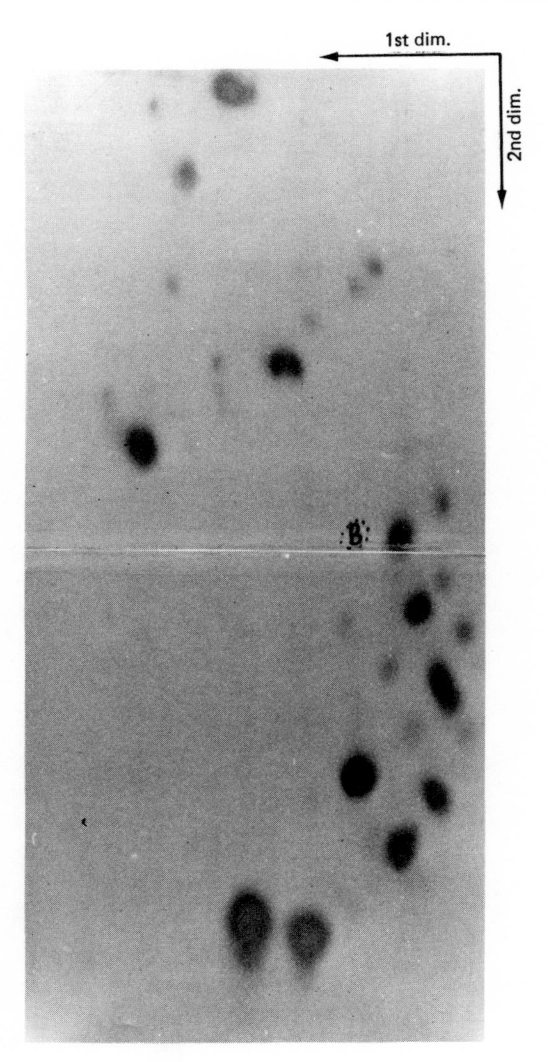

Figure 4.1B

of a mixture of nucleoside-2′-phosphates and nucleoside-3′-phosphates.) Upon digestion with pancreatic RNase it yields two fragments of base compositions (A_2, U) and (A_3, G), respectively. What is the sequence of this oligonucleotide?

As pancreatic RNase digestion products end in C or U and RNase T_1 products in G, the two fragments are AAU and AAAG and the original oligonucleotide has the sequence AAUAAAG.

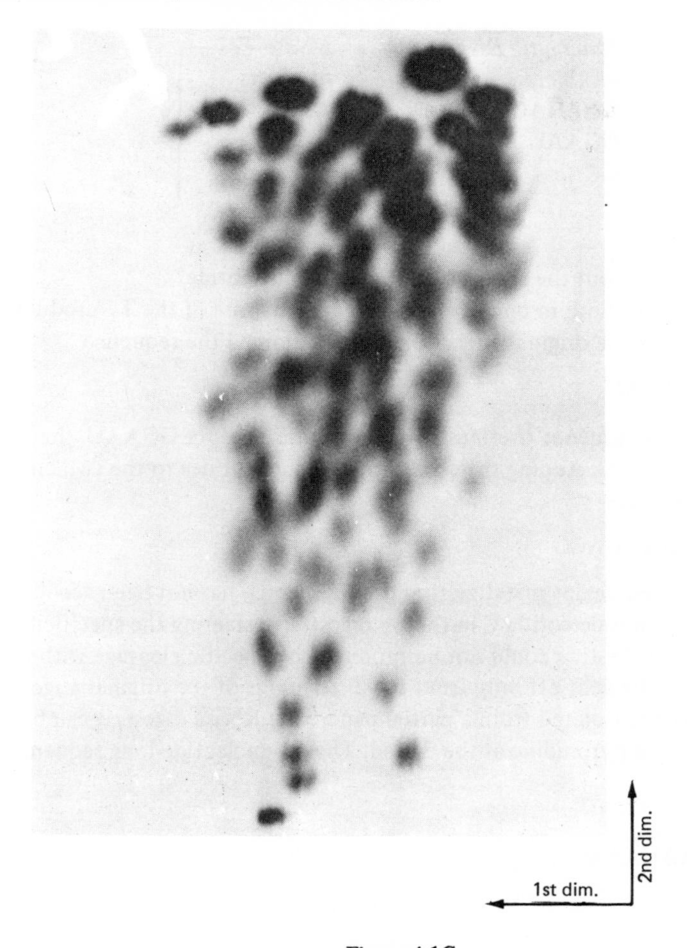

Figure 4.1C

The problem becomes a little more complicated if we are dealing with longer oligonucleotides, for example, with products of partial digestion, when not all specific sites are cleaved by the RNase. The next example shows how the sequence of such longer fragments can be worked out.

Example 2

From a partial pancreatic RNase digest a 13-residue-long oligonucleotide has been isolated which has the base composition (A_6, C_2, G_4, U). On two aliquots of this oligonucleotide, complete digestions have been carried out in parallel, one with pancreatic RNase, the other with RNase T_1. The products of these digestions were the following:

Pancreatic RNase products	*RNase T_1 products*
AAAGAC	AAUAAAG
GGAAU	ACG
GC	2 G
	C

How do we work out the sequence of this oligonucleotide?

AAAG is present both in one of the pancreatic and one of the T_1 products. This indicates that the original oligonucleotide contained the sequence

AAUAAAGAC

The 5′ end of this sequence overlaps with pancreatic product GGAAU, the 3′ end with T_1 product ACG. Adding the corresponding nucleotides to the two ends of the above sequence,

GGAAUAAAGACG

is obtained. There remains one digestion product which has not been accounted for so far, the mononucleotide C in the T_1 digest. Considering the specificity of RNase T_1, this nucleotide could not be produced by specific cleavage with this enzyme. It could be split off only from the 3′ terminus of the original oligonucleotide. As this was isolated from a partial pancreatic RNase digest, it can be expected to have a pyrimidine at the 3′ end. The 13-nucleotide-long sequence can thus be written

GGAAUAAAGACGC

↓ splits by pancreatic RNase

↑ splits by RNase T_1

In essentially the same way, by analysing longer partial digestion products and making use of overlapping sequences, the primary structure of long nucleotide stretches can be established. This kind of analysis is already part of step (4) in the sequencing procedure, and it may in itself be sufficient to work out the complete sequence of small RNA molecules. Over 70 tRNA species and some other small RNAs, like the 5S RNAs and 5.8S RNAs of ribosomes (*see* Chapter 6), have been sequenced in this way.

It can be seen on *figure 4.1* that large RNA molecules give more complex fingerprints than do small ones. In fact, the larger the RNA, the more complex the pattern and the more difficult it becomes to determine the structure of all oligonucleotides and to locate them in the complete polynucleotide chain. These difficulties may be partly overcome by first splitting the RNA into smaller fragments and working out the nucleotide sequences of these fragments individually. This technique was applied with success to R17 RNA, and it yielded the first

CA UGG CGU UCG UAC UUA AAU AUG GAA UUA ACU AUU CCA AUU UUC GCU ACG AAC UCG G

ala trp arg ser tyr leu asn met glu leu thr ileu pro ileu phe ala thr asn ser

Figure 4.2 Nucleotide sequence of a fragment from the coat protein cistron of R17 RNA and amino acid sequence of the corresponding part of the coat protein[6].

sequence of a coded message[5,6] (*figure 4.2*). This 'classical fingerprinting' procedure led to the determination of the complete sequence of MS2 RNA (ref. 7). Novel sequencing methods have also been developed, however, which are more suitable for use on such large nucleic acid molecules.

Copying technique

The basic idea of this technique is to synthesise a complementary copy of a nucleic acid with the aid of a specific RNA or DNA polymerase and to subject this synthetic copy rather than the original molecule to sequence analysis. This approach avoids two difficulties encountered in the classical fingerprinting method: it makes the analytical techniques independent of the size of the RNA or DNA to be sequenced, as complementary stretches of any desired length can be synthesised on large molecular templates, and it makes *in vivo* labelling of the nucleic acid unnecessary as a highly radioactive copy can be produced by using as substrates nucleoside triphosphates labelled in their α-phosphate group. As quite a few nucleic acid species cannot be labelled *in vivo* or at least cannot be labelled to the extent that would be required for fingerprinting, this is of great advantage in extending the applicability of the fingerprinting technique.

As we will see later, new variants of the copying technique open even wider fields for the study of nucleotide sequences: by using different RNA and DNA polymerases, DNA sequencing problems can be converted to RNA sequencing and vice versa. Some years ago, when sequencing techniques for RNA were already being widely used but the sequencing of DNA still presented a more difficult problem, DNA-dependent RNA polymerases were frequently used to transcribe DNA into RNA, as the sequence of the latter was easier to establish. Recently, new methods have been introduced which make the sequencing of DNA the faster and easier task and accordingly, copying techniques have been worked out for sequencing a complementary DNA copied from the RNA molecule. These modern methods will be described in detail later in this Chapter.

A specific transcriptase was first used in 1969 by Weissmann's group in Zurich for studying the nucleotide sequence of the 4600-nucleotide-long RNA of coliphage Qβ (ref. 8). This phage contains a highly specific RNA polymerase which can synthesise a complementary copy ((–)strand) using the viral RNA as template or can copy this (–)strand to produce the (+)strand, a molecule identical to the original viral RNA. The Zurich group used this latter process to produce highly labelled stretches corresponding to the 5′-terminal part of Qβ RNA. These were then subjected either to fingerprint analysis or to one of the special techniques which are applicable only to synthetic transcripts, like nearest-neighbour analysis or kinetic analysis.

Nearest-neighbour analysis

Combined with RNase digestion, nearest-neighbour analysis reveals the structure of oligonucleotides. The way it works can be understood if the mechanism of synthesis and the mechanism of digestion of a polynucleotide are compared.

When a new nucleotide is added to the growing point of a transcript, the α-phosphate group on its 5′ hydroxyl is linked to the 3′ hydroxyl of the preceding nucleotide in the chain. If a NTP labelled in its α-phosphate is used for transcripton, the ^{32}P will form a phosphodiester bond between the 5′ position of the last and the 3′ position of the preceding nucleotide. The phosphodiester bond formed when α-[^{32}P] UTP is added to a transcript ending in G is shown below (the labelled phosphate is marked by an asterisk).

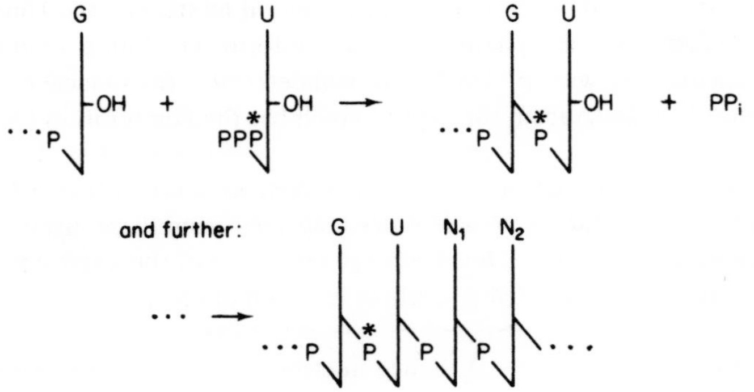

If the RNA is digested with RNase T_1 or hydrolysed by alkali, the phosphodiester bond is split so as to yield 3′ phosphates (in alkaline hydrolysates a mixture of 2′ and 3′ phosphates is produced) and 5′ hydroxyls. As a result, in the above nucleotide, the labelled phosphate group will be attached to the G. In the two steps of transcription and hydrolysis, the labelled phosphate has been transferred to the neighbouring nucleotide.

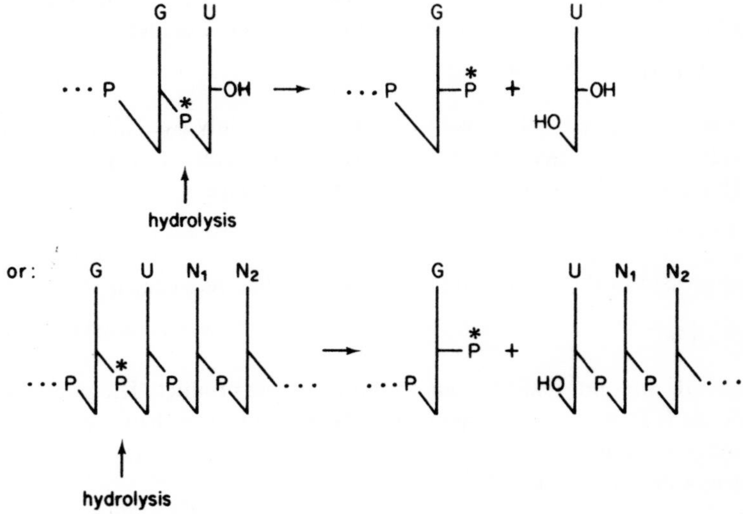

By using labelled UTP in the transcription step and identifying the nucleotides in which the label is found after enzymic digestion or alkaline hydrolysis, the nearest neighbours of U can be determined. In the above example, GU are nearest-neighbour pairs.

The following example shows how the nearest-neighbour technique can be used to work out the structure of oligonucleotides in an RNase T_1 digest.

Example 3

Transcription is carried out in four parallel reaction mixtures, each containing a different nucleoside triphosphate in radioactive form. The four transcripts thus obtained are digested with RNase T_1. We will determine the sequence of one oligonucleotide isolated from the digest, which has the composition (A, C_2, U_2, G).

In the four reaction mixtures the labelled nucleotide was ATP, CTP, GTP and UTP, respectively. In the oligonucleotides recovered from these digests the labelled phosphate group is found in the nearest neighbour of these nucleotides. (The 5′ or 3′ position of the labelled phosphate is shown in the columns below.)

Labelled nucleotide used in transcription	Labelled nucleotide found in digestion product (A, C_2, U_2, G)	Nearest neighbours
pppA	Cp	CA
pppC	Up, Cp	UC, CC
pppG	Ap	AG
pppU	Up, Gp	UU, GU

There is only one G in the molecule and this must be at the 3′ end as this is an RNase T_1 product. AG must therefore be the end-standing dinucleotide. A is also present in CA; these oligonucleotides must overlap in the sequence

. . .CAG

This C may overlap with CC or with UC. But, having only two C's in the oligonucleotide, both these nearest-neighbour pairs can be present only in the form of UCC. This gives us two more nucleotides in the sequence

.UCCAG

U is also present in the dinucleotide UU; the complete sequence is therefore

UUCCAG

(The U in GU does not really belong to this digestion product. The phosphate group of the U from the next oligonucleotide is attached to the G, as can be seen on the diagram on p. 127. Still, this gives us the further information that the next oligonucleotide starts with a U.)

Kinetic analysis

Under the conditions used by the Zurich group, transcription of Qβ RNA proceeded from the 5′ terminus at the speed of five to six nucleotides per second. RNA stretches of different lengths could be synthesised by stopping the reaction at different times. The time required for a given oligonucleotide to appear on the fingerprint of the newly synthesised RNA chain was proportional to its distance from the 5′ terminus. Measuring the 'time of appearance' of each oligonucleotide enabled the exact sequence of the first 175 nucleotides of Qβ RNA to be constructed (*figure 4.3*).

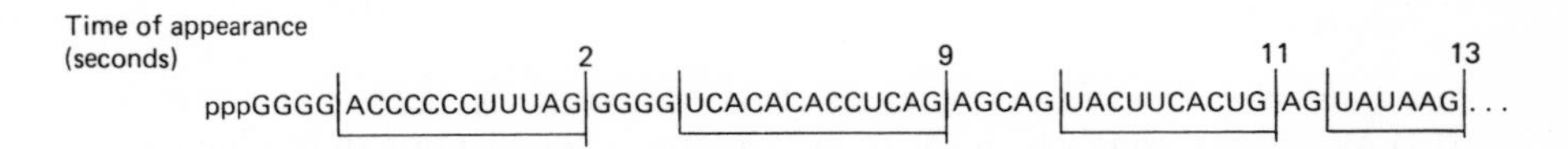

Figure 4.3 Time of appearance of characteristic oligonucleotides in the transcript of the 5′-terminal part of Qβ RNA. (After Billeter *et al.*[8].)

Other variants of the copying technique

As the RNA polymerase of Qβ phage is specific for Qβ RNA as template, it is thus suitable for sequencing this particular RNA molecule only. The principle of copying can, however, also be applied to other RNAs and to DNAs by choosing appropriate polymerising enzymes.

(1) Radioactive copies of nonradioactive DNA can be obtained with the aid of DNA polymerase I. The synthetic DNA can then be subjected to nearest-neighbour and fingerprint analysis[9]. The sequencing of DNA molecules by these techniques is more difficult than that of RNA because of the lack of base-specific endonucleases which would produce similar digestion products from DNA to those obtained from RNA by RNase T_1 or pancreatic RNase digestion. This difficulty can be overcome, however, by making use of the rather weak specificity of DNA pol I in the presence of Mn^{2+} ions. Under these conditions, the polymerase will also insert ribonucleotides into a deoxyribonucleotide chain, and wherever such ribonucleotides are present, the molecule becomes susceptible to RNases and to alkaline hydrolysis (*figure 4.4*). This is called the ribo-substitution technique[10].

(a) . . . dA . dT . dT . dC . dC ↓ rG . dA . dA . . . .   (b) . . . . dA . dT . dT . rC ↓ rC ↓ dG . dA . dA . . .

Figure 4.4 An oligodeoxyribonucleotide sequence with different ribonucleotides substituted. The arrows show the sites where this molecule will be split by (a) RNase T_1 ; (b) pancreatic RNase. Akaline hydrolysis breaks the molecule at all three sites.

(2) If Mg^{2+} is replaced by Mn^{2+} in the reaction mixture, the specificity of DNA pol I is also reduced in another respect: the enzyme can use an RNA tem-

plate as well as DNA. This offers new possibilities of applying the copying technique to RNA molecules. Proudfoot and Brownlee[11] determined the nucleotide sequence of the 3′-terminal regions of globin mRNA by synthesising a complementary DNA copy (cDNA) with DNA pol I and sequencing this radioactive cDNA. *Figure 4.5* shows a short part of this sequence.

5′	3′	
. . . AAAGGAAAUUUAUUUUCAUUGC–polyA		sequence of mRNA
. . . TTTCCTTTAAATAAAAGTAA CG–oligo dT		sequence of cDNA
3′	5′	

Figure 4.5 Part of the 3′-terminal sequence of rabbit β-globin mRNA (Proudfoot and Brownlee[11].)

(3) Another enzyme which has found widespread application in producing a complementary DNA copy from an RNA template is reverse transcriptase (*see* p. 71). It can be used in the same way as DNA pol I to synthesise a radioactive cDNA copy which can be directly sequenced[11a]. A more indirect way of using this enzyme for mRNA sequencing was applied by Marotta and coworkers[12]. In the first step they prepared a nonradioactive cDNA copy from globin mRNA, with the aid of reverse transcriptase. In the second step they transcribed this cDNA molecule into a complementary RNA, using RNA polymerase enzyme and radioactive ribonucleoside triphosphates as substrates. The final product so obtained was an RNA identical to the original mRNA, but it was highly radioactive, and so suitable for sequencing.

Both DNA polymerase I and reverse transcriptase need primers to start DNA synthesis (Chapter 2). In order to provide such primers, the sequence of at least a short stretch of the template molecule must be known. Complementary oligonucleotides can then be synthesised and annealed to the template and these can be extended by the polymerases. Although this means an additional step in the sequencing procedure, it has the advantage that the start-point of copying is thereby exactly defined. In this way a short complementary stretch can be copied and sequenced from a functionally interesting part at the inside of a large template molecule. The copying technique has often been applied to sequencing the 3′-terminal part of eukaryotic mRNAs. These molecules usually carry a poly(A) tail (Chapter 5); therefore, an oligo(dT) primer can be used to synthesise a DNA copy starting from the 3′ end.

Synthesis of a cDNA molecule proved most useful in the sequencing of mammalian messenger RNAs. Direct sequencing of these RNA molecules presents great difficulties because it is not possible to obtain by *in vivo* labelling such highly radioactive preparations as would be needed for the classical sequencing techniques. Eukaryotic mRNAs are biologically much more stable molecules than the mRNAs of prokaryotes; the half life of some mammalian messengers varies between a few hours or even a few days as opposed to half lives often of only a few seconds for the mRNAs of prokaryotic cells. This stability is of advantage in the isolation of eukaryotic mRNAs, but proves a great drawback in the *in vivo* labelling of these molecules. There have been early attempts to label nucleic acids *in vitro* for sequence studies by introducing a radioactive phosphate group into each oligonucleotide in the digest[13] or by iodinating the molecule with radioactive iodine[14]. These techniques could be applied with success to some sequencing problems, but they did not give as satisfactory results as the synthesis of a complementary molecule of high specific radioactivity.

(4) Studying the structure of prokaryotic mRNAs involved difficulties of a different kind. As these molecules are very short-lived, there is no appreciable pool present in the cell; they are synthesised and degraded at a very high rate and, especially those for which synthesis can be induced, can acquire very high specific activity if the cells are grown on radioactive medium. But their lability makes the isolation of intact mRNA molecules an extremely difficult task. Yanofsky's group[15] obtained large fragments of tryptophan mRNA from *E. coli* cells, and they worked out the sequence of a 200-nucleotide-long stretch of it, comprising the 5′-terminal control region, and the first coded message. But so far, no comlete polycistronic mRNA has been isolated from prokaryotic cells and no complete primary structure has been established.

RNA polymerases are of great use in overcoming this problem also, but they are applied in a different way: DNA-dependent RNA polymerase is used to synthesise *in vitro* a mRNA or part of a mRNA which can then be sequenced by any of the RNA sequencing techniques. Modern techniques often make it easier to isolate a gene for a given protein than the corresponding mRNA. *In vitro* transcription provides the radioactive mRNA from these genes, in a form suitable for sequencing. In this way, the first bacterial mRNA sequence, that of the 5′-terminal part of the lac messenger, was determined by Maizels[16] (*figure 4.6*), and the same method was used later in studying the nucleotide sequence of the galactose mRNA[17].

```
5′                         3′
 TGGAATTGTGAGCGGATAACAATT
 ACCTTAACACTCGCCTATTGTTAA
3′                         5′

 pppAAUUGUGAGCGGAUAACAAUU ......
```

Figure 4.6 The nucleotide sequences of the lac operator (above) as determined by Gilbert and Maxam[18] and of the 5′-terminal part of lac mRNA (below), determined by Maizels[16].

(5) Depending on the conditions of transcription, DNA-dependent RNA polymerase may start copying the template at different sites. It is possible to produce copies which cover a complete DNA fragment, or transcripts which start from a specific promoter within this fragment. The former method of transcription provides a way of converting DNA sequencing problems into RNA sequencing tasks.

Gilbert and Maxam[18] isolated the operator of the lac operon and determined its nucleotide sequence by transcribing it into RNA and sequencing this RNA molecule. Later, Dickson *et al.*[19] extended the nucleotide sequence to the whole control region of this operon.

Comparison of a nucleotide sequence obtained from a complete DNA stretch with that of a transcript starting at a specific site within this DNA region may yield information on which parts of control regions are transcribed and which are not. Comparison of the nucleotide sequence of the lac operator as determined by Gilbert and Maxam, and of the 5′-terminal sequence of lac mRNA, determined on a transcript which had a specific start-point at the lac promoter site, showed unequivocally that a major part of the operator region is, in fact, transcribed. As can be seen in *figure 4.6*, there is a 21-nucleotide-long overlap in the two sequences.

Pribnow[20] used a similar technique to locate the start-site of transcription within a promoter region of T7 phage DNA. The promoter site was isolated by binding the RNA polymerase to the DNA molecule and digesting off all the free DNA by a mixture of DNases. The DNA stretch to which the polymerase was bound was protected from digestion. This complete DNA fragment was transcribed into RNA and sequenced. Comparison with a specific transcript which started at a specific site within this DNA fragment showed that the start-site was in the middle of the RNA polymerase binding site (*see also* p. 88 and *figure 3.3*).

Direct sequence readout techniques

Recently, a new principle has been introduced in nucleic acid sequence analysis, first for DNA and later also for large RNA molecules, providing a way of sequencing which avoids the time-consuming analysis of small digestion products and the reconstitution of long sequences from overlapping oligonucleotides. There are several variants of this new method, all based on the production of a series of fragments, each being one nucleotide longer than the preceding one, and on separation of these fragments according to size by gel electrophoresis or chromatography. From the pattern obtained in the gel or on the thin layer plate, the nucleotide sequence can be directly read out. This makes sequencing much more rapid: the latest techniques allow determination of a 100–200-nucleotide-long sequence in a single analysis. A further advantage is that *in vitro* labelling is used in all these methods, thus facilitating work with nucleic acids which are difficult to label *in vivo*.

The new techniques have been originally worked out for DNA, which was rather difficult to approach with the conventional sequencing methods, partly because of the much larger size of these molecules, and partly because of the lack of base-specific endonucleases. However, as the modern methods proved very successful, they have all been adapted since to RNA also.

(1) Wandering spot technique

The first and simplest technique which applied the 'direct readout' principle, was the wandering spot technique[21]. This is suitable for sequencing relatively short, preferably single-stranded DNA or RNA fragments which carry a labelled phosphate group only at their 5′ terminus. Labelling is achieved *in vitro* with the aid of the enzyme, polynucleotide kinase, which transfers the γ phosphate of [^{32}P] ATP to the 5′ hydroxyl of ribo- or deoxyribonucleotides. The nucleic

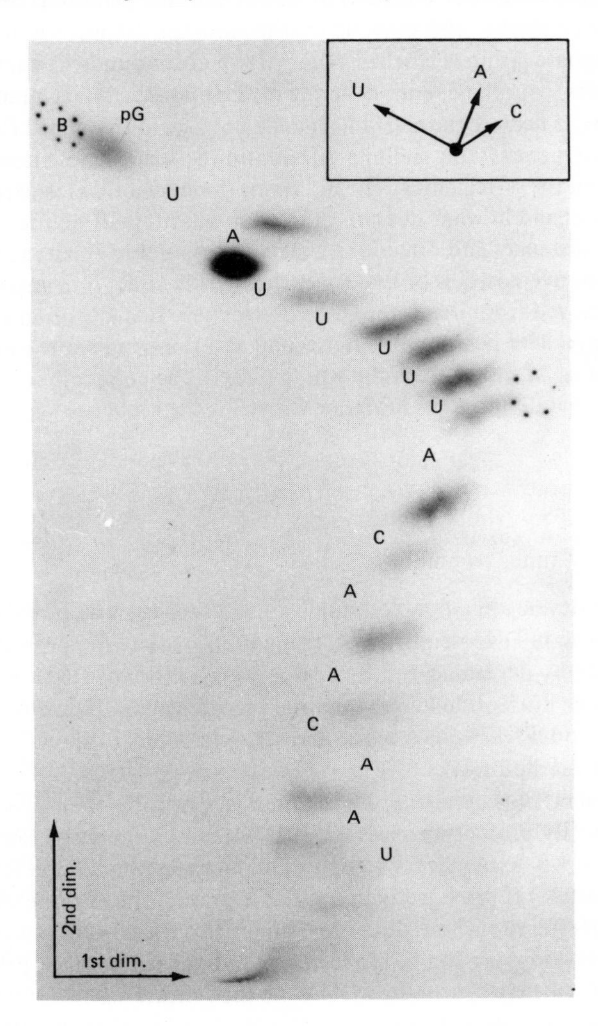

Figure 4.7 Analysis of the 5′-terminal nucleotide sequence of tobacco mosaic virus RNA by the wandering spot technique. For explanation see text. (By Richards *et al.*[21]. Reproduced with permission from *Nature*, **264**, 548 (1977).)

acid fragment is subjected to partial digestion by a nuclease which shows no base specificity, so that digestion products of every possible length are produced. The digest is fractionated in two dimensions, combining electrophoresis in the first dimension with thin layer chromatography in the second. This means that fractionation of the digestion products is according to composition in the first and also according to size in the second dimension: the shorter an oligonucleotide, the farther will it move from the start-line.

Figure 4.7 shows the fractionation of an RNA digest. Only oligonucleotides containing the original 5′ end show up on the autoradiograph, as only the 5′-terminal phosphate group is labelled. The pattern corresponds to a series of oligonucleotides from the 5′ end, differing in their length. The longest, the intact RNA fragment, is nearest the start-line, at the bottom of the plate. The next oligonucleotide moves faster and in a different direction, because it has lost one nucleotide from the 3′ terminus. The nature of the nucleotide lost determines how much faster and in what direction this spot will move. The insert in *figure 4.7* shows the distances and directions which correspond to the loss of an A, C, G or U. The relative position of two spots thus yields direct information on the nucleotide removed from the 3′ terminus. In this case it is a U residue. In the same way, from the relative positions of the second and third spots, the next nucleotide can be defined as A, and so from the whole pattern the complete sequence can be deduced. The 'wandering spot' in *figure 4.7* yields the sequence

$$^{5'}\text{pGUAUUUUUACAACAAU}^{3'}$$

(2) The plus and minus technique

Sanger's group developed a new technique a few years ago which brought about a real breakthrough in DNA sequence determination[22]. Here the problem is approached, not by degrading, but by synthesising a series of DNA fragments, each of them one nucleotide longer than the preceding one. Data from Sanger and Coulson's work on ΦX174 DNA sequences are used below to show on a simplified model how the method works.

DNA polymerase I is used to produce a radioactive copy of single-stranded DNA template. (By proteolytic cleavage of DNA pol I a fragment can be obtained which has no 5′ → 3′ exonuclease activity. This enzyme preparation is usually used in sequence studies.) If provided with a suitable primer, the enzyme will start copying at a specific site where the primer anneals to the complementary sequence in the template*. Copying can be stopped after a brief period, thus producing short stretches of newly synthesised DNA. As the synthesis is not synchronised, a

*Restriction fragments are often used for priming. If the restriction map of the DNA is known, well defined parts of the genome can thus be sequenced.

heterogenous mixture of DNA fragments of different sizes will be produced. It is possible to carry out the DNA polymerase reaction under appropriate conditions so that oligonucleotides of every length between 20 and 30 nucleotides should be produced. The 3′-terminal sequence of such a series of oligonucleotides is shown below.

Nucleotide sequence of DNA template:

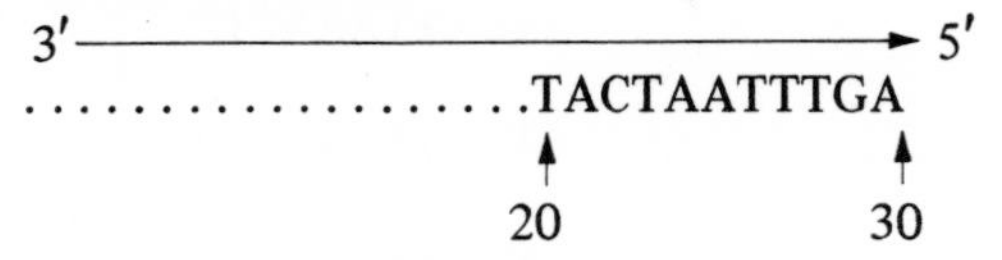

Nucleotide sequences of a series of fragments synthesised:

...................ATGATTAAACT

...................ATGATTAAAC

...................ATGATTAAA

...................ATGATTAA

...................ATGATTA

...................ATGATT

...................ATGAT

...................ATGA

...................ATG

...................AT

...................A

The mixture of these fragments is isolated and re-incubated with the polymerase, but this time in the presence of three nucleoside triphosphates only. The enzyme extends each fragment as far as it can go in the absence of one nucleotide. It stops copying when it arrives at the nucleotide missing in the reaction mixture. If the missing nucleotide is dATP (this is called the minus A system), the fragments will be extended until the next A is reached in the sequence. From the above series the following products will be obtained in the minus A system:

```
                    20        30
.................... ATGATTAAACT
.................... ATGATTAAACT
.................... ATGATTAAACT
.................... ATGATTAA
.................... ATGATTA
.................... ATGATT        minus A system
.................... ATGATT
.................... ATGATT
.................... ATG
.................... ATG
.................... ATG
```

It follows from the principle of the technique that the sizes of the fragments thus extended are directly related to the positions of the A residues in the sequence. If we consider only fragments shorter than 30 nucleotides, we find four distinct sizes, 22, 25, 26 and 27 nucleotides long. This is because in each case the DNA polymerase stopped short of the next A residue, and the positions of A residues are 23, 26, 27 and 28.

Similarly, in the minus C, minus G and minus T systems, in which the DNA polymerase reaction is carried out in the absence of dCTP, dGTP and dTTP, respectively, the extended DNA fragments will be as shown on p. 137.

The fragments in each of the four minus systems are separated according to size by polyacrylamide gel electrophoresis. A diagram of the gel pattern obtained is shown in *figure 4.8*. The sequence can be read from the four patterns directly in the following way. The shortest fragment (20 nucleotides long) is found in the minus T system; this shows that there must be a T in position 21. The next fragment (21 nucleotides long) is found in the minus G system, which indicates that there is a G at position 22; a 22-nucleotide-long fragment is present in the minus A system, which indicates that A is in position 23, etc.

The sequence can be confirmed by using the plus system, in which *E. coli* DNA polymerase is replaced by the enzyme from T4 bacteriophage and the DNA fragments are incubated in four reaction mixtures, each of which contains only one

```
                    20        30
..................ATGATTAAACT
..................ATGATTAAACT
..................ATGATTAAA
..................ATGATTAAA
..................ATGATTAAA
..................ATGATTAAA    minus C system
..................ATGATTAAA
..................ATGATTAAA
..................ATGATTAAA
..................ATGATTAAA
..................ATGATTAAA
```

```
                    20        30                          20        30
..................ATGATTAAACT        ..................ATGATTAAACT
..................ATGATTAAACT        ..................ATGATTAAAC
..................ATGATTAAACT        ..................ATGATTAAAC
..................ATGATTAAACT        ..................ATGATTAAAC
..................ATGATTAAACT        ..................ATGATTAAAC
..................ATGATTAAACT        ..................ATGATTAAAC
..................ATGATTAAACT        ..................ATGAT
..................ATGATTAAACT        ..................ATGA
..................ATGATTAAACT        ..................ATGA
..................AT                 ..................ATGA
..................AT                 ..................A
        minus G system                      minus T system
```

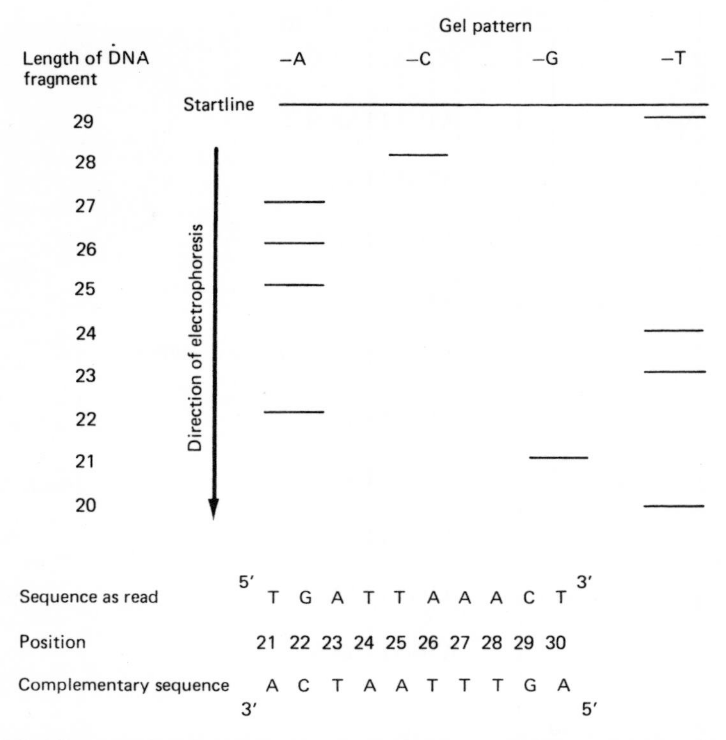

Figure 4.8 Diagram of a gel pattern obtained by the plus and minus method. As the fragments synthesised have been copied from the template DNA by DNA polymerase, the sequence as read from the gel is complementary to the sequence of the original DNA molecule.

nucleoside triphosphate. The T4 phage enzyme has both exonuclease and DNA polymerase activities. As an exonuclease, it removes nucleotides from the 3′ end, but as a DNA polymerase, it can replace these nucleotides if the appropriate dNTPs are present. In the presence of dATP (in the plus A system), for example, the enzyme will remove nucleotides from the 3′ terminus until it arrives at an A residue. Here its DNA polymerase activity will outweigh the exonuclease activity, the A will be removed and replaced continuously, and the apparent action of the enzyme will stop at this nucleotide. However different the enzyme reactions, the result will be similar to that of the minus system: DNA fragments will be produced which end, this time not before, but at the A residues in the sequence. Gel electrophoresis of the four 'plus' mixtures will yield bands which correspond to the positions of the four nucleotides in the sequence, just as those shown in the diagram in *figure 4.8*. The sequence can be read off the gel in the same way as described there.

Introduction of the plus and minus technique dramatically reduced the time required for sequencing large DNA molecules. Sequencing of a 100-nucleotide-long DNA stretch can be accomplished today in a few days. This method enabled

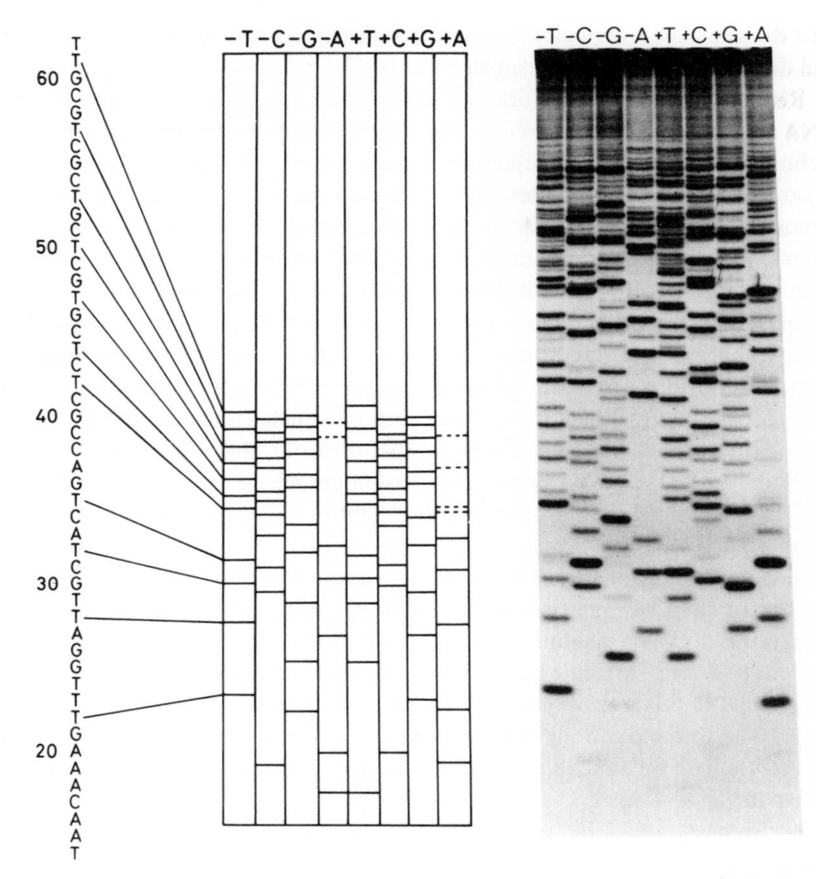

Figure 4.9 Nucleotide sequence determination of a 48-nucleotide-long stretch of ΦX DNA by the plus and minus technique. (Barrell *et al.*[24]. Reproduced with permission from *Nature*, **264**, 34 (1976). Photograph by courtesy of Dr B. G. Barrell.)

the Cambridge group to establish the first complete sequence of a DNA genome, that of ΦX174 phage[23]. *Figure 4.9* shows how from a gel electrophoretogram a 48-nucleotide-long sequence was determined. The total length of the viral DNA is 5386 nucleotides, and all the above described sequencing techniques have been applied in working out the primary structure of different parts of this DNA molecule. The plus and minus technique and the chain termination technique (see below) have been used to establish the complete sequence of the DNA of a related bacteriophage, G4 (ref. 35).

In the course of this work, many interesting new features of the organisation of these viral genomes have been revealed. The amino acid sequence of some viral proteins could be deduced and even the presence of hitherto unknown proteins could be predicted. The most amazing discovery was that the same stretch of DNA can serve different functions: the messages for two proteins can be encoded into the same DNA sequence in two different reading frames (*see* p. 26 and 41) and different control signals can also overlap[23,24,35].

Recently, Sanger and coworkers[25] published a new direct readout method for DNA sequencing which is based on the same principle as the plus and minus technique, but is quicker, simpler and more accurate. DNA pol I is used to copy the template strand in the presence of low concentrations of chain-terminating inhibitors. These are analogues of the dNTPs, usually dideoxynucleoside triphosphates, which, when incorporated into the DNA chain, stop further extension. If a deoxyribonucleoside triphosphate substrate and its chain terminator analogue are present in the right proportion (e.g. 1:100), only partial incorporation of the latter will occur, and fragments of different length will be synthesised, ending at different positions where this nucleotide occurs. The result is a similar mixture of fragments as that obtained in the 'plus' systems which can be fractionated in a gel and interpreted in the same way as those shown in *figure 4.9*. However, this method does not require a two-step extension procedure. A gel, used by Shaw *et al.*[24] in establishing the sequence of a fragment of G4 phage DNA, is shown in *figure 4.10*.

A somewhat different principle has been applied by Barnes[36]. He combined copying of DNA with partial ribo-substitution (*see* p. 129). Under appropriate conditions, only one ribonucleotide is incorporated into each DNA molecule, at random positions. Alkaline hydolysis breaks each molecule at the site of this nucleotide. In this way, a series of fragments is produced the lengths of which correspond to the positions of this nucleotide in the sequence. Fractionation on polyacrylamide gel yields a pattern which can be interpreted in the same way as shown in *figure 4.8*.

Brownlee extended the plus and minus technique to the sequencing of RNA[26]. The principle is the same in DNA and RNA sequencing, but different enzymes are used: reverse transcriptase for the synthesis of complementary DNA fragments and for their extension in minus systems, *E. coli* DNA pol I for the plus systems. Baralle[26a] used this technique for sequencing the 5′-terminal non-coding regions of globin mRNAs. This led to the completion of the sequence of human β-globin mRNA.

Chain termination by specific inhibitors has also been applied in RNA sequencing. dNTP analogues (dideoxynucleoside triphosphates) inhibit reverse transcriptase in the same way as DNA polymerase. They cause termination at a specific base in the cDNA synthesised by reverse transcriptase on an RNA template. The same principle can thus be used for sequencing DNA and RNA; the difference is only in the enzyme used for producing the cDNA fragments. Zimmern and Kaesberg worked out the technique of sequencing RNA with the aid of reverse transcriptase and specific chain terminators[38]. They applied this method to the 3′-terminal nucleotide sequence of encephalomyocarditis virus RNA. At the same time, Hamlyn *et al.*[39] determined the nucleotide sequence of immunoglobin light-chain mRNA by using the chain terminator technique.

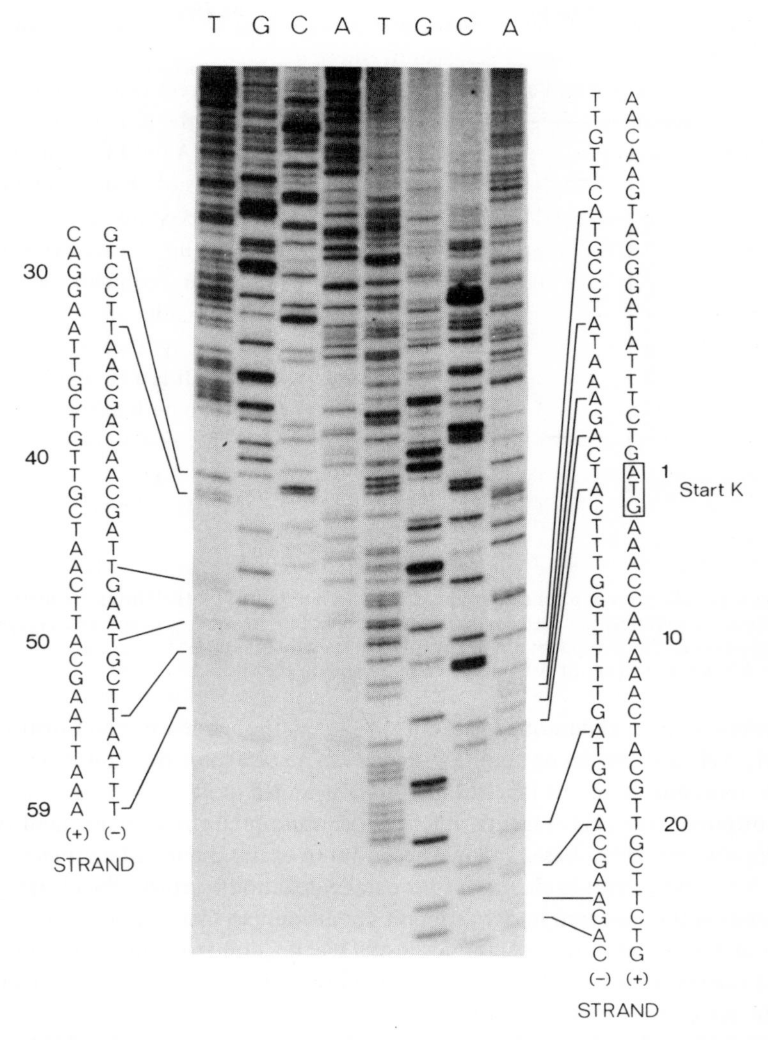

Figure 4.10 Gel electrophoretogram, obtained by Shaw *et al.*[24] from G4 phage DNA, sequenced by the chain termination technique[25]. As in *figure 4.8*, the sequence read directly from the gel is complementary to that of the viral DNA. In this figure the sequence of the viral (+)strand is also shown. (Reproduced with permission from *Nature*, **272**, 511 (1978).)

(3) The Maxam-Gilbert technique

Another recently developed method which has found very widespread application is that of Maxam and Gilbert[27]. This is the only method at present which can also be applied directly to double-stranded DNA.

The DNA is labelled *in vitro* in its 5′-terminal phosphate with the aid of poly-

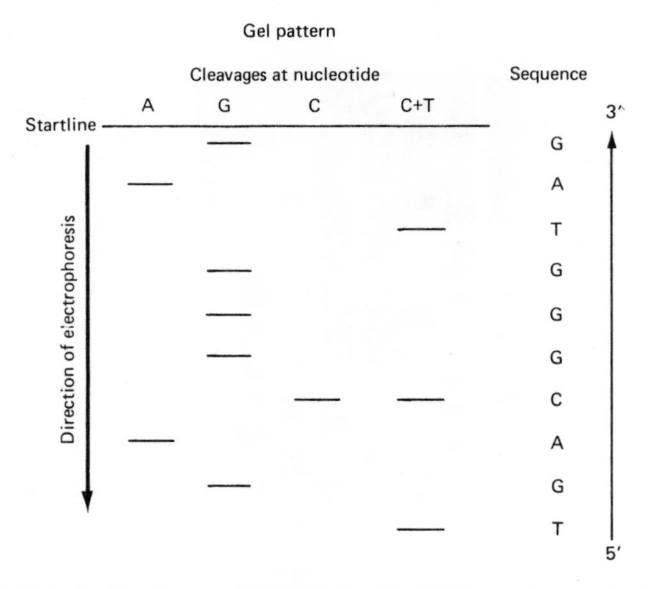

Figure 4.11 Diagram of a gel electrophoretic pattern obtained by the Maxam–Gilbert technique. In practice, the patterns are more complex (*see figure 4.12*) because cleavages at A residues and at G residues are not absolutely specific. Faint G-bands often also appear in the A track and traces of A-bands can also be seen in the G track.

nucleotide kinase, as described earlier (p. 133). (Although the method works equally well with single- or double-stranded DNA, care must be taken that only one 5′ terminus should be labelled in double-stranded molecules.) It is then treated with different chemical reagents which, under appropriate conditions, modify specifically one of the bases and in a second step excise the modified nucleotide from the polynucleotide chain. In four parallel reaction mixtures the chemical degradation is conducted under different conditions, so that cleavages should occur at A residues in the first, at G's in the second, and at C's in the third mixture. (The specificity of cleavages at A and G residues is limited. This difficulty can be overcome, as shown in *figure 4.12*.) Specific cleavage at T's cannot be achieved, but a fourth mixture can be set up in which excisions take place at both C and T residues. Comparison of this system with the one specific for C also gives information on the T-cleavage sites. By carrying out the reaction under such mild conditions that only partial modification should occur, it is possible to excise only

Figure 4.12 (opposite) Part of a gel used for nucleotide sequence determination of polyoma DNA by the Maxam–Gilbert technique. (By courtesy of Dr B. E. Griffin.) As compared to the somewhat simplified diagram in *figure 4.11*, here an additional reaction mixture has been included in which cleavages occur at A and C residues. This is necessary because the specificity of the chemical reactions is limited; under the same conditions which lead mainly to cleavages at A residues, some G residues will also react and vice versa. Additional information is therefore needed to confirm which band represents an A and which a G residue. Different authors use different reaction mixtures to overcome this ambiguity. In this gel, comparison of the A, G, C and A+C tracks defines unequivocally the position of A-bands and G-bands.

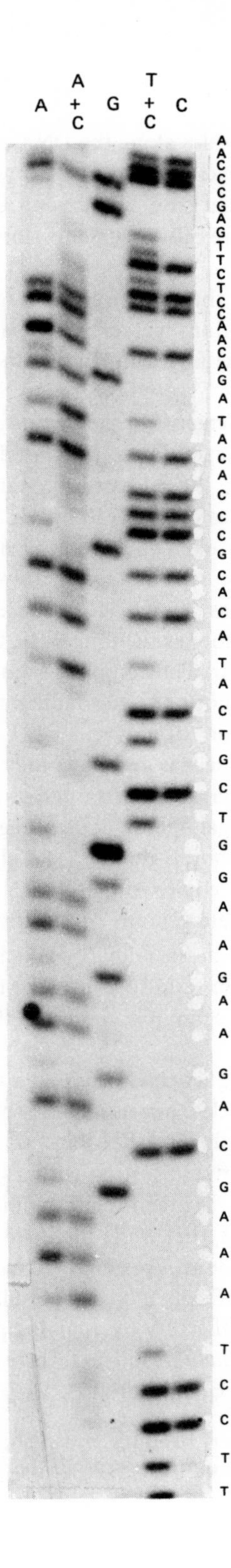
A
A
+
C
G
T
+
C
C
A
A
C
C
C
G
A
G
T
T
C
T
C
C
A
A
C
A
G
A
T
A
C
A
C
C
C
G
C
A
C
A
T
A
C
T
G
C
T
G
G
A
A
G
A
A
G
A
C
G
A
A
A
T
C
C
T
T

one in 50 or one in 100 nucleotides along the DNA chain. As the reaction takes place randomly, fragments of every possible length will be produced, ending at any position where this particular base was present. Only the 5′-terminal phosphate of the DNA is labelled, therefore only fragments starting from the 5′ terminus will be detectable.

In each reaction mixture we thus have a set of fragments which start at the 5′ terminus and end at the positions where a given base is present in the sequence. It follows that the lengths of the fragments in the first mixture correspond to the distances of A residues from the 5′ terminus, i.e. to the positions of A's in the sequence; the lengths of the fragments in the second mixture correspond to the positions of G's, etc. The four reaction mixtures are fractionated according to size by gel electrophoresis, and from the gel pattern obtained, the positions of all four nucleotides in the DNA sequence can be read off, as shown in *figure 4.11*.

The shortest fragment was broken at a T, so this nucleotide is nearest the 5′ terminus. The next cleavage occurred at a G, the third at an A. This gives the sequence from the 5′ terminus as TGA. The sequence is read here from the bottom of the gel towards the top, as the longer the fragment, the farther from the 5′ end did the cleavage occur. The complete sequence read from this gel is TGACGGGTAG. *Figure 4.12* shows the autoradiograph of a gel from which the nucleotide sequence of a fragment of polyoma DNA has been deduced.

The Maxam–Gilbert technique was described only recently, but it has already had very widespread use. Some applications of great interest are: nucleotide sequence determination of the gal operon control region[28]; establishment of the sequences of the mRNAs for rabbit β-globin[29] and chicken ovalbumin[30], based on sequencing the duplex DNA copies of these mRNAs; and determination of the primary structure of the complete genome of SV40 virus[37].

The chemical reactions used in the Gilbert–Maxam technique work on DNA molecules. Conditions have been found recently for base-specific degradation of RNA and an RNA sequencing technique, similar to that described above, has been worked out[40].

Another method of RNA sequencing, based on a similar principle of partial degradation at specific nucleotides, but using base-specific enzymes instead of chemical reagents, has been applied to tRNA and to rRNA[31,41]. If carried out under very mild conditions, partial digestion with a specific enzyme leads to the production of differently sized fragments each ending at a position where a specific base is present. The difficulty is, that while there are base-specific enzymes which split at G residues (RNase T_1) or at A residues (RNase U_2)*, we know of no RNases which would split specifically at C or U residues only. Brownlee's group succeeded in overcoming this difficulty by applying pancreatic RNase and an RNase from *Physarum polycephalum*[31]. The former enzyme splits at C and U residues, the latter at A, G and U, but not at C. The four enzyme digests are

*RNase U_2 splits at both A and G residues, but it shows a preference for the former; under very mild digestion conditions it will therefore act as an enzyme specific for A.

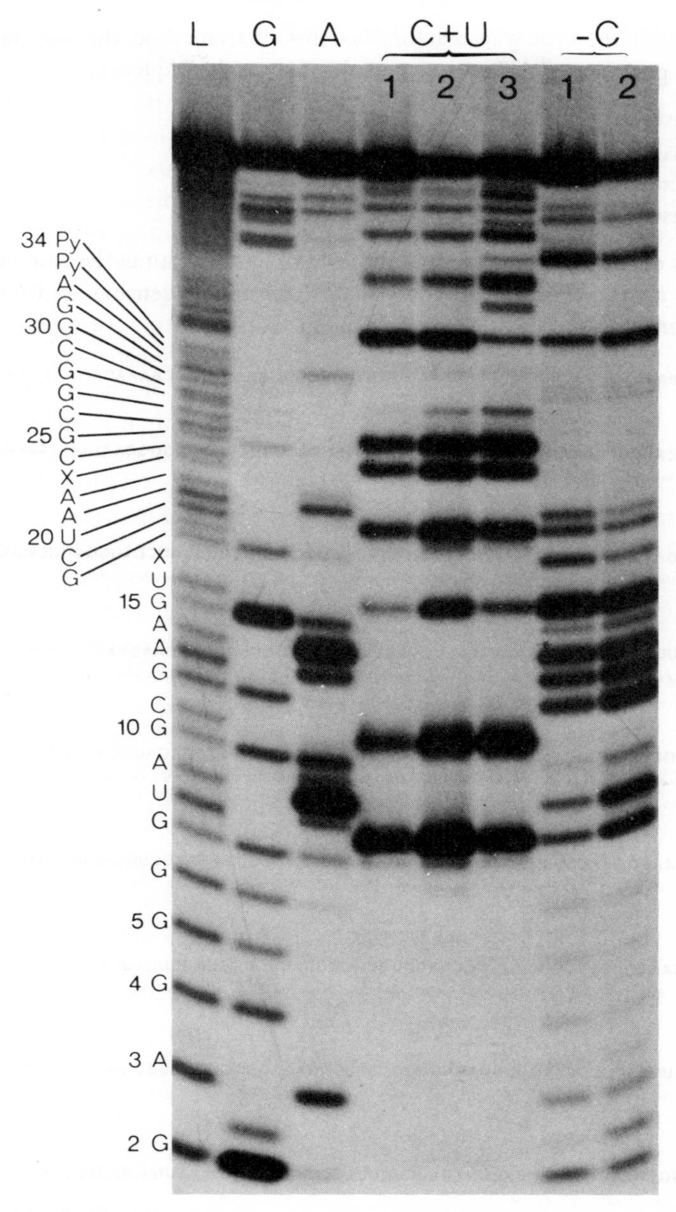

Figure 4.13 Nucleotide sequence determination of $tRNA^{tyr}$ by the technique of Simoncsits *et al.*[31]. L: completely random splits, producing every possible size of oligonucleotide, used for reference, to define the position of the different size bands. G, A, C+U and –C: fragments produced by digestion with RNase T_1, RNase U_2, pancreatic RNase and RNase from *P. polycephalum*, respectively, as described in the text. The three C+U tracks and two –C tracks were obtained under milder and stronger digestion conditions. (Reproduced with permission from *Nature*, **269**, 833 (1977). Photograph by courtesy of Dr G. G. Brownlee.)

fractionated in the same way as in the Maxam–Gilbert method; the four gel patterns define the positions of the following nucleotides in the sequence:

RNase T_1	G
RNase U_2	A
pancreatic RNase	C+U
RNase from *Physarum polycephalum*	A+G+U

Deduction of the sequence is a little more complicated than in the case of DNA, but all the necessary information is available, and the sequence of an RNA can be read off the gel as shown in *figure 4.13*.

mGpppACACUUGCUUUUGACACAACUGUGUUUACUUGCAAUCCCCCAAAACAGACAGAAUG

GUGCAUCUGUCCAGUGAGGAGAAGUCUGCGGUCACUGCCCUGUGGGGCAAGGUGAAUGUG
Val His Leu Ser Ser Glu Glu Lys Ser Ala Val Thr Ala Leu Trp Gly Lys Val Asn Val

GAAGAAGUUGGUGGUGAGGCCCUGGGCAGGCUGCUGGUUGUCUACCCAUGGACCCAGAGG
Glu Glu Val Gly Gly Glu Ala Leu Gly Arg Leu Leu Val Val Tyr Pro Trp Thr Gln Arg

UUCUUCGAGUCCUUUGGGGACCUGUCCUCUGCAAAUGCUGUUAUGAACAAUCCUAAGGUG
Phe Phe Glu Ser Phe Gly Asp Leu Ser Ser Ala Asn Ala Val Met Asn Asn Pro Lys Val

AAGGCUCAUGGCAAGAAGGUGCUGGCUGCCUUCAGUGAGGGUCUGAGUCACCUGGACAAC
Lys Ala His Gly Lys Lys Val Leu Ala Ala Phe Ser Glu Gly Leu Ser His Leu Asp Asn

CUCAAAGGCACCUUUGCUAAGCUGAGUGAACUGCACUGUGACAAGCUGCACGUGGAUCCU
Leu Lys Gly Thr Phe Ala Lys Leu Ser Glu Leu His Cys Asp Lys Leu His Val Asp Pro

GAGAACUUCAGGCUCCUGGGCAACGUGCUGGUUAUUGUGCUGUCUCAUCAUUUUGGCAAA
Glu Asn Phe Arg Leu Leu Gly Asn Val Leu Val Ile Val Leu Ser His His Phe Gly Lys

GAAUUCACUCCUCAGGUGCAGGCUGCCUAUCAGAAGGUGGUGGCUGGUGUGGCCAAUGCC
Glu Phe Thr Pro Gln Val Gln Ala Ala Tyr Gln Lys Val Val Ala Gly Val Ala Asn Ala

CUGGCUCACAAAUACCACUGAGAUCUUUUUCCCUCUGCCAAAAAUUAUGGGGACAUCAUG
Leu Ala His Lys Tyr His

AAGCCCCUUGAGCAUCUGACUUCUGGCUAAUAAAGGAAAUUUAUUUUCAUUGC-polyA

Figure 4.14 The complete nucleotide sequence of rabbit β-globin mRNA. The complete sequence has been established by combining the results obtained by several scientists in different laboratories, as described on p. 147. (Reproduced, with permission, from *Cell*, **10** (1977).)

An interesting example from which the application of practically all sequencing techniques can be seen is the sequence determination of rabbit β-globin mRNA. This was the first messenger isolated in pure form, and since then, elucidation of its structure has always attracted keen interest. It has taken years from the first attempts to arrive at the stage at which the complete structure has been disclosed. The sequence of this mRNA, shown in *figure 4.14*, was the first complete nucleotide sequence obtained from an eukaryotic mRNA.

A number of research groups, using a variety of sequencing techniques, contributed to the success of this work. Proudfoot and Brownlee[11] established the 3′-untranslated sequences using the DNA pol I copying technique, and adjacent stretches of the coding region by copying with reverse transcriptase[11a]. The ribosome binding site was determined by Legon[32], on *in vitro* labelled, iodinated mRNA[14]. The cap structure and a short sequence at the 5′ end were established by Lockard and RajBhandary[33], who made use of the wandering spot technique. Baralle[34] completed the sequence of the 5′-terminal untranslated region, applying Brownlee and Cartwright's variant of the plus and minus technique[26]. The major part of the coding sequence was established by Efstratiadis and coworkers[29], who synthesised a double-stranded DNA copy on rabbit β-globin mRNA and sequenced this DNA by the Maxam–Gilbert technique. This led to the completion of the nucleotide sequence of this mRNA molecule.

REFERENCES

1 Holley, R. W. *et al. Science,* **147**, 1462 (1965).
2 Sanger, F., Brownlee, G. G. and Barrell, B. G. *J. Mol. Biol.,* **13**, 373 (1965).
Barrell, B. G. in *Procedures in Nucleic Acid Research* (eds Cantoni, G. L. and Davies, D. R.) Harper & Row, Vol 2, 751 (1971).
3 Brownlee, G. G. and Sanger, F. *Eur. J. Biochem.,* **11**, 395 (1969).
4 de Wachter, R., Merregaert, J., Vanderberghe, A., Contreras, R. and Fiers, W. *Eur. J. Biochem.,* **22**, 400 (1971).
Fon Lee, Y. and Wimmer, E. *Nucl. Acids. Res.,* **3**, 1647 (1976).
5 Sanger, F. *Biochem. J.,* **124**, 833 (1971).
6 Adams, J. M., Jeppesen, P. G. N., Sanger, F. and Barrell, B. G. *Nature,* **223**, 1009 (1969).
7 Min Jou, W., Haegeman, G., Ysebaert, M. and Fiers, W. *Nature,* **237**, 82 (1972).
Fiers, W. *et al. Nature,* **256**, 273 (1975).
Fiers, W. *et al. Nature,* **260**, 500 (1976).
8 Billeter, M. A., Dahlberg, J. E., Goodman, H. M., Hindley, J. and Weissmann, C. *Nature,* **224**, 1083 (1969).
9 Sanger, F., Donelson, J. E., Coulson, A. R., Kossel, H. and Fischer, D. *PNAS,* **70**, 1209 (1973).
10 Sanger, F., Donelson, J. E., Coulson, A. R., Kossel, H. and Fischer, D. *J. Mol. Biol.,* **90**, 315 (1974).
11 Proudfoot, N. J. and Brownlee, G. G. *Nature,* **252**, 359 (1974).
11a Proudfoot, N. J., Cheng, C. C. and Brownlee, G. G. *Progr. Nucl. Acid Res. Mol. Biol.,* **19**, 123 (1976).

12 Marotta, C. A., Forget, B. G., Cohen-Solal, M. and Weissman, S. M. *Progr. Nucl. Acid Res. Mol. Biol.*, **19**, 165 (1976).
13 Szekely, M. and Sanger, F. *J. Mol. Biol.*, **43**, 607 (1969).
Szeto, K. S. and Soll, D. *Nucl. Acids Res.*, **1**, 171 (1974).
14 Robertson, H. D., Dickson, E., Model, P. and Prensky, W. *PNAS*, **70**, 3260 (1973).
15 Squires, C., Lee, F., Bertrand, K., Squires, C. L., Bronson, M. J. and Yanofsky, C. *J. Mol. Biol.*, **103**, 351 (1976).
Bronson, M. J., Squires, C. and Yanofsky, C. *PNAS*, **70**, 2335 (1973).
16 Maizels, N. M. *PNAS*, **70**, 3585 (1973).
17 Musso, R. E., de Crombrugghe, B., Pastan, I., Sklar, J., Yot, P. and Weissman, S. *PNAS*, **71**, 4940 (1974).
18 Gilbert, W. and Maxam, A. *PNAS*, **70**, 3581 (1973).
19 Dickson, R. C., Abelson, J., Barnes, W. M. and Reznikoff, W. S. *Science*, **187**, 27 (1975).
20 Pribnow, D. *PNAS*, **72**, 784 (1975).
21 Jay, E., Bambara, R., Padmanabhan, R. and Wu, R. *Nucl. Acids Res.*, **1**, 331 (1974).
Richards, K., Jonard, G., Guilley, H. and Keith, G. *Nature*, **267**, 548 (1977).
22 Sanger, F. and Coulson, A. R. *J. Mol. Biol.*, **94**, 441 (1975).
23 Sanger, F. *et al. Nature*, **265**, 687 (1977); *J. Mol. Biol.*, **125**, 227 (1978).
24 Barrell, B. G., Air, G. M. and Hutchison, C. A. III *Nature*, **264**, 34 (1976).
Smith, M., Brown, N. L., Air, G. M., Barrell, B. G., Coulson, A. R., Hutchison, C. A. III and Sanger, F. *Nature*, **265**, 702 (1977).
Shaw, D. C., Walker, J. E., Northrop, F. D., Barrell, B. G., Godson, G. N. and Fiddes, J. C. *Nature*, **272**, 510 (1978).
25 Sanger, F., Nicklen, S. and Coulson, A. R. *PNAS*, **74**, 5463 (1977).
26 Brownlee, G. G. and Cartwright, E. M. *J. Mol. Biol.*, **114**, 93 (1977).
26a Baralle, F. E. *Cell*, **12**, 1085 (1977).
27 Maxam, A. M. and Gilbert, W. *PNAS*, **74**, 560 (1977).
28 Musso, R., DiLauro, R., Rosenberg, M. and de Crombrugghe, B. *PNAS*, **74**, 106 (1977).
29 Efstratiadis, A., Kafatos, F. C. and Maniatis, T. *Cell*, **10**, 571 (1977).
30 McReynolds, L., O'Malley, B. W., Nisbet, A. D., Fothergill, J. E., Givel, D., Fields, S., Robertson, M. and Brownlee, G. G. *Nature*, **273**, 723 (1978).
31 Simoncsits, A., Brownlee, G. G., Brown, R. S., Rubin, J. R. and Guilley, H. *Nature*, **269**, 833 (1977).
32 Legon, S. *J. Mol. Biol.*, **106**, 37 (1976).
33 Lockard, R. E. and RajBhandary, U. L. *Cell*, **9**, 747 (1976).
34 Baralle, F. E. *Cell*, **10**, 549 (1977).
35 Godson, G. N., Barrell, B. G., Staden, R. and Fiddes, J. C. *Nature*, **276**, 236 (1978).
36 Barnes, W. M. *J. Mol. Biol.*, **119**, 83 (1978).
37 Fiers, W. *et al. Nature*, **273**, 113 (1978).
Reddy, V. B. *et al. Science*, **200**, 494 (1978).
38 Zimmern, D. and Kaesberg, P. *PNAS*, **75**, 4257 (1978).
39 Hamlyn, P. H., Brownlee, G. G., Cheng, C., Gait, M. J. and Milstein, C. *Cell*, **15**, 1067 (1978).
40 Peattie, D. A. *PNAS*, **76**, 1760 (1979).
41 Carbon, P., Ehresmann, C., Ehresmann, B. and Ebel, J. P. *FEBS Lett.*, **94**, 152 (1978).
Ross, A. and Brimacombe, R. *Nucl. Acids Res.*, **5**, 241 (1978).

5 The Structure of Messenger RNA

THE PRIMARY STRUCTURE OF MESSENGER RNA

Interest in the primary structure of mRNA dates back to the time when the messenger concept was first introduced by Jacob and Monod. Although their hypothesis was based rather on the functional requirement for a mediator between DNA and protein than on direct evidence for the existence of such a substance, it was strongly supported by the work of Volkin and Astrachan[1], who characterised an RNA fraction which seemed to fit the above requirements. This RNA fraction, synthesised in *E. coli* cells after phage infection, was shown to have a base composition similar to that of DNA. This finding directed attention to the 'rapidly labelled' or 'DNA-like' RNA and encouraged further research[2] into the characterisation of this RNA fraction.

This line of research, while bringing some positive results, was also misleading in so far as it implied that mRNA, being a copy of DNA, should have a DNA-like base composition. With the elucidation of the mechanism of transcription, it emerged that mRNA is synthesised by copying only one strand of the DNA and cannot therefore be expected to show any special base composition. The message carried by mRNA is encoded in the sequence of the nucleotides; the structural characteristics relevant to mRNA function are therefore represented by the base sequence and not by the base composition of the molecule. In 1961, Hall and Spiegelman[3], using hybridisation techniques, had detected complementary sequences in T2 phage DNA and in an RNA synthesised in *E. coli* upon infection with this phage. These early results provided proof for the existence of RNA species with functional and structural characteristics predicted for mRNAs but they contributed little to our knowledge of the actual structure of these RNAs. Progress in this direction was made possible some years later by the development of new techniques for determining nucleotide sequences in RNA.

In the meantime, determination of the code alphabet posed new problems which could be solved only by establishing the primary structure of a mRNA. The genetic code had been established mainly on the basis of *in vitro* assays. Only the comparison of the amino acid sequence of a protein and of the nucleotide sequence of the corresponding message could provide direct proof that it was operational *in vivo*. Such studies were also expected to shed light on the practical implications of some special characteristics of the code: on the use of degenerate codons and on the structural features which unambiguously define an initiation site, in spite of the ambiguity of the initiation codons.

The technique for determining nucleotide sequences in large RNA molecules was worked out in the late 'sixties, but the isolation of pure mRNAs from eukaryotic cells was achieved only years later, and mRNAs (or rather fragments of mRNAs) from bacterial cells have become available for sequence studies only in recent years. The first studies on the primary structure of messengers were carried out on RNAs from bacteriophages. These molecules were very suitable for sequence work as they have the same messenger function as mRNAs from the host cells: they carry the information for specific viral proteins. At the same time, they are easily obtained in pure form with a high specific radioactivity.

Much of our present knowledge on the structure of messengers comes from studies on RNA viruses. However, as these RNAs constitute a rather special group, some caution is indicated in generalisations as to structural characteristics of mRNA molecules. Recent progress in studies on RNA structure made it possible to compare some characteristic features of viral and non-viral mRNAs in both prokaryotes and eukaryotes. Results on the structures of globin, ovalbumin and immunoglobulin mRNAs as well as on the bacterial lactose, galactose and tryptophan messengers suggest that viral and non-viral mRNAs have some basic characteristics in common. At the same time, however, significant differences have been observed in control regions of eukaryotic and prokaryotic mRNAs.

(1) Prokaryotic mRNAs

The first investigation of the primary structure of mRNA was started by Sanger's group in Cambridge. They determined the nucleotide sequences of different stretches in the RNA of a small coliphage, R17. Their first results yielded information on the structure of the coded message and of the initiation and termination sites, and revealed the presence of untranslated sequences in viral RNA[4,5,6]. Within an amazingly short time, similar information was provided on the structure of the RNA from the coliphage Qβ by research groups in Geneva and Zurich[7]. In the next few years, major parts of the RNA from MS2 virus were sequenced by Fiers' group in Ghent[8]. The study of nucleotide sequences has been extended since to a great number of viral and also non-viral messengers.

As all the RNAs concerned are rather large molecules (at least several hundred but often several thousand nucleotides long), in many cases the structure of only part of the molecule or of only some fragments is known. The first mRNA to be completely sequenced was the RNA of MS2 virus[8,9,10]. This is a 3569-nucleotide-long molecule with a structure almost identical with that of R17 RNA. It codes for three proteins: the maturation or A-protein, coat protein and a virus-specific RNA polymerase which is also called replicase or synthetase. The sequences of two long stretches in this RNA are shown in *figures 5.1* and *5.2*. These parts of the RNA contain the coat protein cistron and parts around the initiation site (130–132) and termination site (1309–1311) of the A-protein cistron.

New results in DNA sequencing allow the deduction of mRNA sequence from the sequence of the DNA of bacteriophages. The nucleotide sequences of mRNAs from ΦX174 phage are identical with those of the single-stranded DNA

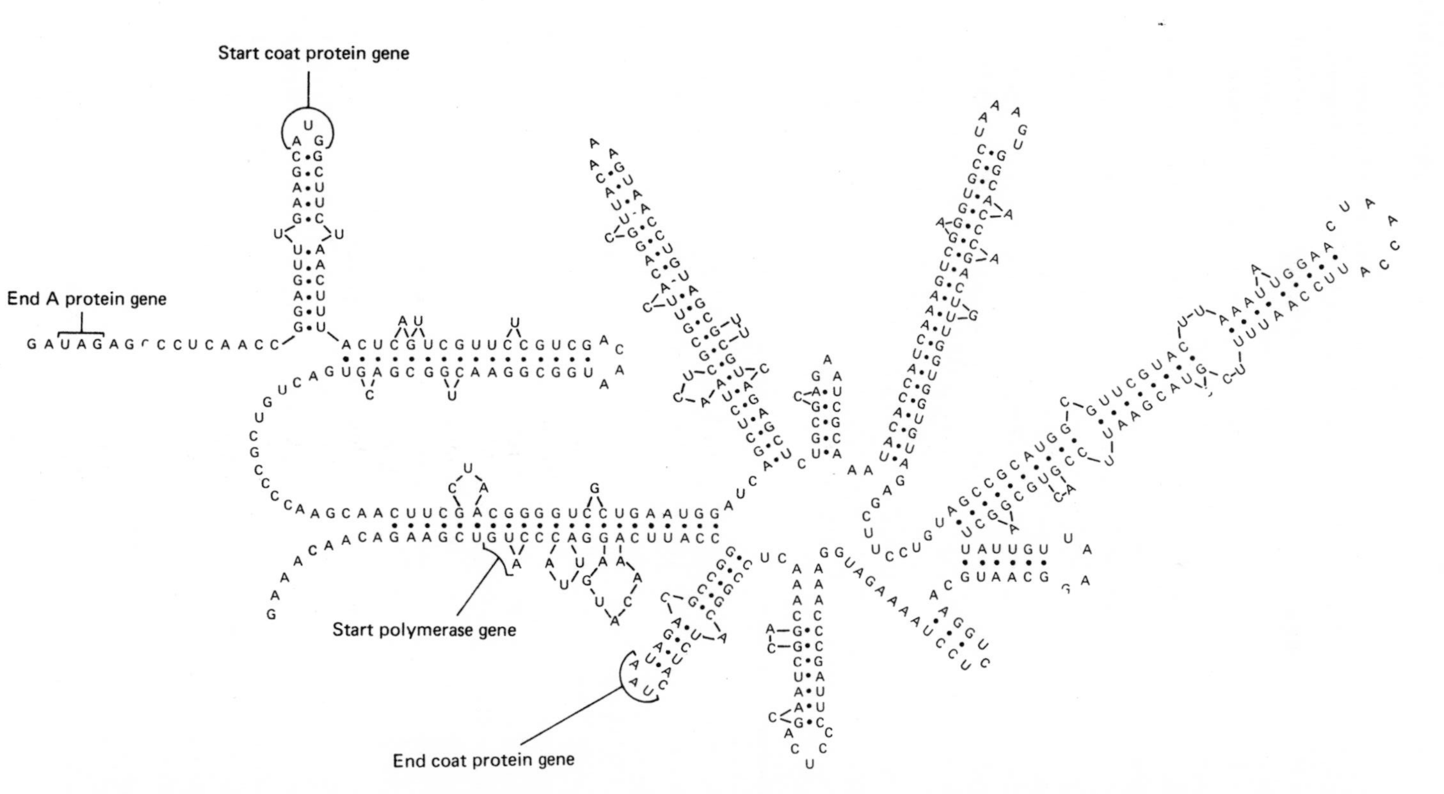

Figure 5.1 Nucleotide sequence of the coat protein region of MS2 RNA (Min Jou *et al.*[8]). (Reproduced with permission from *Nature*, **237**, 82 (1972).)

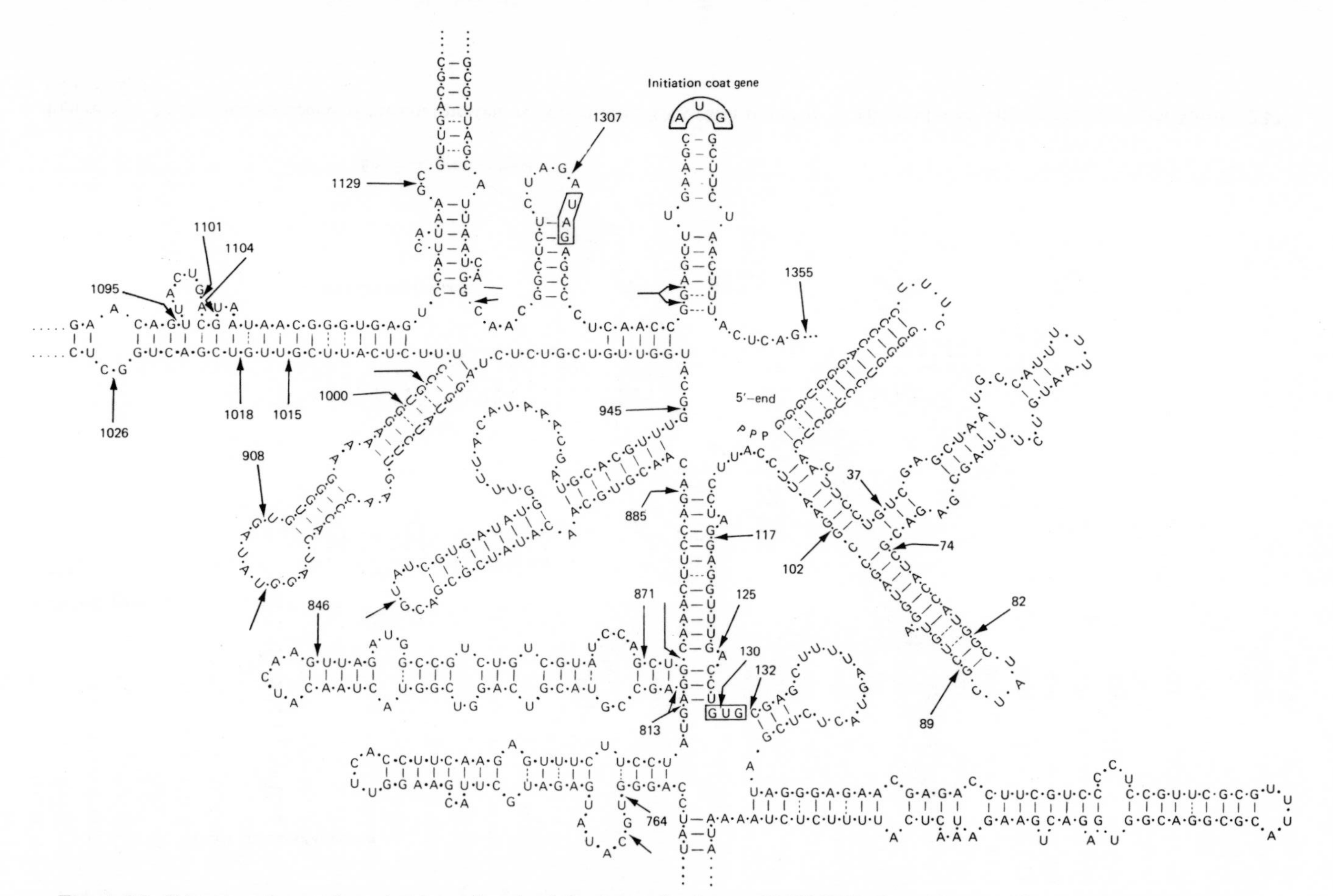

Figure 5.2 Primary and secondary structure of parts of the A-protein cistron of MS2 RNA (from Fiers *et al.*[9]). Arrows show the sites cleaved by mild RNase T_1 treatment. (Reproduced with permission from *Nature*, **256**, 273 (1975).)

of this phage[11]. As mRNAs can also be obtained by *in vitro* transcription of the viral genome[12], elucidation of the primary structure of viral mRNAs shows fast progress.

Greater difficulties had to be overcome in studying the primary structure of bacterial mRNAs. The extreme lability of these molecules (their half life may be in the range of a few seconds or a few minutes) makes it a very difficult task to isolate an intact mRNA molecule. Their high turnover means also that any individual mRNA species is always present only in very low amounts; consequently, the isolation and identification of any specific mRNA requires special techniques. The possibility of inserting well defined DNA fragments, containing a specific gene, into transducing phages (*see* Chapter 1) provides a way of picking out one mRNA species which hybridises to this DNA from the mixture of RNAs in a bacterial extract. Yanofsky's group isolated the mRNA of the tryptophan operon in this way, and determined the nucleotide sequence of a long fragment from the 5′ terminus, which comprised an untranslated sequence containing control signals and the coded message for the first enzyme of this operon[13]. *In vitro* transcription of DNA has been used to synthesise and sequence the 5′-terminal parts of the mRNAs of the lactose[14] and galactose[15] operons. Comparison of the primary structure of these mRNAs revealed a number of common characteristics which shed light on different aspects of mRNA function.

(a) The use of different codewords

Confirmation of the code alphabet was obtained in the first studies on small fragments of the RNA of R17 virus. This RNA codes for three viral proteins: A (or maturation) protein, coat protein and synthetase. The structures of the proteins and RNA of this virus are almost identical with those of MS2 virus. The amino acid sequence of the coat protein was determined by Weber[16] in 1967. The nucleotide sequence of a stretch of R17 RNA coding for 19 amino acids of this protein is shown in *figure 4.2*. This was the first information ever obtained on the structure of a message. Comparison of the amino acid and nucleotide sequences shows perfect agreement with the code alphabet and also confirms that more than one codon is used to specify the same amino acid (*see* codons for thr, ser, asn, ala).

Nucleotide sequences determined in other viral and host cell mRNAs in recent years provide further proof for the validity of the genetic code and for the use of degenerate codons. In the case of MS2 RNA, it was shown that all degenerate codewords are used in the message for the replicase and that only UGU is missing from the message for A-protein. However, the frequency of the different codewords differs.

Figure 5.3 shows how many times each codon is used in the A-protein cistron to specify the different amino acids. There is a striking difference in the frequency of different degenerate codewords: asparagine is coded for much more frequently by AAC than by AAU; ACU stands about twice as often for threonine as any of the other threonine codons; GGU is used very frequently, GGA only twice, to

	U		C		A		G		
U	Phe	6	Ser	5	Tyr	4	Cys	0	U
		10		6		12		3	C
	Leu	8		8		–		–	A
		6		10		–	Trp	12	G
C	Leu	6	Pro	5	His	2	Arg	7	U
		9		5		3		6	C
		5		4	Gln	9		6	A
		2		3		9		3	G
A	Ileu	1	Thr	11	Asn	2	Ser	4	U
		8		5		15		3	C
		7		5	Lys	5	Arg	3	A
	Met	7		6		9		4	G
G	Val	8	Ala	6	Asp	8	Gly	15	U
		7		12		5		6	C
		7		7	Glu	5		2	A
		8		10		12		5	G

Figure 5.3 Frequency of different codewords in the A-protein cistron of MS2 RNA (initiation and termination codons not included). (Data from Fiers *et al.*[9].)

code for glycine, etc. It is the exception rather than the rule when all degenerate codons occur with about equal frequency, as in the case of proline and valine. The preferential use of one or the other codon may be characteristic for a given message: in the coat protein cistron of the same RNA some codewords are absent which are used quite frequently in the A-protein cistron, e.g. AUA and CGA. In the replicase cistron AAU is used more frequently for asn than AAC and the most frequent codeword for thr is ACC.

Although the significance of these preferences is not known, some interesting suggestions have been put forward which connect the use of different degenerate codewords with a specific control mechanism in protein synthesis. Fiers suggested that rare codons may have a modulating role in translation if their cognate tRNA species are also rare[10]. In the synthesis of proteins of bacteriophage T4, the codeword CUG seems to be used less frequently for leucine than in the synthesis of the protein of the host cell. Upon infection of *E. coli* cells with this phage, the $tRNA^{leu}$I species which reads the CUG codon becomes degraded and a new type of $tRNA^{leu}$ is synthesised[17]. Adaptation of the tRNA profile to the special needs of viral protein synthesis may be part of the mechanism by which viruses control protein synthesis in the infected cell. A parallelism has been confirmed between the tRNA profile and the frequency of different codewords in some types of cells which synthesise mainly one protein species, e.g. fibroin or globin[18]. Another interesting observation on the non-random use of codons was made in the ΦX174 genome, in which a very marked preference is shown for codewords with T in the third position[11].

The use of degenerate codons leaves a considerable degree of freedom for the primary structure of mRNA. In theory, a great number of different nucleotide sequences could carry the same message: on average, almost every third nucleotide could be altered in the mRNA without affecting the structure of the protein it codes for. The importance of this flexibility has been recognised only recently, when the multi-functional character of mRNAs as well as of the DNA genome was revealed. It was first observed by Platt and Yanofsky[19], in the case of the tryptophan messenger, that the same RNA stretch which carries the coded message for one protein also provides the binding site for ribosomes and thus the specific initiation signal for another protein. Very similar structures have been detected at two sites in ΦX174 virus, where mRNAs carry out the same multiple functions[11].

Flexibility of the nucleotide sequence of mRNA may be needed to fulfil the structural requirements for different functions. The multi-functional character of some small DNA viruses creates even greater demands for a flexible primary structure of the message: the genomes of bacteriophages ΦX174 and G4 contain overlapping genes for different proteins which are translated in different reading frames (*see* Chapter 1). This phenomenon is not restricted to prokaryotic genomes: a partial overlap, about 110 nucleotides long, coding in different reading frames for the C terminus of one viral protein and for the N terminus of another has also been reported for the genome of the small monkey virus SV40 (ref. 20). Only by non-random use of degenerate codons can two different messages be encoded into the same nucleotide sequence.

(b) Initiation and termination sites

The structure at the initiation site of the message attracted special interest in view of the highly specific process by which these sites are selected in the translation of different mRNAs. There is one ambiguity in the genetic code: the initiator codons are identical with the codewords for methionine and valine. In MS2 RNA, AUG is used 18 times as an internal codon for methionine and twice as an initiator codon. GUG is the initiator codon for A-protein and it also occurs 18 times as an internal codon for valine. The same triplets also occur quite frequently out of phase.

The mechanism of elongation of a polypeptide chain includes a step which ensures that no formylmethionine (which initiates a polypeptide chain) can be inserted in place of a methionine or valine residue. At the stage of initiation of translation, however, we know of no such built-in safeguards which would prevent the initiator formylmethionyl-tRNA from interacting with an internal or an out-of-phase AUG or GUG triplet. Yet translation is carried out in an unambiguous way; there is no evidence for faulty initiation *in vivo.* Ribosomes do not bind to the 'wrong' AUG or GUG codons. There has been much controversy as to how the right initiation sites are recognised, how ribosomes discriminate between the initiator codons and the other AUG or GUG triplets. The structure of initiation sites from a great number of mRNAs has been studied to find out whether they contain some additional information which could act as an 'initiation signal'. Such a signal could be present either in the primary or in the secondary structure.

First, the nucleotide sequence of the different initiation sites was compared in the hope of finding some common characteristic feature. The technique to isolate initiation sites of different messages was first developed by Steitz[5] and later applied by several authors. It was based on the observation that when ribosomes bind to mRNA they protect a 25-30-nucleotide-long stretch of this RNA molecule against nuclease action. Ribosomes were bound to mRNA in the presence of all the protein factors and other components required for the initiation of protein synthesis. Under these conditions ribosomes bind almost exclusively to the genuine initiation sites. The free RNA was then degraded with RNase, and the

Table 5.1 Nucleotide sequences at the initiation sites of prokaryotic mRNAs

mRNA	Sequence of initiation site
MS2 A-protein	GAU UCC UAG GAG GUU UGA CCU GUG CGA GCU UUU AGU G
R17 A-protein	GAU UCC UAG GAG GUU UGA CCU AUG CGA GCU UUU AGU G
MS2 coat	CC UCA ACC GGA GUU UGA AGC AUG GCU UCU AAC UUU
R17 coat	CC UCA ACC GGG GUU UGA AGC AUG GCU UCU AAC UUU
MS2, R17 replicase	AA ACA UGA GGA UUA CCC AUG UCG AAG ACA ACA AAG
Qβ A-protein	UCA CUG AGU AUA AGA GGA CAU AUG CCU AAA UUA CCG CGU
Qβ coat	AAA CUU UGG GUC AAU UUG AUC AUG GCA AAA UUA GAG ACU
Qβ replicase	AGUAA CUA AGG AUG AAA UGC AUG UCU AAG ACA G
f1 coat	UUU AAU GGA AAC UUC CUC AUG AAA AAG UCU UU
f2 coat	CC UCA ACCG(AG)GUU UGA AGC AUG GCU UCC AAC UUU ACU
trp mRNA, leader	CAC GUA AAA AGG GUA UCG ACA AUG AAA GCA AUU UUC GUG
trp E mRNA	GAA CAA AAU UAG AGA AUA ACA AUG CAA ACA CAA AAA CCG
trp A mRNA	GAA AGC ACG AGG GGA AAU CUG AUG GAA CGC UAC GAA UCU
lac Z mRNA	AAU UUC ACA CAG GAA ACA GCU AUG ACC AUG AUU ACG GAU
gal E mRNA	AUA AGC CUA AUG GAG CGA AUU AUG AGA GUU CUG GUU ACC
ΦX174 A gene*	CA AAT CTT GGA GGC TTT TTT ATG GTT CGT TCT TAT
ΦX174 B gene*	AA AGG TCT AGG AGC TAA AGA ATG GAA CAA CTC ACT

The underlined sequences are complementary to the 3′ end of 16S RNA.
*DNA sequences of the viral genome.

fragment protected by the ribosome was isolated and sequenced. *Table 5.1* shows the sequence of initiation sites in a number of viral and non-viral mRNAs. The AUG (or, in the case of MS2 A-protein, GUG) is always found in the middle of the RNA fragment which is protected by the ribosomes. It is followed by the codons specifying the first four amino acids from the N-terminus of the protein. Any specific initiation signal would be expected to be present at the 5′ side of the fragment, in the nucleotide stretch preceding the AUG codon. When comparing the sequences in *table 5.1,* we cannot detect any consistent homology which could be defined as an initiation signal. There is one consistent feature in all the above sequences, although this does not amount to a strict homology: a short purine-rich stretch is present at a distance of about 10 nucleotides from the initiator codon. At the initiation sites of the coat protein, A-protein and replicase cistrons of MS2 RNA, these purine-rich sequences are GGAG, AGGAGGU and AGGA, respectively.

The importance of purine-rich stretches in mRNA was recognised when a pyrimidine-rich sequence was detected in one of the ribosomal RNA molecules. The 16S RNA from *E. coli* ribosomes contains at the 3′ terminus a sequence consisting mainly of C and U residues which is complementary to the purine sequences at the above mRNA initiation sites. It seems probable that this pyrimidine-rich region of 16S RNA has a functional role. The 3′-terminal sequence of 16S RNA from *E. coli* ribosomes is shown in *figure 5.4*. The underlined pyrimidine-rich part contains complementary sequences to all the known initiation sequences in mRNAs which are translated by these ribosomes.

$$\mathrm{UCGU^{m}AACAAGGUAACCGUAGGG(G)GA^{m_2^6}A^{m_2^6}CGUGCGGUUGG\underline{AUCACCUCCUUA}_{OH}}$$

Figure 5.4 Nucleotide sequence at the 3′ terminus of 16S ribosomal RNA.

Shine and Dalgarno[21] put forward the idea that the complementary sequences may be involved in the mechanism of recognition of the initiation sites. The 3′ end of the ribosomal RNA could anneal to the complementary stretches in the mRNA and could thus fix the ribosome in the proper position, near the initiator codon. The formylmethionyl-tRNA which binds to the ribosome could then easily interact with the initiating AUG or GUG. Steitz and Jakes[22] found direct experimental proof for the annealing of the 3′ tail of 16S RNA to an initiation site in R17 RNA. They isolated the A-protein initiation site from ^{32}P-labelled R17 RNA and incubated it with ^{32}P-labelled ribosomes under conditions of initiation complex formation. The complex was treated with colicin E3, a bacterial protein which acts as a very specific RNase and splits off a 45-nucleotide-long tail from the 3′ end of the 16S RNA. The ribosomes were then removed from the mRNA and the initiation site was re-isolated and fingerprinted. It was found that the initiation site was associated with the 3′ tail of the 16S RNA. Two sets of oligonucleotides were detected on the fingerprints, those from the A-protein initiation site and those from the 3′ terminus of the ribosomal RNA.

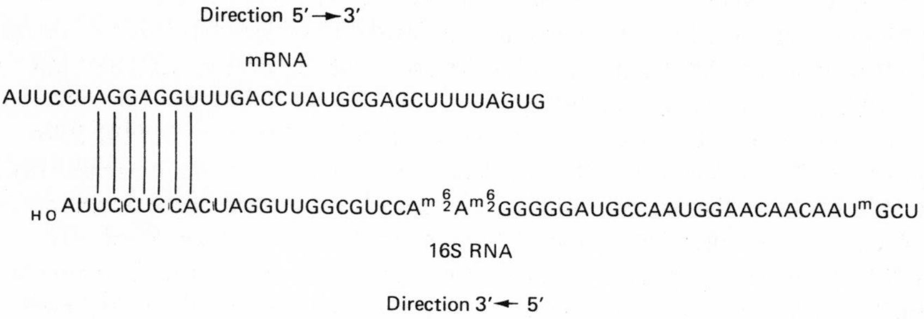

Direction 5′→3′

mRNA

AUUCCUAGGAGGUUUGACCUAUGCGAGCUUUUAGUG

$_{HO}$AUUCCUCCACUAGGUUGGCGUCCA$^{m}{}^{6}_{2}$A$^{m}{}^{6}_{2}$GGGGAUGCCAAUGGAACAACAAU^{m}GCU

16S RNA

Direction 3′← 5′

Figure 5.5 Annealed structure of A-protein initiation fragment and 3′ tail of 16S RNA, split off by colicin E3.

Figure 5.5 shows the base-paired structure that can be formed between the 3′ tail of 16S RNA and the initiation fragment of the A-protein cistron. There are seven complementary bases in the two RNA fragments which can form a very stable annealed structure.

The conditions of base pairing are often less favourable with other cistrons in R17 RNA or in other mRNAs. As can be seen in *table 5.1*, some initiation sites, e.g. at the coat protein cistron of R17, Qβ or f1 RNA, or the leader sequence of the tryptophan messenger, contain only a trinucleotide sequence complementary to the 3′ terminus of 16S RNA. Although very stable annealing should not be expected at this stage, as this would prevent the ribosome from moving along the mRNA during translation, it is still questionable whether the pairing of three bases is sufficient to bring about the binding of ribosomes to mRNA without a further stabilising effect, e.g. without the contribution of some specific protein factor.

As far as specific selection of genuine initiation sites is concerned, it also seems improbable that a three- or four-nucleotide-long purine tract should be sufficient for recognition. The probability is rather high that in a long polynucleotide chain with any random sequence, a short purine stretch will be present in the neighbourhood of an AUG triplet. In fact, in MS2 RNA there are more than a dozen AUG triplets (some of them out of phase) near to trinucleotide sequences which are complementary to the tail of 16S RNA, but which do not act as ribosome binding sites. Even longer polypurine sequences do occur in such positions without specifying an initiation site, e.g. in the replicase cistron of MS2 RNA the sequence UAAGGAGCCUGAUAUG occurs in the middle of the message[10]. The first seven nucleotides in this sequence are complementary to the 3′ terminus of 16S RNA and this stretch is at a distance of seven nucleotides from AUG. Still, this AUG acts as an internal methionine codon. Furthermore, there seems to be no parallelism between the efficiency of a genuine initiation site and the extent of base pairing that can be achieved at this site[23]. It seems therefore that, although annealing of ribosomal RNA to mRNA may play an important role in initiation complex formation, it may represent merely one of several determining factors in the selection of the specific initiation site. The actual initiation process is probably much more complex than that outlined above. As will be discussed in more detail in Chapter 8, several hypotheses have been put forward implicating the role of protein factors, or that of the secondary structure of mRNA, but it is still an open question exactly what additional information is built into the structure of mRNA, specifying the site where translation starts.

The termination codon for the A-protein cistron is UAG both in MS2 RNA (position 1309–1311) and in R17 RNA, and it is the same for the replicase cistron in MS2 RNA. At the terminus of the coat protein cistron two nonsense codons, UAA and UAG, follow each other. Whether a double termination signal is an additional precaution against 'readthrough', i.e. against an ambiguous interpretation of a nonsense codon, is not known. It is not used frequently. Nevertheless, it is interesting to note that such readthrough can indeed occur in the translation of another coliphage, Qβ. The products of translation of Qβ RNA are three major viral proteins, A_1 protein, coat protein and replicase, and a minor component, A_2 protein, which has the same N-terminal amino acid sequence as coat protein but is some 200 amino acids longer. A_2 protein is thought to be the product of translation beyond the UGA terminus of the coat cistron[24]. UGA is less efficient as a termination codon than UAG or UAA; it has been characterised as 'leaky'.

(c) Untranslated sequences

All mRNAs, whatever their origin, contain some stretches which do not belong to the message itself and are never translated. The presence of such untranslated sequences could be predicted even before they were actually detected in R17 RNA. The molecular weights of the three proteins of this virus were known and the length of the coded messages could thus be calculated: about 1200 nucleotides for the A-protein, 387 for the coat protein, and 1500 to 1600 for the synthetase. This gives a total length of approximately 3100 or 3200 nucleotides. However, R17 RNA contains over 3500 nucleotides, an excess of more than 300 nucleotides over those required by the triplet code. (The lengths of the messages are exactly known today: A-protein cistron, 1176 nucleotides; coat protein cistron, 387 nucleotides; synthetase, 1632 nucleotides. Total length of the RNA molecule, 3569 nucleotides.) A similar redundancy has been observed in all mRNAs characterised so far, although the actual number of untranslated nucleotides varies from one mRNA to the other.

Nucleotide sequence studies revealed the location of untranslated sequences in the different mRNAs. An untranslated 'leader sequence', preceding the actual message, is always present at the 5′ terminus of prokaryotic mRNAs*. The length of this leader sequence varies in different molecules: in MS2 RNA and R17 RNA it is 129 nucleotides long, in Qβ RNA, 61 nucleotides. In the mRNA of the lactose operon the initiation codon is preceded by 38 nucleotides only. Another untranslated region has been found at the 3′ end, after the termination codon; in MS2 RNA its length is 174 nucleotides.

*So far only one exception has been found: an mRNA of bacteriophage λ has no leader sequence but begins with the AUG initiation codon[24a].

In mRNAs carrying information for more than one protein (polycistronic mRNAs), in addition to these untranslated sequences at the ends of the molecule, the cistrons are usually, but not always, separated by untranslated 'intercistronic regions'. The leader sequence of MS2 RNA (nucleotides 1-129) and the intercistronic region between the A-protein and coat protein cistron (nucleotides 1309-1334) can be seen in *figure 5.2*. Another intercistronic region, 36 nucleotides long, exists between the coat protein and synthetase cistron; *figure 5.1* shows both intercistronic regions. In the tryptophan mRNA of *E. coli*, however, there is no intercistronic region between the messages for proteins B and A. The termination codon for protein B, UGA, overlaps with the AUG initiation codon for protein A, forming a pentanucleotide structure, UGAUG, of which only the first two nucleotides are not translated (*figure 5.6*.) The two messages are out of phase; the reading frame for protein B is different from that for protein A. This is

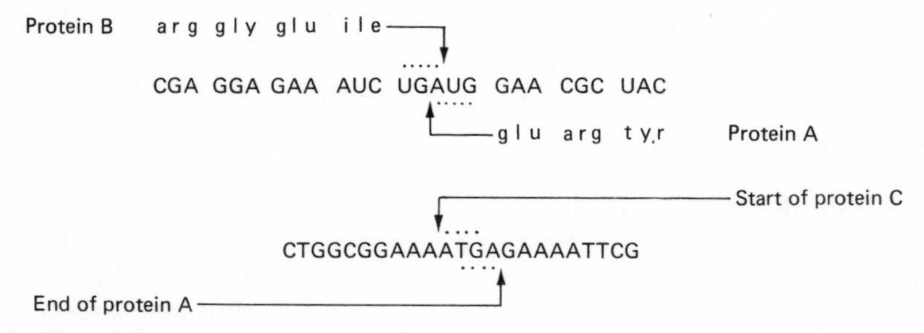

Figure 5.6 Overlapping termination and start signals. Above: sequence of the B–A region of the tryptophan messenger (Platt and Yanofsky[19]). Below: nucleotide sequences from the ΦX174 genome (Sanger *et al.*[11]). Initiation and termination codons are marked by dotted lines.

not a unique structure; in the ΦX174 and G4 genomes similar sites have been detected, where termination codons for one protein overlap with initiation codons for another protein (*figure 5.6*).

As described above, the region preceding the initiation codon contains the recognition site for ribosomes. In these overlapping cistrons a stretch of RNA must therefore fulfil two functions: in *figure 5.6* the stretch at the left provides the ribosome binding and recognition site for the A-protein cistron in the tryptophan operon and at the same time contains the codewords for the last amino acids of the B protein. A polypurine sequence, GAGGAGAAA, similar to those described on p. 157–158 is indeed present in this RNA stretch. AGGAG is complementary to the 3′ end of 16S RNA.

There are indications that a double function of an RNA stretch may not be a rare phenomenon. Part of a coded message may also serve as binding site for protein factors, or an RNA stretch which can bind a protein may also anneal to another RNA molecule.

The untranslated sequences are thought to take part in control functions.

They may contain control signals, as the initiation signal, discussed above, or binding sites for proteins which can regulate translation. In the leader sequence of some bacterial mRNAs, AUG is present, followed by a few codons and eventually by a termination codon[26a]. These sequences may code for a short peptide, but it is not known whether such peptides are actually synthesised *in vivo*. Lee and Yanofsky[25] assume that translation of this 'leader peptide' may serve the control of gene expression, in conjunction with the attenuator, described below. At the 3′ terminus some mRNAs contain a series of U residues, a sequence which is supposed to provide the signal for the termination of transcription (*see* p. 93). In viral RNAs, the 3′-terminal untranslated sequence also contains the site of interaction with RNA polymerase, the enzyme responsible for the replication of the RNA of the virus. Untranslated sequences also contribute to the secondary and tertiary structure of the mRNA molecule which in turn may control the efficiency of translation. Furthermore, the leader sequence may have a stabilising effect: it may protect the mRNA against some RNases. An exonuclease present in *E. coli* ribosomes acts in the 5′ → 3′ direction, starting at the 5′ terminus. If the ribosome binding site is rather distant from the 5′ end, the ribosomal enzyme cannot degrade the mRNA.

In the leader sequence of tryptophan mRNA, a CUUUUUUU sequence is present which serves transcriptional control. Synthesis of mRNA may be terminated at this site, preventing transcription and translation of the coded message (*see* p. 93). Bertrand and coworkers[26] consider this 'attenuator' as a new type of regulatory mechanism which controls the expression of the operon. Attenuators have since been detected also in the leader regions of the histidine and phenylalanine operons[26a].

(2) Eukaryotic mRNAs

A number of eukaryotic mRNAs have been isolated in recent years, some of them in sufficiently pure form to carry out nucleotide sequence analysis of the molecules. Their relative ease of isolation is due partly to their stability, which is much greater than that of their prokaryotic counterparts, and partly to some common structural characteristics: most eukaryotic mRNAs contain a poly(A) tail at their 3′ end and this provides a simple way of separating the messenger fraction from other RNA species of the cell by annealing it to a complementary oligo- or polynucleotide. All cytoplasmic RNAs which contain poly(A) can thus be retained on a membrane filter to which poly(U) has been bound or on a column made of a cellulose derivative containing oligo(dT) groups. Avid and Leder[27] worked out a technique for the isolation of globin mRNA, using this principle; Brownlee *et al.*[28] purified immunoglobulin mRNA for sequence studies in this way. The technique is now widely used for the isolation of different mRNAs from eukaryotic cells.

Another approach is based on immunoprecipitation of the nascent polypeptide chain which is being synthesised on the mRNA template. A complete polysome complex (*see* Chapter 8) is precipitated, which comprises the mRNA template, several ribosomes engaged in translation and nascent polypeptide chains of different length. The other components can be removed and the specific mRNA isolated from this complex. This method was first applied to ovalbumin mRNA by Schimke *et al.*[29] and several modifications of the method have been used since to obtain specific mRNA preparations.

Identification of a specific mRNA species is usually based on *in vitro* translation and analysis of the protein synthesised under the direction of this RNA. It is now also possible to use hybridisation techniques to identify eukaryotic mRNAs, as great progress has been made in the isolation of genes from eukaryotic DNAs. Synthetic cDNA, made on an authentic mRNA template, has also been used to identify other mRNA preparations by hybridisation.

All these methods produce mRNA of high purity and often also in high enough yield, but the preparations are not radioactive. The lack of highly labelled preparations delayed investigations of the primary structure of eukaryotic mRNAs. Only the latest sequencing techniques succeeded in overcoming this difficulty (*see* Chapter 4), with the spectacular result of producing the first complete mRNA sequence, that of rabbit β-globin mRNA[30], and shortly afterwards the complete sequences of the human β-globin mRNA[31], of rabbit α-globin mRNA[51], of chicken ovalbumin mRNA[52] and of the constant and 3′-untranslated regions of immunoglobulin light-chain mRNA[55]. Major parts of the nucleotide sequences of other eukaryotic mRNAs have also been established, including the mRNAs of viruses (e.g. SV40, ref. 56). In this latter case, some mRNAs have been directly sequenced; the structure of others can be deduced from the known nucleotide sequence of the genome.

Comparison of the primary structure of prokaryotic and eukaryotic mRNAs reveals similarities in the non-random use of degenerate codons, in the preference for AUG as initiation codon, and in the presence of untranslated sequences preceding and following the coded message. However, eukaryotic mRNAs show very special structures at their 5′ as well as their 3′ termini and it seems that great differences may also exist between the two groups of mRNAs with respect to the nature of some control signals.

Untranslated sequences in eukaryotic mRNAs

Different eukaryotic mRNAs contain untranslated sequences of widely varying length at both their 5′ and their 3′ termini. We have to distinguish here between (a) sequences which are copied from the DNA, i.e. which are produced by transcription of the DNA and splicing of the transcript; and (b) additional sequences which are not present in the DNA template, but are added to the termini of the mRNA molecule in the course of post-transcriptional modifications.

(a) Untranslated sequences transcribed from DNA

The leader sequence The length of the 5′-terminal untranslated sequences shows great variability. The shortest stretch has been reported for brome mosaic virus mRNA: it is only 10 nucleotides long. In the different reovirus mRNAs the coded region starts at the 15th to 33rd nucleotide from the 5′ end. Rabbit β-globin mRNA has a 5′-terminal leader sequence of 53 nucleotides, ovalbumin mRNA a leader sequence of 64 nucleotides. In the late mRNAs of SV40 virus, a stretch more than 200 nucleotides long precedes the initiation codon. It is interesting that this latter leader sequence contains a potential peptide coding sequence[56].

It has been mentioned in Chapter 3 that a number of eukaryotic viruses (SV40, polyoma, adenovirus type 2) produce spliced mRNAs in which the leader sequence and the coded message are copied from separate, remote parts of the viral genome. In adenovirus the leader sequence itself is transcribed from three different parts of the DNA. In all these viruses more than one mRNA contains the same leader sequence: the same long, untranslated RNA stretch is attached to different coding sequences. This suggests that some structural feature of the leader sequence may be of importance for the translation of these messages[32].

As in prokaryotes, in eukaryotic mRNAs it is assumed that the untranslated sequences have some role in regulatory functions. As the essential steps in translation of prokaryotic and eukaryotic mRNAs are analogous, it was expected that the mechanism of ribosome binding and thus the structure of ribosome binding sites would show some similarities in the two kinds of RNAs. The nucleotide sequences around the initiation codon were therefore determined in a number of eukaryotic mRNAs, to see if purine tracts, complementary to the well preserved 3′-terminal sequence of eukaryotic 18S rRNAs, were present in the neighbourhood of the initiation codons.

The first results were negative: no complementary sequences were detected in globin, reovirus or brome mosaic virus mRNAs in the expected positions. In β-globin mRNA a complementary hexanucleotide could be found but at an improbable site, overlapping with the initiation codon itself[33,34]. Recently, however, Hagenbuchle *et al.*[35] determined longer nucleotide sequences at the 3′ termini of different eukaryotic 18S rRNA molecules. They discovered a polypurine tract, a few nucleotides farther from the 3′ end, which shows complementarity to four- to six-nucleotide-long stretches in the leader sequence of a number of eukaryotic mRNAs. This indicates that mRNA:rRNA annealing may also take place in eukaryotes when ribosomes bind to the initiation sites of mRNAs. Base pairing may occur here between a purine tract in the rRNA and a pyrimidine tract in the mRNA.

The variations in the structures of different leader sequences, listed in *table 5.2*, suggests, however, that this interaction is not as simple as in prokaryotic systems. The position of the complementary stretches is very variable; in some mRNAs (e.g. reovirus) it is quite close to the initiation codon, in others (histone, human β-globin) it is so far from it that it is difficult to envisage how 18S RNA and initiator tRNA, which are at a fixed distance on the ribosome, can anneal simultaneously to these sequences. Some mRNAs do not contain a complement-

Table 5.2 Nucleotide sequences at the initiation sites of eukaryotic mRNAs

mRNA	Sequence
α-globin (rabbit)	m⁷Gpppm⁶AmCACUUCUGGUCCAGUCCGACUGAGAAGGAACCACCAUG
β-globin (rabbit)	m⁷Gpppm⁶AmCACUUGCUUUUGACAACUGUGUUUACUUGCAAUCCCCCAAAAGAGACAGAAUG
α-globin (human)	m⁷GpppA*CUCUUCUGGUCCCCACAGACUCAGAGAGAACCCACCAUG
β-globin (human)	m⁷GpppA*CAUUUGCUUCUGACACAACUGUGUUCACUAGCAACCUCAAACAGACACCAUG
Histone 114 (sea urchin)	. . .CACAGAACUCGCUCUCAACUAUCAAUCAUCAUCAUG
Reovirus s54	m⁷GpppCmCUAUUUUGCCUCUUCCCAGACGUUGUCGCAAUG
s45	m⁷GpppCmCUAAAGUCACGCCUGUCGUCCUCACUAUG
s46	m⁷GpppGmCUAUUCGCUGGUCAGUUAUG
m30	m⁷GpppGmCUAUUCGCGGUGAUG
m44	m⁷GpppGmCUAAAGUGACCGUGGUCAUG
m52	m⁷GpppGmCUAAUCUGCUGACCGUUACUCUGCAAAGAUG
SV40 VP1	. . .ACGGAAGTGTTACTTCTGCTCTAAAAGCTTATG
BMV coat protein	m⁷GpppGUAUUAAUAAUG

The underlined regions indicate sequences which may anneal to the 3′-terminal region of 18S rRNA[35]. U•G base pairs are marked by single dots. An 18-nucleotide-long sequence from the 3′ end of 18S RNA is shown below. The complementary purine-rich tract is underlined.

3′ HO-AUUACUAGGAAGGCGUCC. . . . 5′

ary pyrimidine stretch at all. The interpretation of these data is at present rather equivocal, and the role of annealing between eukaryotic mRNAs and rRNA is a matter of controversy. According to Baralle and Brownlee[53], there is no signal, apart from the initiation codon, which would be consistently present in all eukaryotic mRNAs and which could be considered as a recognition site. This idea is in agreement with recent results of Kozak and Shatkin[54], who also emphasise the role of the AUG codon as the essential feature required for ribosome binding.

There are reasons to suppose that the leader sequence may be in a folded structure or may form a loop when ribosome binding occurs. This would imply that the actual distance between the pyrimidine tract and the AUG codon can be much shorter than would be expected from the linear sequence. This possibility is suggested by the variable size of ribosome binding sites as determined in β-globin mRNA[33] and in reovirus mRNAs[36].

The length of the RNA stretch protected by the particles depends on whether 40S ribosomal subunits or 80S ribosomes are bound to the mRNA. The latter protect a 25–30-nucleotide-long fragment, similar in size to the ribosome binding sites in prokaryotic mRNAs and corresponding well to the dimensions of the ribosomal particles. The 40S subunit protects a much longer fragment, up to 54 nucleotides long, which can be accommodated on the ribosome only in a folded structure. McReynolds *et al.*[52] propose a possible hairpin loop structure in the leader sequence of ovalbumin mRNA.

3′-terminal untranslated sequences The length of the untranslated sequences at the 3′ end of eukaryotic mRNAs also varies widely. In rabbit β-globin mRNA a 95-nucleotide-long untranslated region follows the coded message, the human β-globin mRNA contains an additional 39 nucleotides at the 3′ end, while the 3′-terminal untranslated sequence is 211 nucleotides long in immunoglobulin light-chain mRNA[55] and 637 nucleotides in ovalbumin mRNA[52].

There are only assumptions as to the function of these untranslated sequences. The very length of some untranslated regions makes doubtful the functional significance of these sequences: no control signals, protein binding sites, etc. could require the presence of an RNA stretch over 600 nucleotides long. One protein binding site can certainly be expected in this position in eukaryotic mRNAs: these molecules interact with poly(A) polymerase, an enzyme which in the course of post-transcriptional processing synthesises a long poly(A) sequence joined to the 3′ terminus.

(b) Untranslated sequences resulting from post-transcriptional modification

Poly(A) tail Post-transcriptional modifications affect both the 3′ and the 5′ terminus of eukaryotic mRNAs. With the exception of histone mRNAs, almost all mature eukaryotic mRNA species contain a 3′-terminal poly(A) sequence the average length of which may be estimated as 200 nucleotides, although the exact length is difficult to define as it varies not only from one RNA species to another but also during the lifetime of a single mRNA species.

No clearly defined function has so far been ascribed to the poly(A) tail, although suggestions include that it is involved in the transport of mRNA from the nucleus to the cytoplasm, that it increases the stability of cytoplasmic mRNA, and that it may be required for efficient translation. Its involvement in the transport process is indicated by the effect of 3′-deoxy-adenosine, an inhibitor of poly(A) polymerase. In the presence of this inhibitor, no poly(A) tails are synthesised and at the same time there is a decrease in the amount of mRNA appearing in the cytoplasm, although the synthesis of RNA is not affected. The strongest evidence for a stabilising effect of poly(A) was described by Huez *et al.* and Marbaix *et al.*[37]. Globin mRNA can be translated *in vivo* in *Xenopus* oocytes and its decay can be followed by the decline of globin synthesis. The above authors found that the half-life of globin mRNA injected into oocytes was considerably reduced if the poly(A) tail had been removed from the molecule. Experiments with poly(A)-containing mRNAs, aimed at establishing the effect of the length or of the removal of this tail on the efficiency of translation, yielded rather controversial results. It adds to the uncertainty of attributing a specific function to the poly(A) tail that most but not all mRNAs possess this structure. There are thus several possible roles for the poly(A) tail, but it is still an open question which of these can be considered as its main function[38].

Capped 5′ termini At the 5′ end of eukaryotic mRNAs a rather unusual structure has been detected. In most mRNAs studied so far (whether they originate from prokaryotes or eukaryotes), the first nucleotide is a purine, more often G than A. It was first observed by Perry and Kelley that the 5′ terminus of mRNA from mouse L cells contains methylated nucleotides. The exact structure of the 5′ termini of a great number of viral and non-viral eukaryotic mRNAs has been studied in several laboratories with the result that a unique structure has been established in all these molecules[39,40]. It is called the ‘capped structure’. The terminal G (or A) is methylated in the ribose moiety and is linked to a 7-methyl guanosine, through a pyrophosphate or triphosphate group

m^7G (HO–) –P–O–P– $\overset{m}{G}$ –P– N_1 –P– N_2 ... or m^7G (HO–) –P–O–P–O–P– $\overset{m}{G}$ –P– N_1 –P– N_2 ...

The pyrophosphate linkage involves the 5′ hydroxyls of both the 7-methyl G and the ribose-methylated G. These two nucleotides are thus linked to each other in an ‘inverse’ way, different from all the other internucleotide linkages in RNA. All other nucleotides are joined by 5′, 3′ phosphate bonds, while this terminal structure contains a 5′, 5′ bond. As a result, there is no free 5′ hydroxyl or 5′ phosphate at this so-called 5′ end. This 5′, 5′ pyrophosphate bond cannot be split by nucleases but is susceptible to pyrophosphatase, an enzyme which does not attack other bonds in RNA. It was, in fact, the lack of a free 5′ group and the resistance to nucleases which led to the discovery of this terminal structure. There is some variation in the structure of different mRNAs, according to whether both methylated nucleotides are G's and whether they are linked by two or three phosphate groups. The general formula can be written as follows:

$$\begin{array}{lll} m^7G & \overset{m}{N} & N \\ HO- & & \\ P{-}O({-}P{-}O){-}P & P & P\ldots \end{array}$$

There is still some controversy about the function of these capped structures. Shortly after their detection it was claimed that they were required for efficient translation and took part in the formation of the initiation complex[40,41]. Caps have been found to constitute part of the ribosome binding site in different eukaryotic messengers. The longer RNA fragment, which is protected by the 40S ribosomal subunit, includes in most cases (but not always, *see* ref. 33) the capped termini. In the shorter sequence, protected by 80S ribosomes, the capped end is often not present[36]. As binding of 40S subunits to mRNA takes place at an early stage of initiation of translation and the 80S initiation complex is formed in a later step, these data suggest that caps may play a role in an early event in the initiation of protein synthesis, probably by providing a recognition site for the ribosome.

Another approach to exploring the function of the capped termini has also been followed by several authors: the caps have been modified or removed and the effects of such modifications on the translation of the message have been studied in *in vitro* systems. The translation of reovirus mRNAs, globin mRNA and a number of other eukaryotic mRNAs is greatly reduced if the cap structure is missing[41,42,54]. All this evidence strongly suggests that the capped structure at the 5′ termini of eukaryotic mRNAs plays an essential role in translation. The controversial fact remains, however, that not all eukaryotic mRNAs contain this structure, and mRNAs which possess no capped termini (e.g. poliovirus mRNA) can still be efficiently translated.

SECONDARY STRUCTURE OF mRNA

mRNAs are single-stranded molecules which contain many base-paired regions but do not form a regular double-helical structure. Completely double-helical RNA cannot be translated; some viruses which contain double-stranded RNA as genetic material (e.g. reovirus) produce a single-stranded transcript which can serve as messenger.

In a long mRNA molecule there are several stretches which contain complementary bases. The polynucleotide chain can fold back on itself, forming a hairpin loop in which the complementary sequences are annealed. A number of hairpin loops can be seen in the structure of MS2 RNA in *figures 5.1* and *5.2*. The annealed regions may not always consist of perfectly matched base pairs. If we consider the three hairpin loops formed at the 5′-terminal part of this RNA, from

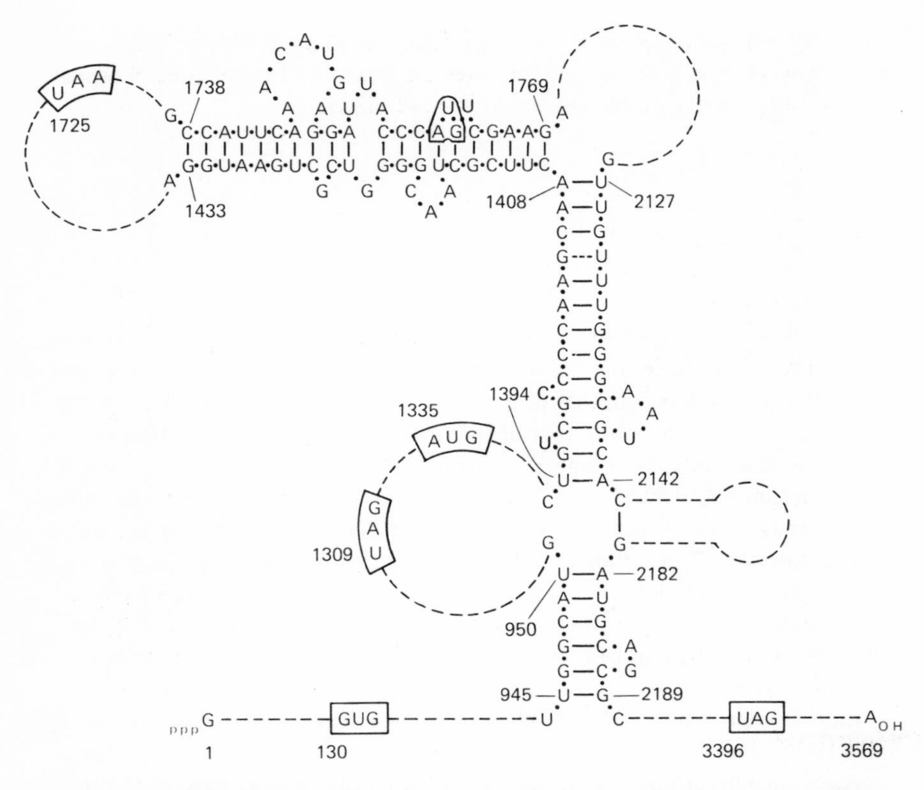

Figure 5.7 Possible long-range interactions in the MS2 RNA molecule, according to Fiers *et al.*[10]. (Reproduced with permission from *Nature*, **260**, 500 (1976).)

nucleotides 1 to 109, we find several imperfections in the base-paired structure. Some nucleotides (e.g. C in position 31, A in position 94) are not matched by any nucleotide in the opposite strand. In the long hairpin loop formed by nucleotides 36–73, base-paired regions alternate with short unpaired stretches. This structure is quite characteristic of single-stranded RNA. As the two strands do not form a regular double-helical structure, the base pairs in the annealed stretches are not necessarily of the Watson–Crick type. G and U can also be linked by hydrogen bonds, although this base pair is not as stable as the G–C or A–U type. In the three hairpin loops there are five G–U pairs and several more can be found in other loops.

Annealing of complementary sequences does not always lead to the formation of hairpin loops. A different structure results if complementary stretches are present at distant parts of the molecule. Such structures can also be seen in *figure 5.2*: a long, double-stranded stretch is formed near the initiation codon of the A-protein cistron by the annealing of nucleotides 115–129 and 874–885, two base pairs being contributed by nucleotides 814–815.

Annealing of distant stretches may determine the shape and size of the whole molecule; such long-range interactions are essential in the formation of the tertiary structure. Little is known today about the exact secondary structure, and even less about the tertiary structure of mRNAs. Fiers and coworkers[10] proposed a

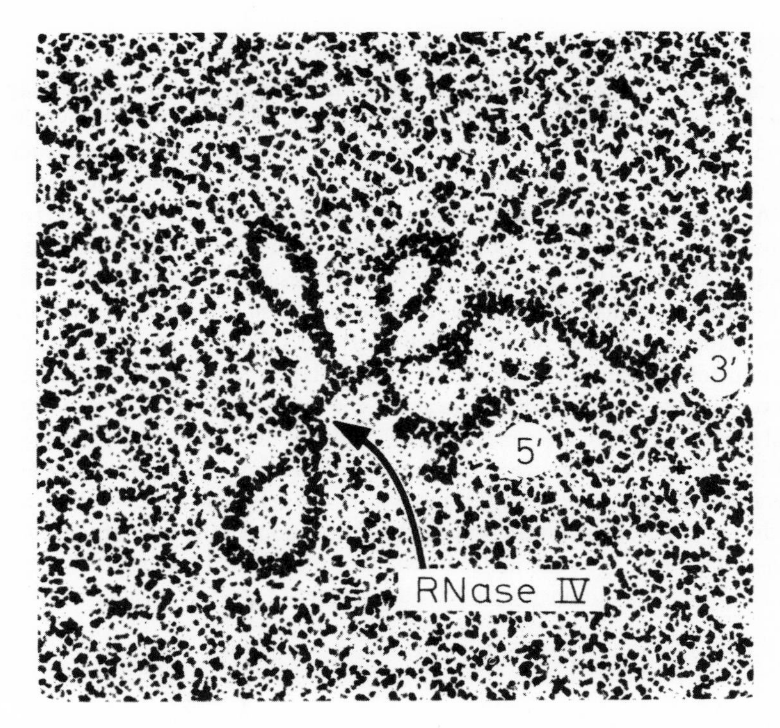

Figure 5.8 Electron micrograph of MS2 RNA by Jacobson and Spahr[43]. The initiation sites of the coat protein cistron and of the replicase cistron are located close to each other, near the base of the central loop. (Reproduced with permission from *J. Mol. Biol.*, **115**, 279 (1977). Copyright by Academic Press Inc. (London) Ltd. Photograph by courtesy of Dr A. Jacobson.)

model for long-range interactions in MS2 RNA based on possible base pairing between stretches in different parts of the molecule the complementarity of which could be detected only after the complete nucleotide sequence had been determined (*figure 5.7*). Jacobson and Spahr[43] observed native MS2 RNA as well as fragments of this molecule under the electron microscope and obtained a clear map of the major loops (the small hairpin loops do not show up in the electron microscope). So far, this is the only mRNA for which the conformation has been determined by direct physical methods; it is shown in *figure 5.8.* Knowledge of the position of the three cistrons in the nucleotide sequence enabled the authors to map approximately the initiation sites of the proteins in their RNA model. As can be seen in the figure, the conformation of MS2 RNA, as determined by electron microscopy, confirms that complex interactions do occur between parts of the molecule which are far apart in the linear sequence. Since the determination of the complete nucleotide sequence of several eukaryotic mRNAs, some speculative proposals have been put forward concerning their secondary structure, but no models have so far been produced. Baralle[34] called attention to the possibility that the 5′-terminal and 3′-terminal parts of the rabbit β-globin mRNA could anneal to each other. This may be of importance for some regulatory mechanisms which control translation of this mRNA.

Determination of the secondary structure

The fine details of the secondary structure of mRNA are not accessible today to direct physicochemical analysis. The smaller tRNA molecules have been subjected to X-ray crystallography, and exact data on the secondary and tertiary structure have been obtained. This approach, however, cannot yet be applied to large mRNAs. Jacobson and Spahr's electron microscopic technique enables us to visualise the overall conformation of MS2 RNA. If similar investigations are carried out in the future on other mRNA molecules, knowledge of the conformation together with that of the nucleotide sequence will provide a very valuable way of mapping mRNAs. As we will see later, this should also help to elucidate some control mechanisms which influence translation. The overall conformation, however, does not yield information about the individual small hairpin loops, the actual sequences which are annealed. For the function of different parts of the RNA it is of great importance which bases are freely accessible and which are present in base-paired form. In this respect we can rely only on hypothetical models, constructed on a rather speculative basis, combining the data on the total degree of base pairing in the molecule which can be determined by direct measurements and the data obtained by nucleotide sequence analysis which show the

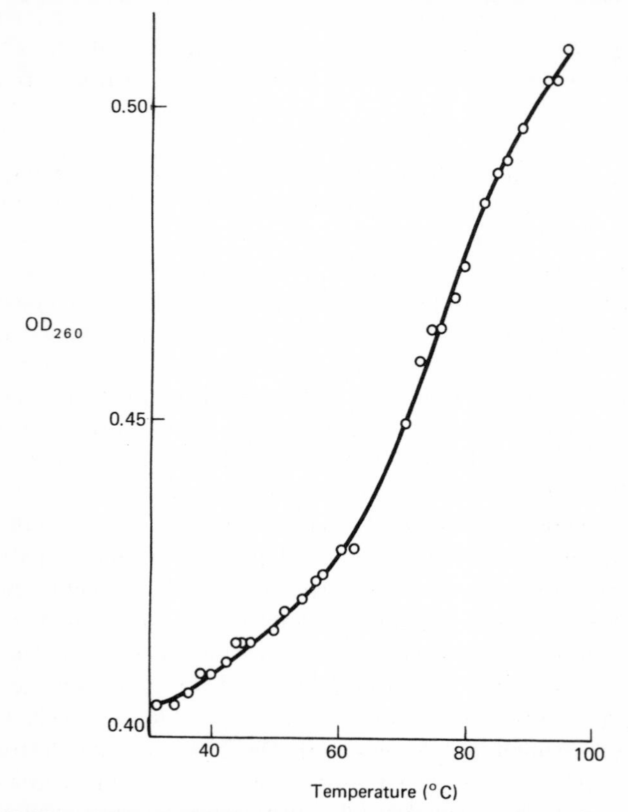

Figure 5.9 Melting profile of MS2 RNA. (By courtesy of Mr M. J. F. Fowler.)

position of complementary regions.

The average content of base-paired regions in an RNA molecule can be studied by classical physicochemical techniques. As a result of imperfect base pairing, the melting profiles of RNA molecules are different from those of DNA. The different base-paired regions melt at different temperatures, the OD_{260} increases steadily over a wide temperature range, 25–30°C (*figure 5.9*). From the transition mid-point, the total amount of base-paired structures in the molecule can be calculated. The Tm of MS2 and R17 RNAs is rather high, 60–70°C, depending on the salt concentration of the medium. This indicates a highly hydrogen-bonded structure. A compact secondary structure is also suggested by their slow reaction with formaldehyde and their small radius of gyration. An overall estimate of the degree of base pairing in viral RNAs has also been obtained by infra-red spectral analysis[44]. The total amount of base-paired structures in viral mRNAs has been estimated to be 60–85%. These techniques do not reveal, however, where these base-paired regions are located in the molecule.

In the course of nucleotide sequence determination, useful information is also obtained concerning the most probable secondary structure of the molecule. It is possible to calculate the stability of different base-paired structures which can be formed with the established sequences and to select the most stable model[45]. The results of partial digestion also reveal some characteristics of the secondary structure. Different sites of the RNA show different susceptibilities towards RNases. Thus, in *figure 5.2*, the arrows indicate the sites where even mild digestion with RNase T_1 causes a split. As RNase T_1 acts specifically on single RNA strands, it can be assumed that these G residues form part of a single-stranded stretch of the molecule. G residues, at which RNase T_1 acts only under more drastic conditions, are probably present in base-paired form. Other RNases which act differently on single- and double-stranded RNA structures (e.g. pancreatic RNase, S_1 nuclease, RNase III) can also be used to obtain similar information. Because of the preference of some RNases for structures which are not base paired, hair-pin loops can often be recovered in the digest of RNA in the form of intact large fragments.

Compilation of the results obtained by these different methods allows the construction of probable models, but the data are usually not sufficient to define a unique secondary structure. Often, alternative structures are proposed for the same RNA. The uncertainties of this approach are caused by the vast number of possible interactions between distant parts of the molecule. No reliable comparison can be made of the stability of different base-paired structures until the total sequence of the RNA is known. The interpretation of the susceptibility of different sites to RNases also becomes somewhat ambiguous if the structures concerned are neither completely single stranded nor completely double stranded, but contain imperfectly base-paired stretches. The different mRNA models are therefore helpful in visualising some characteristics of the conformation and also in interpreting the role of conformation in some functions of mRNA, but many details of these structures should be regarded as hypothetical until direct physicochemical analysis can be applied to confirm or disprove the suggested structures.

In the following we will use mainly the MS2 RNA models, constructed by Fiers *et al.*[8,9,10] and by Jacobson and Spahr[43], to illustrate how the conformation of mRNA can influence its function.

The role of secondary structure in mRNA function

Many observations point to an involvement of the secondary structure in the control of protein synthesis. Most of the evidence comes from studies on viral RNAs, mainly on MS2, R17 and f2 RNA. As the structures of these three RNAs are very similar, the results are valid for all three of them and the interpretation of the data in terms of the role of the secondary structure is probably also valid for the RNAs of other viruses. We cannot be certain, however, how much farther we can go with generalisations. There is too little information available at present on the secondary structure of eukaryotic mRNAs to compare them with these viral RNAs. It can also be expected that the controls which work at translational level and which are influenced by the secondary structure, differ in several respects in these two groups of mRNAs.

As will be discussed below, the secondary and tertiary structure of viral RNAs affects the relative efficiency of translation of different cistrons. This problem is irrelevant in the case of eukaryotic mRNAs which are monocistronic. Still, a strong probability exists that the efficiency of an initiation site or the recognition of a specific site depends on the secondary structure of this RNA region in eukaryotic mRNAs in much the same way as in viral RNAs. Eukaryotic mRNAs interact with ribosomes and with different protein factors involved in protein synthesis in a way analogous to the interactions of MS2 RNA with ribosomes and factors of *E. coli*. Annealing between eukaryotic mRNA and eukaryotic ribosomal RNA has also been assumed, although it has not yet been confirmed. Protein–RNA interactions as well as RNA–RNA interactions usually depend very strongly on the secondary and tertiary structure. It therefore seems probable that accessibility of a functional site, or recognition of a specific tertiary structure at a functional site, will be of importance for the translation of all the above mRNA species. We have to keep in mind, however, that eukaryotic mRNAs also possess another structural characteristic which is supposed to play a role in translational control, though by an as yet unknown mechanism: protein molecules are attached to the mRNA in the cytoplasm (*see* Chapter 8).

The phenomena described below which emphasise the importance of the conformation of viral RNAs for the process of translation, therefore seem relevant for eukaryotic mRNAs, even though the actual results obtained with MS2 RNA cannot be directly applied to globin or ovalbumin mRNA. It is doubtful, however, if we can go any further with generalisation. In bacterial mRNAs, the formation and the role of the secondary and tertiary structure may be very different. In bacterial cells transcription and translation occur in a coupled process: translation of mRNA starts near the 5′ terminus while the rest of the RNA molecule is still being synthesised. The mRNA is thus functional before the complete RNA molecule is present, before its final secondary and tertiary structure, as determined by the complete nucleotide sequence, could have been formed. The secondary structure of these mRNAs may change in the course of transcription, because further complementary stretches are produced, or it may change because attachment of ribosomes to the nascent RNA stretches disrupts or abolishes annealed structures. Under these conditions it is impossible to define what, if any, kind of secondary structure is present at a given stage of the transcriptional-translational process, and whether the secondary structure has any role in controlling the binding of ribosomes, the initiation of translation or any other step in bacterial protein synthesis.

The following results refer mainly to viral RNAs, which contributed much, in this as well as in other respects, to our present knowledge on protein synthesis. The data obtained with these RNAs do not enable us to describe exactly the structure required for efficient translation, but they show how the conformation of mRNA can play an important role in the regulation of translation.

(a) Disruption of the secondary and tertiary structure alters the efficiency of translation of different cistrons

In MS2, R17 and f2 viruses which code for three viral proteins, translation of the three cistrons occurs with different efficiency both *in vivo* and *in vitro*. Coat protein is produced in great excess over A-protein and replicase. It was shown by Lodish and Robertson[46] that in an *E. coli* extract, synthesis of the two latter proteins could be greatly increased if the secondary structure of the viral RNA was at least partly abolished by mild formaldehyde treatment. When using native f2 RNA as messenger, A-protein and replicase constituted only a minor fraction of the proteins produced. If the RNA was denatured by mild formaldehyde treatment, their yield was increased 4–20-fold. This suggests that in the native RNA molecule the secondary and tertiary structure prevents translation of the A-protein and replicase cistrons. Thus, synthesis of both these proteins seems to be under the negative control of RNA conformation.

(b) Changes in the secondary structure alter the efficiency of different ribosome binding sites

The three initiation sites in the above RNAs also show very different efficiencies in ribosome binding. The binding of ribosomes and formation of an initiation complex occur with greatest efficiency at the coat protein cistron and with least efficiency at the A-protein cistron. Steitz[47] has found that the efficiency of these sites in R17 RNA is drastically altered if the RNA is fragmented by mild alkaline degradation. With fragmentation, the activities of the coat protein and of the synthetase initiation sites decrease while the efficiency of the A-protein initiation site increases. If the short initiation fragments shown in *table 5.1* are isolated from the A-protein and from the coat protein cistron, the former will bind ribosomes very efficiently, while the latter will not, quite contrary to their behaviour in the intact mRNA.

The effect of fragmentation can be interpreted in terms of a change in the secondary structure, as small RNA fragments may take up a very different conformation from the one they displayed in the intact molecule. This suggests that formation of an initiation complex does not depend solely on the primary structure of the ribosome binding site but is strongly influenced by the conformation of this site which, in turn, is determined by the rest of the RNA molecule.

Using the MS2 RNA models (*figures 5.1, 5.2* and *5.8*), we can try to find an explanation for both the above phenomena. As can be seen in *figure 5.2*, at the coat protein initiation site, the untranslated sequence anneals to the sequence at the beginning of the message, forming a hairpin loop with the initiation codon exposed at the top of the loop. Similar structures have been proposed for a number of other messages. It has been suggested that a hairpin loop structure may facilitate the recognition of the initiation codon, although later results are against the general validity of this idea. It is nevertheless certain that in such a structure the AUG sequence is easily accessible. At the A-protein initiation site, on the other hand, a strongly base-paired stretch is formed between the untranslated leader sequence and a distant part of the mRNA. The initiation codon is buried in the secondary and tertiary structure. This conformation seems to prevent attachment of ribosomes and translation of the message. There are indications that the A-protein cistron is actively translated only while the RNA is in the nascent state and that translation becomes inhibited once this double-stranded stretch has been formed[9].

The regulation of translation of the replicase cistron is a more complex

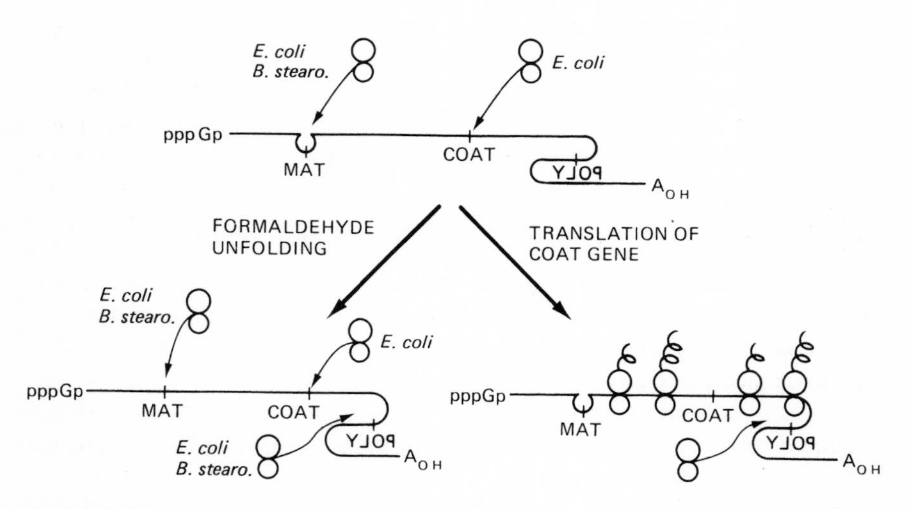

Figure 5.10 Translation of the RNA of f2 phage. Model of Lodish[48]. The initiation site of the polymerase (replicase) cistron is inaccessible to ribosomes because of the strong secondary and tertiary structure. If the RNA is unfolded by formaldehyde treatment, this initiation site becomes accessible. Ribosomes moving along the RNA molecule, translating the coat cistron, also cause an unfolding of RNA and make it possible for other ribosomes to attach to the polymerase initiation site. (Reproduced with permission from *J. Mol. Biol.*, **50**, 689 (1970). Copyright by Academic Press Inc. (London) Ltd.)

phenomenon which may be connected with the tertiary structure and also with long-range interactions in the RNA molecule. A very interesting aspect of this regulatory mechanism has been revealed in genetic studies with amber mutants in the coat protein cistron, viz. that translation of the coat protein enhances translation of the replicase *in vivo*. Lodish[48] proposed that in the tertiary structure, the coat protein and replicase cistrons interact in a way which makes the initiation site of the latter inaccessible to ribosomes (*figure 5.10*). Disruption of the tertiary structure releases this initiation site and makes ribosome binding, and thus translation, possible. He assumed that during translation of the coat protein cistron, a conformational change takes place in the RNA which, as shown in *figure 5.10*, has much the same effect as chemical disruption of the tertiary structure: the above interaction is abolished and the initiation site of the replicase cistron becomes accessible for ribosome binding. According to the more recent model of Jacobson and Spahr (*figure 5.8*) it is the physical proximity of the coat protein and the replicase initiation sites which prevents the binding of ribosomes to the latter. Jacobson and Spahr suggest a similar mechanism to explain the dependence of replicase synthesis on the translation of the coat protein cistron: during translation a conformational change would occur, and this would result in a physical separation of the coat protein and the replicase initiation codons[43].

(c) Abolishing the tertiary structure alters the requirement for a specific protein factor

One of the ribosomal proteins of *E. coli*, protein S1, is required for the binding of ribosomes to mRNA. If this protein is removed, no initiation complex is formed, and no translation takes place. The activity of the ribosomes can be restored by the addition of small amounts of S1 protein. Van Dieijn *et al.*[49], in studies on the translation of MS2 RNA in an *E. coli* system, found that this effect of S1 protein can only be observed when the mRNA is in its native conformation. If the tertiary structure is abolished by mild formaldehyde treatment, the requirement for S1 protein is also lost. This result can be interpreted by assuming that S1 protein acts by altering the secondary or tertiary structure of MS2 RNA and thus making it more suitable for ribosome binding. It is probably relevant to this function of S1 protein that it has been shown to possess an activity to unwind double-stranded stretches in RNA molecules[50].

(d) In eukaryotic mRNAs initiation complex formation is probably accompanied by conformational changes

Binding sites for 80S ribosomes as well as for 40S ribosomal subunits have been determined in several eukaryotic mRNAs. As mentioned earlier (p. 165 and p. 167), the 40S subunit protects a longer RNA stretch which often also includes the cap at the 5′ terminus (*table 5.3*). To account for the length of some of the 40S ribosome binding sites (about 50 nucleotides in globin mRNA, 52 and 54 nucleotides in reovirus mRNAs), Kozak and Shatkin[57] proposed that the ribosomal particle

Table 5.3 Ribosome binding sites on reovirus mRNAs

Small-size class mRNAs	$m^7Gppp\overset{m}{G}$CUAUUUUGCCUCUUCCCAGACGUUGUCGCAAUGGAGGUGUGCUUGCCCAACG
	$m^7Gppp\overset{m}{G}$CUAAAGUCACGCCUGUCGUCGUCACUAUGGCUUCCUCACUCAG
	$m^7Gppp\overset{m}{G}$CUAUUCGCUGGUCAGUUAUGGCUCGCUGCGCGUUCCUAUUCAAG
Medium-size class mRNAs	$m^7Gppp\overset{m}{G}$CUAAUCUGCUGACCGUUACUCUGCAAAGAUGGGGAACGCUCUUCCUAUCG
	$m^7Gppp\overset{m}{G}$CUAAAGUGACCGUGGUCAUGGCUUCAUUCAAGGGAUUCUCCG
	$m^7Gppp\overset{m}{G}$CUAUUCGCGGUCAUGGCUUACAUCGCAG

The complete sequences shown are protected by 40S ribosomal subunits. The shorter stretches marked by arrows are protected by 80S ribosomes. (Data from Kozak[36].)

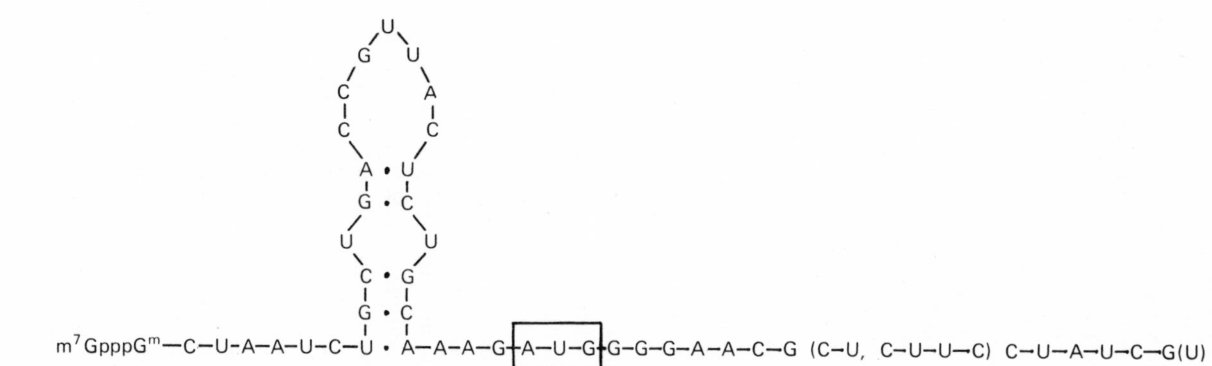

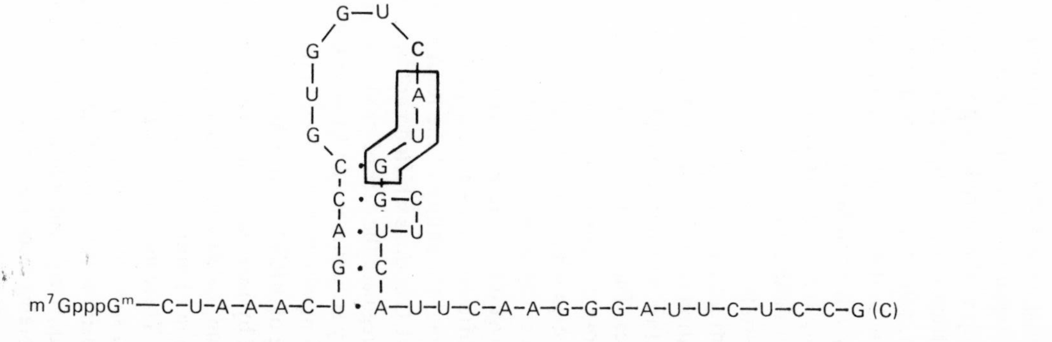

Figure 5.11 Possible folded structures of two reovirus mRNAs at the ribosome binding site. From Kozak and Shatkin[57]. (Reproduced with permission from *J. Mol. Biol.*, **112**, 75 (1977). Copyright by Academic Press Inc. (London) Ltd.)

interacts with a folded structure rather than a linear RNA stretch (*figure 5.11*). Binding of the small ribosomal subunit is the first step in initiation complex formation; it is followed by the attachment of the large subunit, leading to the formation of the 80S initiation complex. This latter process seems to involve some conformational change in the mRNA. The RNA stretch protected by the 80S particle is smaller and often does not include the capped termini (*table 5.3*). The effect of the change in the length of the protected RNA and the release of the terminal nucleotides upon formation of the 80S complex can thus be interpreted in terms of a rearrangement of the secondary structure, probably an opening up of the hairpin loop within the ribosome-associated region. As the capped structures are supposed to play an important role at the early stage of initiation, a flexible secondary structure which enables these nucleotides to participate in the first step of initiation complex formation, but allows abolition of a hairpin loop to facilitate the further steps of translation when initiation has been completed, may ensure high efficiency of both the early and the later steps in the translation process[41].

The results outlined above all point to a strong influence of the secondary and tertiary structure on the efficiency of translation, especially on the efficiency of the initiation step. The information available is not sufficient to define a secondary structure which is required or best suited for the formation of initiation complexes. There is, in fact, some controversy in the conclusions which can be drawn from the above data. On the one hand, mild denaturation or the presence of S1 protein which can disrupt the secondary structure do enhance initiation, although at the same time they decrease the specificity of the process. This would suggest that the secondary structure exercises a negative control.

The different annealed structures, hairpin loops, etc. reduce to a greater or lesser extent the ability of the RNA to bind ribosomes. This may prevent initiation at AUG codons which do not belong to genuine initiation sites and may serve to regulate the rate of production of different proteins. Such a role of the secondary structure would fit in well with the requirements for RNA–RNA interactions which are involved in the binding of ribosomes to mRNA and in the decoding of mRNA by tRNAs. Annealing of two RNA molecules may only be hindered by base-paired structures existing in any of the components. On the other hand, specific secondary and tertiary structures seem to be involved in the recognition of initiation sites by ribosomes, both in viral RNAs and in eukaryotic mRNAs. There are also indications that for the interaction of mRNA with protein factors, a specific secondary or tertiary structure may be required, the nucleotide sequence alone is not sufficient for recognition.

This apparent contradiction can be resolved if we assume a two-step mechanism: some proteins recognise a nucleotide sequence within a given secondary structure and thus select the specific site to which the ribosome can bind. In order to obtain a sufficiently stable association of ribosomes with mRNA, however, the secondary structure has to be abolished in this region so as to allow the interaction of mRNA with ribosomal RNA. S1 protein may play a specific role here, in opening up locally the annealed structures, and thus enhance ribosome binding and also allow subsequent movement of the ribosome along the mRNA molecule.

The idea that while the secondary structure of mRNA influences its interactions with ribosomes and proteins, these same interactions also have an effect on the structure of the mRNA, is supported by direct observations made on smaller RNA molecules. Transfer RNAs, as well as the small RNA components of ribosomes, have been shown to undergo conformational changes when interacting with proteins or with other RNA molecules (*see* Chapters 6 and 7). The structure of a large RNA molecule can be expected to show even greater fluctuations. Flexibility of the structure is certainly required for the above described control functions. We have seen that translation of a cistron can be prevented or triggered off by changes in the conformation of the mRNA. If RNA molecules had a fixed, rigid secondary structure, such control mechanisms would not be feasible.

REFERENCES

1 Volkin, E. and Astrachan, L. *Virology,* **2**, 149 (1956).
2 Gros, F., Hiatt, H., Gilbert, W., Kurland, C. G. and Risebrough, R. W. *Nature,* **190**, 581 (1961).
3 Hall, B. D. and Spiegelmann, S. *PNAS,* **47**, 137 (1961).
4 Adams, J. M., Jeppesen, P. G. N., Sanger, F. and Barrell, B. G. *Nature,* **223**, 1009 (1969).
5 Steitz, J. A. *Nature,* **224**, 957 (1969).
6 Sanger, F. *Biochem. J.,* **124**, 833 (1971).
7 Billeter, M. A., Dahlberg, J. E., Goodman, H. M., Hindley, J. and Weissmann, C. *Nature,* **224**, 1083 (1969).
8 Min Jou, W., Haegeman, G., Ysebaert, M. and Fiers, W. *Nature,* **237**, 82 (1972).
9 Fiers, W. *et al. Nature,* **256**, 273 (1975).
10 Fiers, W. *et al. Nature,* **260**, 500 (1976).
11 Sanger, F., Air, G. M., Barrell, B. G., Brown, N. L., Coulson, A. R., Fiddes, J. C., Hutchison, C. A. III, Slocombe, P. M. and Smith, M. *Nature,* **265**, 687 (1977).
12 Ravetch, J. V., Model, P. and Robertson, H. D. *Nature,* **265**, 698 (1977).
13 Squires, C., Lee, F., Bertrand, K., Squires, C. L., Bronson, M. J. and Yanofsky, C. *J. Mol. Biol.,* **103**, 351 (1976).
Bronson, M. J., Squires, C. and Yanofsky, C. *PNAS,* **70**, 2335 (1973).
14 Maizels, N. M. *PNAS,* **70**, 3585 (1973).
15 Musso, R. E., de Crombrugghe, B., Pastan, I., Sklar, J., Yot, P. and Weissman, S. *PNAS,* **71**, 4940 (1974).
Sklar, J., Weissman, S., Musso, R. E., Di Lauro, R. and de Crombrugghe, B. *J. Biol. Chem.,* **252**, 3538 (1977).
16 Weber, K. *Biochemistry,* **6**, 3144 (1967).
17 Yudelevich, A. *J. Mol. Biol.,* **60**, 21 (1971).
Scherberg, N. H. and Weiss, S. B. *PNAS,* **67**, 1164 (1970).
18 Garel, J. P., Hentzen, D. and Daillie, J. *FEBS Lett.,* **39**, 359 (1974).
Garel, J. P. *J. Theor. Biol.,* **43**, 225 (1974).
Smith, D. W. E. *Science,* **190**, 529 (1975).
19 Platt, T. and Yanofsky, C. *PNAS,* **72**, 2399 (1975).

20 Contreras, R., Rogiers, R., Van der Voorde, A. and Fiers, W. *Cell,* **12**, 529 (1977).
21 Shine, J. and Dalgarno, L. *Nature,* **254**, 34 (1975).
22 Steitz, J. A. and Jakes, K. *PNAS,* **72**, 4734 (1975).
23 Sprague, K. U., Steitz, J. A., Grenley, R. M. and Stocking, C. E. *Nature,* **267**, 462 (1977).
24 Moore, C. M., Farron, F., Bohnert, D. and Weissmann, C. *Nature NB,* **234**, 204 (1971).
Weiner, A. M. and Weber, K. *Nature NB,* **234**, 206 (1971).
24a Walz, A., Pirrotta, V. and Ineichen, K. *Nature,* **262**, 665 (1976).
25 Lee, F. and Yanofsky, C. *PNAS*, **74**, 4365 (1977).
26 Bertrand, K., Korn, L., Lee, F., Platt, T., Squires, C. L., Squires, C. and Yanofsky, C. *Science,* **189**, 22 (1975).
26a Di Nocera, P. P., Blasi, F., Di Lauro, R., Frunzio, R. and Bruni, C. B. *PNAS,* **75**, 4276 (1978).
Barnes, W. M. *PNAS,* **75**, 4281 (1978).
Zurawski, G., Brown, K., Killingly, D. and Yanofsky, C. *PNAS,* **75**, 4271 (1978).
Lee, F., Bertrand, K., Bennett, G. and Yanofsky, C. *J. Mol. Biol.,* **121**, 193 (1978).
27 Aviv, H. and Leder, P. *PNAS,* **69**, 1408 (1972).
28 Brownlee, G. G., Cartwright, E. M., Cowan, N. J., Jarvis, J. M. and Milstein, C. *Nature NB,* **244**, 2 (1973).
29 Schimke, R. T., Palacios, R., Sullivan, D., Kiely, M. L., Gonzales, C. and Taylor, J. M. *Methods in Enzymology,* **30F**, 631 (1974).
30 Proudfoot, N. J. *Cell,* **10**, 559 (1977).
Efstratiadis, A., Kafatos, F. C. and Maniatis, T. *Cell,* **10**, 571 (1977).
31 Baralle, F. E. *Cell,* **12**, 1085 (1977).
32 Hsu, M. and Ford, J. *PNAS,* **74**, 4982 (1977).
Fiers, W. *et al. Nature,* **273**, 113 (1978).
Klessig, D. F. *Cell,* **12**, 9 (1977).
Chow, L. T. and Broker, T. R. *Cell,* **15**, 497 (1978).
33 Legon, S. *J. Mol. Biol.,* **106**, 37 (1976).
34 Baralle, F. E. *Cell,* **10**, 549 (1977).
35 Hagenbuchle, O., Santer, M. and Steitz, J. A. *Cell,* **13**, 551 (1978).
36 Kozak, M. *Nature,* **269**, 390 (1977).
37 Huez, G. *et al. PNAS,* **71**, 3134 (1974).
Marbaix, G. *et al. PNAS,* **72**, 3065 (1975).
38 Brawerman, G. *Progr. Nucl. Acid Res. Mol. Biol.,* **17**, 117 (1976).
39 Rottman, F., Shatkin, A. J. and Perry, R. P. *Cell,* **3**, 197 (1974).
Abraham, G., Rhodes, D. P. and Banerjee, A. K. *Nature,* **255**, 37 (1975).
40 Adams, J. M. and Cory, S. *Nature,* **255**, 28 (1975).
Mutukrishnan, S., Both, G. W., Furuichi, Y. and Shatkin, A. J. *Nature,* **255**, 33 (1975).
41 Shatkin, A. J. *Cell,* **9**, 645 (1976).
42 Zan-Kowalczewska, M., Bretner, M., Sierakowska, H., Szczesna, E., Filipowicz, W. and Shatkin, A. J. *Nucl. Acids Res.,* **4**, 3065 (1977).
43 Jacobson, A. B. and Spahr, P. F. *J. Mol. Biol.,* **115**, 279 (1977).
44 Isenberg, H., Cotter, R. I. and Gratzer, W. B. *Biochim. Biophys. Acta,* **232**, 184 (1971).
45 Tinoco, T., Borer, P. N., Dengler, B., Levine, M. D., Ohlenback, O. C., Crothers, D. M. and Gralla, J. *Nature NB,* **246**, 40 (1973).

46 Lodish, H. F. and Robertson, H. D. *Cold Spring Harbor Symp.*, **34**, 655 (1969).
Lodish, H. F. *J. Mol. Biol.*, **50**, 689 (1970).
47 Steitz, J. A. *PNAS*, **70**, 2605 (1973).
48 Lodish, H. F. in *RNA phages* (ed. Zinder, N. D.) Cold Spring Harbor Laboratory, New York, 301 (1975).
49 vanDieijn, G., Van Knippenberg, P. H. and Van Duin, J. *Eur. J. Biochem.*, **64**, 511 (1976).
50 Bear, D. G., Ng, Ray, Van Derveer, D., Johnson, N. P., Thomas, G., Schleich, T. and Noller, H. F. *PNAS*, **73**, 1824 (1976).
Szer, W., Hermoso, J. M. and Boublik, M. *Biochem. Biophys. Res. Comm.*, **70**, 957 (1976).
51 Heindell, H. C., Liu, A., Paddock, G. V., Studnicka, G. M. and Salser, W. A. *Cell*, **15**, 43 (1978).
52 McReynolds, L. *et al. Nature*, **273**, 723 (1978).
53 Baralle, F. E. and Brownlee, G. G. *Nature*, **274**, 84 (1978).
54 Kozak, M. and Shatkin, A. J. *Cell*, **13**, 201 (1978).
55 Hamlyn, P., Brownlee, G. G., Cheng, C., Gait, M. J. and Milstein, C. *Cell*, **15**, 1067 (1978).
56 Ghosh, P. K., Reddy, V. B., Swinscoe, J., Choudary, P. V., Lebowitz, P. and Weissman, S. M. *J. Biol. Chem.*, **253**, 3643 (1978).
Bina-Stein, M., Thoren, M., Salzman, N. and Thompson, J. A. *PNAS*, **76**, 731 (1979).
57 Kozak, M. and Shatkin, A. J. *J. Mol. Biol.*, **112**, 75 (1977).

PART 3

The Synthesis of Proteins

6 The Site of Protein Synthesis: The Ribosome

THE FUNCTION OF THE RIBOSOME

It has been known since the earliest studies on the biosynthesis of proteins that the ribosomal particles are the sites of protein synthesis in the cell. These nucleoprotein particles, present in every cell which carries out protein synthesis, vary somewhat in size but are of basically similar structural build-up in different organisms and fulfil the same function. They bring together the different components of the translation system, in the correct spatial arrangement and in the right conformation required for the highly specific interactions between these components. They bring about the formation of peptide bonds between the amino acids which are specified in the message.

Ribosomes contain 50 to 70 proteins and 3 or 4 RNA molecules. In spite of extensive studies in this field, the exact arrangement of all these components in the particles has not yet been clarified. Neither do we know the exact role of each individual component in bringing about the synthesis of a polypeptide chain. Years of research have been spent in many laboratories to find the answer to the question: how does the ribosome fulfil its important function? Elucidation of structural problems had to go hand in hand with studies on functional properties, as the topographical arrangement of the different molecules is closely connected with the mechanism of the different translational events which take place on the particle. Our image of the ribosome has changed very markedly in both these aspects since the first oversimplified ideas which depicted the proteins as small spheres attached to a long string of RNA and which assumed that the different ribosomal functions could be ascribed to single, well defined molecules. At different periods of this work the importance of either ribosomal proteins (r-proteins) or ribosomal RNAs (rRNAs) was emphasised until it eventually transpired that a better understanding of the structural as well as the functional organisation of the particles can be achieved only if these components are not considered separately but as parts of a structural–functional entity, much more complex than envisaged before. At present, we regard the ribosome as a particle in which strongly folded RNA, in a tight secondary and tertiary structure, forms a kind of matrix into which are embedded the r-proteins, many of them elongated molecules, which are intertwined with the rRNAs.

Proteins and RNAs have many sites of interaction as a result of which the function of each molecule is strongly affected by neighbouring molecules. The different steps of translation are carried out not by single molecules but by groups of molecules which influence each other's action and coordinately bring about a translational event. Kurland defines different 'domains' on the ribosome[1] which are responsible for the different functions; others refer to 'functional sites'. What is meant by this is a region of the particle which is comprised of a number of proteins and a stretch of RNA and which carries out in a coordinated way one step in the synthesis of protein. It follows from the above considerations that in order to understand the mechanism of these events it is not enough to collect information on the structure, on the molecular environment and on the functional activities of each ribosomal component; we must also study their interactions with each other and with the extra-ribosomal components of the translation system. A large amount of data are available today in this field; unfortunately, it is still difficult to organise them so as to obtain a clear picture. In the following, some characteristic properties of the ribosomal particle and of its components will be described and an attempt made to fit these data into an up to date structural and functional model.

Before considering the components individually, it is necessary to summarise the different functions fulfilled by the ribosome as an entity. For the translation of the message an adaptor molecule is required, transfer RNA, which recognises both the codon and the corresponding amino acid. The different tRNAs carry their respective amino acids to the site of protein synthesis where specific interaction with the mRNA must occur to select the aminoacyl-tRNA specified in the message. The ribosome has to accommodate both the mRNA and the aminoacyl-tRNA so as to allow this specific interaction. Furthermore, the amino acid will enter into a peptide bond with the C-terminal amino acid of the growing polypeptide chain; thus, the ribosome also accommodates another tRNA, that which carries the peptide chain (peptidyl-tRNA), in such a position that the reaction between the aminoacyl and peptidyl groups, i.e. the formation of the peptide bond, should readily occur.

We will thus try to locate and characterise the functional sites on the ribosome at which

(a) the mRNA is accommodated
(b) the aminoacyl-tRNA is bound (called A-site)
(c) the peptidyl-tRNA is bound (P-site)
(d) peptide bond formation takes place (peptidyl transferase centre)

In addition to these functions, the ribosome recognises and binds to the specific initiation sites on mRNA; it moves along the mRNA to 'read' the complete message, and drops off when translation is terminated. We will consider the ribosomal components which are known today to have a role in these events and will attempt to explain these specific interactions on the basis of the structural characteristics of these components as far as they are known.

Even if we could explain exactly how the above processes are carried out (which is certainly not the case), this would still leave open an important and much argued question. Do ribosomes provide only an exact machinery for polypeptide synthesis without any specificity for the message or can they select specifically the mRNA they translate? Numerous studies on heterologous systems suggest that ribosomes from one organism can translate an mRNA from a different organism, both *in vitro* and *in vivo*. A variety of mRNAs have been translated in cell-free systems from wheat germ, e.g. rabbit globin mRNA[2] and histone mRNA from sea urchin[3]; myosin mRNA could be translated in reticulocyte lysates[4]. Globin has been synthesised in *Xenopus laevis* oocytes into which globin mRNA has been injected[5]. A more detailed discussion of the results obtained in heterologous systems will be given in Chapter 8, in relation to the different factors which may determine the specificity of translation.

The ability of ribosomes to translate a variety of mRNAs does not necessarily imply a complete lack of specificity. The efficiency of translation of different messages by the same ribosomes or of the same message by different ribosomes may vary widely[6]. Several observations also indicate that a preference for one message above another may prevail within the cell[7]. The control mechanisms which operate at the translational level are not very well known and often it cannot be decided whether some inherent property of the ribosomes is responsible for the higher or lower efficiency of synthesis of a given protein or if the efficiency is entirely under the control of extra-ribosomal factors. In spite of the rather significant results with heterologous systems, it seems that the question, how far ribosomes may contribute to the specificity of translation, cannot be answered unambiguously.

SIZE AND SUBUNIT STRUCTURE OF RIBOSOMES

As ribosomes fulfil the same function in every organism, it can be expected that particles of different origin should resemble each other in their essential structural characteristics. There is indeed a basic similarity in the structural organisation of ribosomes, although particles from eukaryotic organisms show a greater complexity and a less pronounced conservation of structural features of the individual components than can be observed in ribosomes of prokaryotes. Prokaryotic ribosomes are smaller, they have a sedimentation coefficient of 70S and are built of two subunits with sedimentation coefficients of 30S and 50S. Their protein: RNA ratio is 1:2. Bacterial ribosomes show a remarkable homology in their protein pattern. The size of eukaryotic ribosomes is larger, 80S, due to an increase in both the number and size of protein and RNA components. The ribosomal proteins contribute more to the increased size; the protein:RNA ratio in eukaryotic particles is about 1:1. There is a slight variation in the size of eukaryotic ribosomes from different species, reflecting variations in the individual components. 80S ribosomes also consist of two subunits: a smaller 40S and a larger 60S subparticle.

A functional difference seems to exist between the two ribosomal subunits. The small subparticle is involved in sequence-specific recognition processes, e.g. recognition of the initiation site, interaction of the codon with the complementary anticodon in tRNA. The proteins which were shown to be directly connected with the fidelity of translation, S12 and S4, belong to this subunit in *E. coli*. The large subparticle is responsible for functions which are common to all amino acids and to all tRNAs, i.e. the formation of the peptide bond and the binding of the aminoacyl-tRNA and peptidyl-tRNA in suitable positions to bring about this reaction.

Ribosomes undergo dissociation into subunits and association into 70S or 80S particles both *in vivo* and *in vitro*. *In vivo*, the dissociation cycle is connected with the functional stages through which the ribosome goes in the course of translation. At the early phase of initiation, free ribosomal subunits are required; the later steps of translation are carried out by associated 70S or 80S particles. A specific dissociation, or rather anti-association, factor controls this process: it combines with the smaller subunit and thereby prevents it from associating with the large subparticle. If the factor is released, association can occur. *In vitro*, it depends on the ionic environment whether ribosomes are present in the dissociated or associated state. *E. coli* ribosomes dissociate into subunits at Mg^{2+} concentrations of 10^{-3} M or lower, whereas they form stable 70S particles in 10^{-2} M Mg^{2+}. However, too low a Mg^{2+} concentration, or complete removal of Mg^{2+} by chelating agents like EDTA, affect the conformation of the subunits themselves; the particles unfold, leading to changes in their physicochemical properties, to the

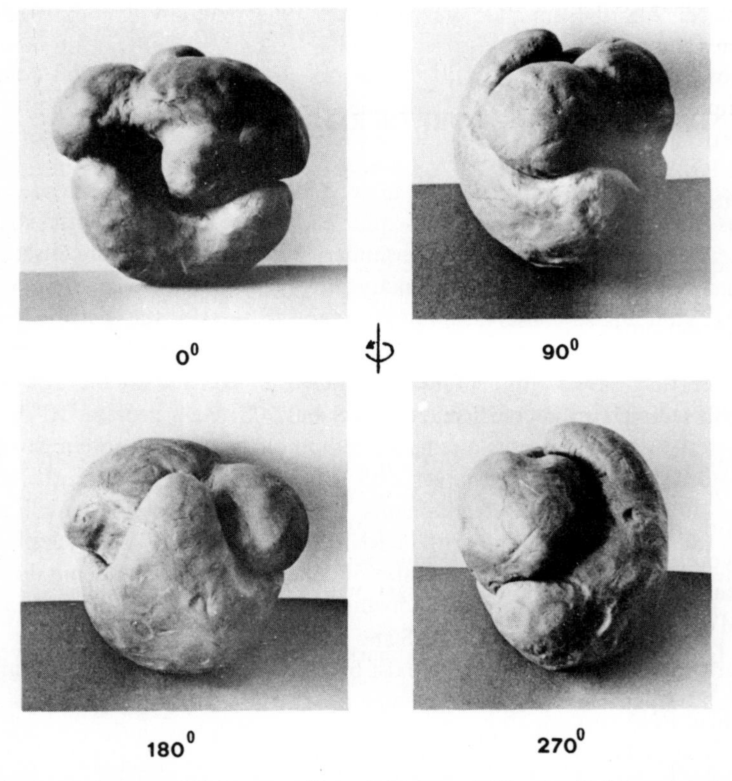

Figure 6.1 Model of the 70S ribosome. From Stöffler and Wittmann[8]. (Reproduced with permission from *Molecular Mechanisms of Protein Biosynthesis*, Academic Press (1977). Photograph by courtesy of Professor H. G. Wittmann and Dr G. Stöffler.)

release of some ribosomal proteins and to a loss of biological activity. The conditions for dissociation of eukaryotic ribosomes are more drastic: either EDTA or high salt concentration has to be used, or dissociation can be brought about by causing termination of polypeptide synthesis and release of the peptide chain.

The shape of the ribosomal subunits of *E. coli* and their mode of association have been studied by electron microscopy[8]. The model of the *E. coli* ribosome constructed from electron microphotographs is shown in *figure 6.1*. It can be seen that as the two subunits associate, a cavity is formed between them, which, like a tunnel, leads through the complete 70S particle. Electron microscopy of eukaryotic ribosomes revealed differences in the fine structure but a basic similarity in the orientation and overall shape of 70S and 80S ribosomes[9].

The distinction between the two types of ribosomes is not quite clear-cut. The main population of ribosomes in both prokaryotic and eukaryotic cells is present in the cytoplasm either in the form of free particles or particles bound to membranes. For these particles the above described differences are strictly valid. Cell organelles, however, have a distinct population of ribosomes which differ in their structural characteristics from the ribosomes outside these organelles. In chloroplasts, 70S ribosomes are present which strongly resemble the prokaryotic ribosomes and have probably evolved from bacterial ribosomes. In mitochondria, 70S ribosomes and even smaller ribosomal particles may be found, depending on the complexity of the organism. These small particles also differ from both eukaryotic and prokaryotic ribosomes in their protein components and in the size of their RNAs. In some respects, 70S mitochondrial ribosomes show more resemblance to prokaryotic than to eukaryotic particles. Still, it seems that they have evolved from eukaryotic ribosomes. Sequence homologies in 3′-terminal regions of ribosomal RNA show a closer relationship between mitochondrial ribosomes and cytoplasmic ribosomes of eukaryotes than between mitochondrial and bacterial ribosomes.

THE RNAs OF THE RIBOSOME

Ribosomal particles comprise only RNA and protein molecules. The RNA:protein ratio is 2:1 in the 70S, prokaryotic ribosomes and about 1:1 in cytoplasmic eukaryotic ribosomes. The smaller subunits contain one RNA molecule which in prokaryotes has a MW of 500,000 and a sedimentation coefficient of 16S, while the eukaryotic rRNA is somewhat larger, with a sedimentation coefficient of 18S and a MW of 700,000.

The larger subparticles differ more markedly in prokaryotes and eukaryotes. The prokaryotic 50S particle contains a 23S RNA molecule, of 1.05×10^6 MW, and a small RNA species, 5S RNA, which contains 120 nucleotides and has a MW of 36 000. The 60S subunit of the eukaryotic cytoplasmic ribosome contains a larger RNA molecule, of a somewhat varying size in different species: its MW is 1.3×10^6 to 1.7×10^6 and the sedimentation coefficient is around 28S. In addition to this high molecular weight RNA, there are two small RNA species in this subunit: 5S RNA of the same size as that of the prokaryotic ribosome, and 5.8S RNA, a somewhat bigger molecule built of about 150 nucleotides and with a MW of about 47,000. The 5.8S RNA is strongly associated with 28S RNA by hydrogen bonding. The base composition of ribosomal RNAs shows that all rRNAs, with the exception of the 5S RNAs, contain some substituted bases.

The primary structure of bacterial rRNAs has been extensively studied, especially that of the rRNAs of *E. coli*. The complete sequence of 16S RNA has been recently established; sequencing of 23S RNA is also under way. Suggestions have also been made for the secondary structures of tRNAs, but the models proposed are still hypothetical. Some structural features could be correlated to ribosomal functions. Less information is available on the rRNAs of eukaryotic cells; here, only the small RNAs and some stretches of the 18S and 28S RNAs have been sequenced.

Bacterial rRNAs

5S RNA

The nucleotide sequence of 5S RNA of *E. coli* has been known since 1967, and it was one of the first nucleic acids to be sequenced by the fingerprinting technique[10]. It is 120 nucleotides long and contains no minor bases. Its nucleotide sequence is shown in *figure 6.2*. Comparison with 5S RNA from other bacteria shows a degree of conservation of primary structures, although the nucleotide sequences are by no means identical[10a]. In *figure 6.2* the highly conserved regions are underlined. One of the strongly conserved sequences in prokaryotic 5S RNAs is the tetranucleotide,

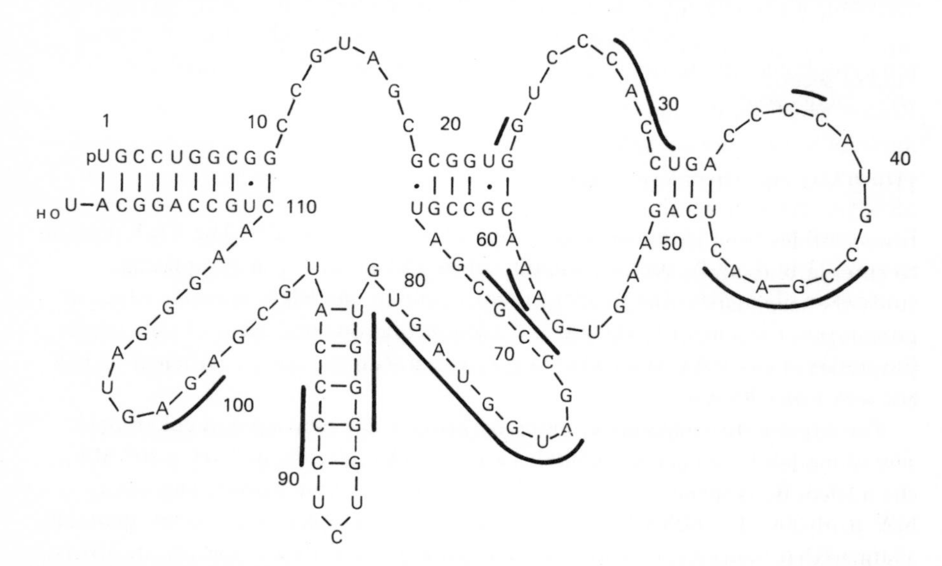

Figure 6.2 The structure of 5S RNA from *E. coli*. Highly conserved sequences underlined. Secondary structure model according to Fox and Woese[15].

CGAA, around positions 40–50. This is an interesting sequence, because it is complementary to a tetranucleotide present in all tRNAs (except eukaryotic initiator tRNAs, *see* p. 233), TψCG. This complementarity was already known when the first 5S RNA sequence was available and it was immediately suggested that annealing of these sequences may play a role in binding the tRNA to ribosome. The occurrence of the same sequence in all prokaryotic 5S RNAs studied so far strongly supports this idea. Direct experimental evidence for the participation of the CGAA sequence in tRNA binding was obtained by Ofengand and Henes[11], who showed that the tetranucleotide, TψCG, has an inhibitory effect in a purified protein synthesising system. The inhibition is brought about by TψCG binding to 5S RNA and preventing its interaction with tRNA.

Another conserved stretch of 12 nucleotides (positions 72–83) is complementary to a dodecanucleotide sequence in 23S RNA[12]. It has been suggested that these two RNAs anneal in the 50S particle. According to a 5S RNA model of Weidner *et al.*[13], the sites complementing tRNA and 23S RNA, respectively, may be near each other in the three-dimensional structure, bringing the tRNA and 23S RNA into close proximity.

On the basis of the nucleotide sequence, several models can be constructed for the secondary structure of 5S RNA[14]. Optical data suggest a strong secondary structure for all rRNAs; in the case of 5S RNA a double-helical content of over 60% can be calculated. As no X-ray crystallography data are available, the models proposed are based on maximal base pairing as deduced from the primary structure, on the susceptibility of different sites to RNases and on other indirect assays which relate to the accessibility of different regions in the RNA molecules. These assays do not define the position of double-stranded regions exactly, but they give information on 'ordered areas', i.e. on areas where nucleotides are involved in some kind of secondary or tertiary interactions. These areas are not susceptible to RNases which show strong specificity for a free single-stranded structure (e.g. nuclease S_1) and are not available for annealing to oligonucleotides with a complementary sequence. If a variety of trinucleotides are allowed to interact with 5S RNA, the ones which do anneal will define the accessible regions, whereas the trinucleotides complementary to the inaccessible, ordered regions will be unable to enter into any form of binding. This assay method has been widely used to study the secondary structure of small RNA molecules. Whilst it does not give unambiguous results, it yields useful information which also reflects the functional properties of the RNA, in that it characterises the regions which are able to interact with other RNA molecules.

Considering the ambiguities of the techniques used, it is not surprising that several models have been proposed for the secondary structure of 5S RNA. From the nucleotide sequences of a number of 5S RNA species, generalised patterns of base pairing could be derived which fit most 5S RNA molecules[13,15,16]. It may be assumed that the secondary structure of all 5S RNAs (and this may also include eukaryotic 5S RNAs) show some basic similarities.

A common secondary structure for different 5S RNAs is also indicated by

Erdmann's results[14]. Using the oligonucleotide binding assay, he found considerable similarity in the distribution of free and ordered areas in 5S RNAs from different prokaryotic and eukaryotic species. A common feature in all 5S RNA models is the double-stranded region formed by annealing of complementary stretches at the 5′ and 3′ ends of the molecule. Although the actual nucleotide sequences at the two termini are different in different 5S RNA species, this complementarity exists in all 5S RNAs studied so far. In *figure 6.2*, the nucleotide sequence of *E. coli* 5S RNA is shown in the secondary structure proposed by Fox and Woese[15]; *figure 6.3* shows the models of Weidner *et al.*[13], referring

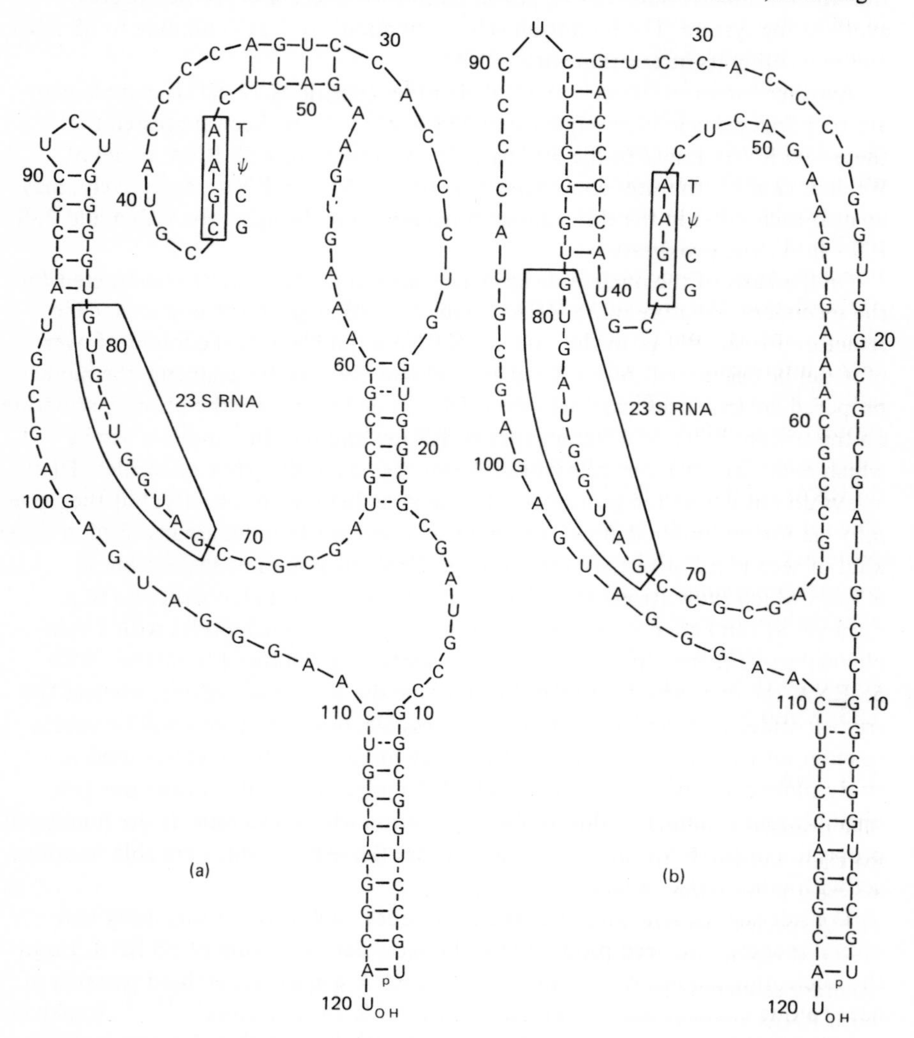

Figure 6.3 Two conformations of 5S RNA from prokaryotic ribosomes, according to Weidner *et al.*[13]. Sequences in boxes: sites of interaction with tRNA and with 23S RNA. For explanation see text. (Reproduced with permission from *Nature,* **266**, 193 (1977).)

to two different conformations of 5S RNA. These are generalised patterns of base pairing; the double-stranded regions represent conserved structures in several prokaryotes. Weidner *et al.* propose that both conformations exist and correspond to different functional states. In conformation *b*, the sites for annealing of tRNA and 23S RNA are adjacent, while in model *a* they are farther apart. A conformational change may thus cause a movement of tRNA relative to 23S RNA. Such movement may have an important role in translation; it might be involved in the migration of the ribosome along the mRNA (p. 269–271).

When studying the secondary structure of ribosomal RNAs, we must also keep in mind that the structure within the ribosomal particle may differ from that found in isolated RNA molecules which are free of proteins. RNA–protein interactions often lead to alterations in the conformation of both partners. In ribosomes where each RNA molecule is surrounded by a number of protein molecules, the effect of such interactions cannot be ignored. Bear *et al.*[17] have, in fact, observed a shift in the circular dichroism spectra of 5S RNA upon interaction with two 5S RNA binding proteins of the 50S ribosomal subunit, L18 and L25. Similar effects have also been detected in the case of other rRNAs. Such conformational changes may be of great importance for RNA function. At different stages of translation, the precision of the process is secured by specific conformational changes arising from protein–RNA or RNA–RNA interactions (*see* Chapter 8).

An interesting observation of Erdmann[14], concerning the annealing of tRNA and 5S RNA, also emphasises the important effect of ribosomal proteins on the properties of rRNA. Erdmann and coworkers used the oligonucleotide binding assay to study conformational changes in *E. coli* and *Bacillus stearothermophilus* 5S RNAs. In the free RNAs, the tRNA-binding sequence, CGAA, was not available for annealing. If, however, proteins L5, L18 and L25 (the 5S RNA binding proteins) of *E. coli* ribosomes or the corresponding ribosomal proteins of *B. stearothermophilus* were added, the CGAA sequence became accessible and annealed to the complementary oligonucleotide. This suggests that within the ribosomal particle, 5S RNA takes up some specific conformation which enables it to bind tRNA. It is interesting that attachment of the same proteins, L18 and L25, is also required for the formation of a 5S RNA–23S RNA complex.

23S RNA

The primary structure of the large RNA component of the 50S subunit is only partially known[18]. Information is still insufficient to correlate structural features to specific functions. Comparison of the known sequences with those of the other rRNA species suggests some sites for interactions within the ribosome. At the 3′-terminal region, the 23S RNA of *E. coli* contains both homologous and complementary sequences to the 3′-terminal region of 16S RNA[19]. As the 3′ termini of both RNAs are near the attachment site of the two ribosomal subunits, these sequences may play a role in the association of the 30S and 50S particles. Complementarity of 12-nucleotide-long sequences in 23S RNA and 5S RNA has been mentioned above. These sequences may represent the site of attachment of the two RNAs in the 50S subunit.

Complementarity also exists between 23S RNA and initiator tRNA. Dahlberg *et al.*[75] detected, in the middle of the 23S RNA molecule, a sequence which could form a 17 base pair-long annealed structure with initiator tRNA. Interaction between these molecules may be of importance for the formation of initiation complexes.

23S RNA was shown to be present in the peptidyl transferase centre in the proximity of the growing peptide chain[20]. It has been suggested that the peptide chain may have several attachment sites on the RNA.

16S RNA

The RNA of the smaller subunit of *E. coli* ribosomes, 16S RNA, has been completely sequenced[21]. The 3′-terminal sequence of 16S RNA has attracted great interest. The first findings which indicated that this RNA stretch had an important function in translation came from studies on ribosomes treated with colicin E3, a bactericidal protein which has the effect of introducing one specific split into 16S RNA, thus removing a 49-nucleotide-long fragment from the 3′ end[22]. This 'colicin fragment' may remain attached to the ribosomal particle even though the continuity of the 16S RNA has been severed. Colicin E3-treated ribosomes lose their ability to carry out protein synthesis. The exact cause of this inactivation is not yet clear: it has been argued that splitting off the 3′ terminus may affect translation at the stage of initiation or of elongation.

It seems probable that more than one function can be attributed to the 3′-terminal region of 16S RNA. In 1975, Shine and Dalgarno[23] called attention to the pyrimidine-rich sequence, $ACCUCCUUA_{OH}$, at the 3′ terminus of 16S RNA, and suggested that this stretch had an essential role in the binding of ribosomes to mRNA. Near the initiation sites of practically all the prokaryotic mRNAs studied so far, purine-rich tracts have been detected which show varying degrees of complementarity to the 3′-terminal sequence of 16S RNA. Shine and Dalgarno assumed that annealing between these complementary stretches forms the basis of the specific attachment of ribosomes to the initiation sites in mRNA (*see* Chapter 5). Comparison of the 3′-terminal sequences in 16S RNAs from *E. coli* and *B. stearothermophilus* shows sufficient homology to support the idea that mRNA–rRNA annealing exists as a general mechanism (at least in prokaryotes), contributing to the binding of ribosomes to the initiation site. The extent of possible base pairing varies with different mRNAs and different ribosomes. The stability of annealing is not directly correlated to the efficiency in ribosome binding. Several other factors are involved in this process and their role is not yet clear[24] (*see* Chapter 8).

There has been some controversy about the secondary structure of the 3′-terminal sequence of 16S RNA. Dahlberg and Dahlberg[25] proposed that this RNA stretch is itself strongly base paired and that it is made available for annealing to mRNA by the action of protein S1 of the 30S ribosomal subunit. This protein attaches to the 3′-terminal part of the RNA and thereby abolishes its secondary structure. *Figure 6.4* shows the assumed secondary structure of this RNA stretch

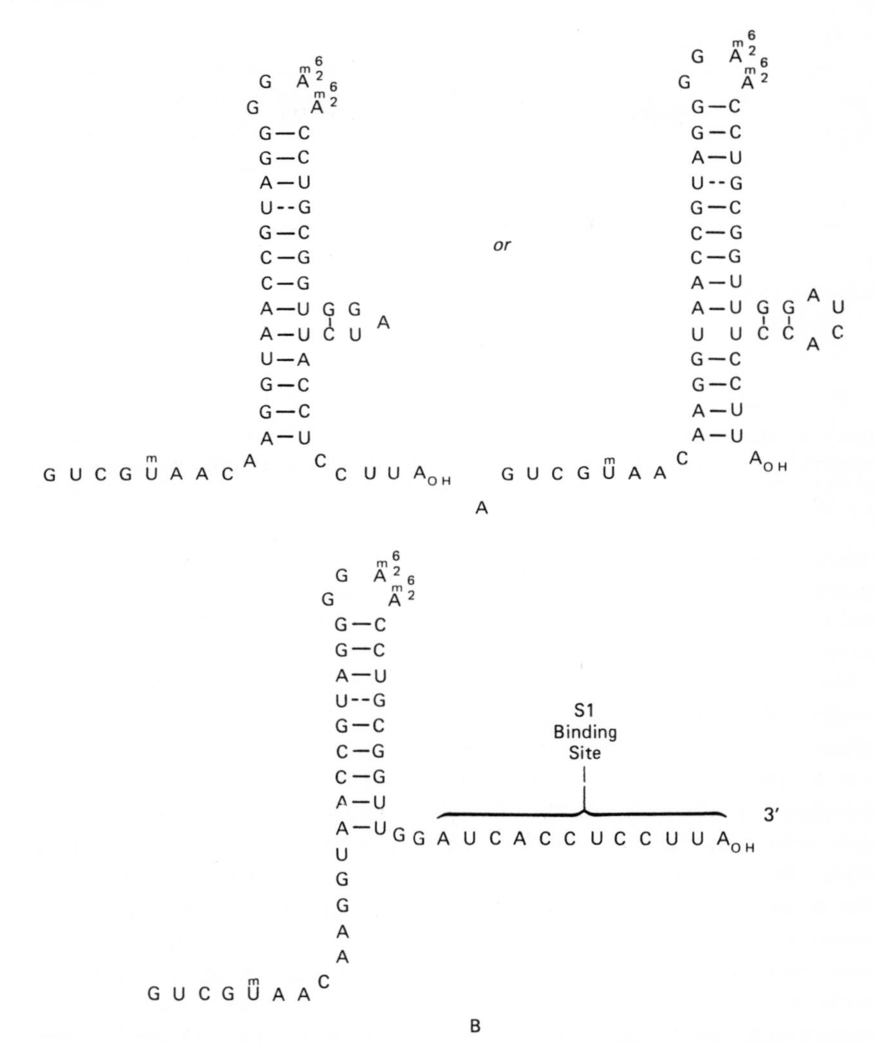

Figure 6.4 The effect of S1 protein on the conformation of 16S RNA. Assumed secondary structure of the 3′-terminal part of 16S RNA in the absence (A) and in the presence (B) of S1 protein. (According to Dahlberg and Dahlberg[25].)

in the absence and in the presence of protein S1. The melting characteristics of the colicin fragment seemed to support the possibility of such an effect[26]. Baan *et al.*[27], on the other hand, measuring proton magnetic resonance of the same fragment, came to the conclusion that there is no necessity to invoke a function of ribosomal proteins in opening up hairpin loops, as the 3′-terminal dodecanucleotide is in any case unpaired in 16S RNA. *Figure 6.5* shows the secondary structure of this 49-nucleotide-long fragment, deduced from these studies.

Near the 3′ terminus there is also a sequence in 16S RNA which shows complementarity to 23S RNA. This stretch may have a role in the association of the two

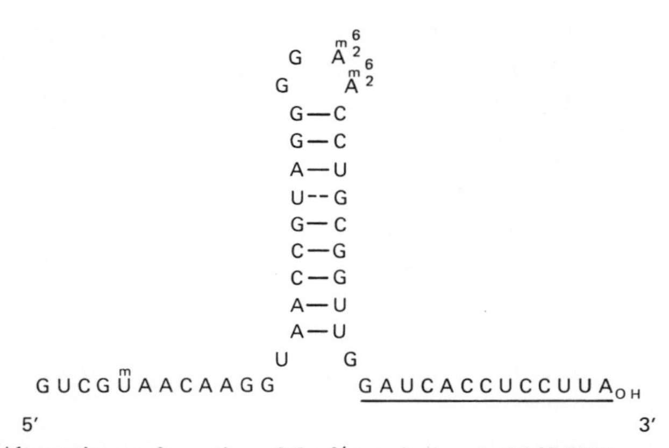

Figure 6.5 Alternative conformation of the 3′-terminal part of 16S RNA, constructed by Baan *et al.*[27] on the basis of NMR spectra. This structure would not require unwinding by S1 protein in order to anneal to mRNA. The pyrimidine-rich sequence taking part in the annealing is underlined.

subunits[19]. As the sequences complementary to 23S RNA and to mRNA are adjacent in the 16S RNA molecule, annealing to one or the other of these RNAs may be mutually exclusive processes. This would explain why only free 30S subparticles can attach to mRNA.

The secondary structure of 16S RNA is not known. Optical properties suggest a highly ordered structure with more than 60% of base-paired and stacked regions[28], but no direct information is available as to the position of hairpin loops or other secondary and tertiary interactions. Susceptibility to RNases and other assay techniques described earlier which can be used with some success to locate ordered and single-stranded regions in small RNA molecules cannot be usefully applied to RNAs of this large size. A model of the arrangement of hairpin loops in the molecule derived mainly from the primary structure, taking into account maximal and most stable base pairing, has been proposed, but has proved rather controversial. The similarity of free 16S RNA and of RNA in the ribosomal particle first led to the assumption of only one stable secondary structure for this RNA. This idea was strongly refuted later[29,30]. Hochkeppel and Craven[31] found that free 16S RNA can exist in at least two different conformations depending on the techniques used for its extraction from the particles. Moreover, they provided unequivocal evidence for the influence of ribosomal proteins on the RNA conformation. It seems probable that all ribosomal RNAs are capable of conformational changes induced by ribosomal proteins or possibly by extra-ribosomal components in the course of translation.

Eukaryotic rRNAs

5S RNA

The primary structure of 5S RNA from a number of different species has been established[10a,14]. Sequence homologies are more extensive within the group of eukaryotic 5S RNAs than between eukaryotic and prokaryotic species. The secondary structure may be less divergent: the distribution of ordered and single-stranded areas seems to follow a similar pattern in prokaryotic and eukaryotic organisms. Recently, a cloverleaf structure has been proposed for yeast 5S RNA by Luoma and Marshall[31a], who suggest that other 5S RNAs and 5.8S RNAs may have a similar secondary structure.

The most interesting difference in the conserved sequences in prokaryotic and eukaryotic 5S RNAs is the absence in the latter of the CGAA sequence around positions 40-50. As this sequence serves to bind tRNA, this difference in the structure also suggests a difference in the function of the two RNA species. Another conserved sequence, GAUC, is present in this position in eukaryotic 5S RNAs. The complementary tetranucleotide sequence, GAUC, has been found to replace the TψCG sequence in initiator tRNAs of eukaryotic cells (p. 233). Annealing can thus occur in eukaryotic ribosomes between 5S RNA and initiator tRNA but not between 5S RNA and tRNAs engaged in elongation. This, together with other evidence, described below, suggests that in spite of their identical size, 5S RNA of the eukaryotic ribosome is not the functional counterpart of bacterial 5S RNA.

5.8S RNA

Various observations suggest that the 5.8S RNA of eukaryotic ribosomes corresponds both genetically and functionally to the 5S RNA of bacterial ribosomes. The extent of sequence homologies is not sufficient to substantiate a relationship between the two RNA species. Nevertheless, the presence of the CGAA sequence in 5.8S RNA strongly suggests that this RNA fulfils the same function in binding tRNA as does bacterial 5S RNA[14,16,28]. Whilst both these RNA species are involved in elongation, the 5S RNA of eukaryotic ribosomes is an additional component probably connected with initiation of translation.

A further indication that 5.8S RNA rather than 5S RNA of the eukaryotic ribosome corresponds to bacterial 5S RNA comes from studies on RNA-protein interactions. 5S RNAs from a number of prokaryotic ribosomes interact with ribosomal proteins L18 and L25 of *E. coli*. 5S RNA from yeast did not bind these proteins while 5.8S RNA proved almost as efficient in binding the bacterial proteins as 5S RNA of the bacterial ribosomes[32].

The same idea is supported by studies on the genetics of rRNAs. The genes for the three RNAs of prokaryotic ribosomes are organised into one transcription unit, as are the genes for the 28S, 5.8S and 18S RNAs of eukaryotic ribosomes[33,34]. However, the parallelism is not perfect, for the order of the three cistrons is different in the two cases; in prokaryotes it is 16S-23S-5S, while in eukaryotes the 5.8S RNA cistron is between the cistrons for 18S RNA and 28S RNA[35]. The primary transcripts contain the sequences for all three rRNAs, in both prokaryotes and eukaryotes, and the mature rRNAs are the products of post-transcriptional processing (*see* Chapter 3). 5S RNA genes, however, are separated from this transcription unit in eukaryotic chromosomes. 5S RNA sequences are not present in the primary transcript of the eukaryotic ribosomal transcription unit. As far as genetic loci and mechanism of biosynthesis are concerned, prokaryotic 5S RNA shows a striking similarity to 5.8S RNA and a distinct dissimilarity to eukaryotic 5S RNA.

18S RNA and 28S RNA

Information on the primary structure of the two high molecular weight components is rather scarce. 3′-terminal sequences of 18S RNAs from a number of eukaryotic organisms have been determined, as it was of special interest whether annealing between ribosomal RNA and mRNA can take place in the same way in eukaryotic organisms as it occurs in bacteria. A highly conserved sequence has indeed been found; the last 18 nucleotides are identical in 18S RNAs from mouse, silkworm and wheat ribosomes: . . .CC$\underline{\text{UGCGGAAGGAU}}$CAUUA$_{OH}$ (ref. 36). It has been suggested that the underlined, purine-rich stretch may anneal to pyrimidine-rich sequences in eukaryotic mRNAs, but it is still uncertain whether such a function can indeed be ascribed to 18S RNA (*see* 163).

THE PROTEINS OF THE RIBOSOME

One-third of the total mass of prokaryotic ribosomes and half of the mass of eukaryotic ribosomes are made up of protein molecules. There are 21 different proteins in the 30S subunit of the bacterial ribosome and 34 in the 50S subunit. Still, the total number of proteins is only 54 in the 70S ribosome, as one protein of the smaller subunit (S20) is identical to one found on the larger subunit (L26). All 54 proteins of the *E. coli* ribosome have been isolated and characterised as to their molecular weight and physicochemical characteristics. The amino acid sequences of a number of ribosomal proteins have also been established[8]. Quite a few of the ribosomal proteins have been found to have a highly elongated shape. Most of the proteins show basic character: they contain a large proportion of basic amino acids and have isoelectric points around *p*H 10. They are usually small molecules in the MW range 6000 to 32 000, with the exception of protein S1, which is much larger, 65 000. S1 is one of the few acidic proteins of the ribosomal particle. (Recently, another protein of similar size and charge has been detected on *E. coli* ribosomes[37]. This might be the 22nd protein of the 30S subunit.)

In eukaryotic ribosomes, the proteins show considerable variations between species[38]. Less ribosomal proteins have been isolated in pure form and subjected to extensive analysis. The data on the number and sizes of different proteins are often only approximate: in different ribosomes the large subunit has been found to contain 36 to 40, the small subunit 29 to 32 proteins and the MW of these proteins ranges from 8000 to 40 000 (ref. 28).

As proper characterisation of the proteins and allocation of some ribosomal functions to individual proteins have been carried out only in bacterial ribosomes, in the following, the properties and functions of these proteins will be discussed in more detail.

Proteins can be extracted from ribosomes by acetic acid extraction and can be fractionated by two-dimensional electrophoresis. In this way, Wittmann and coworkers separated and identified all proteins of both the small subunit (proteins S1 to S21) and the large subunit (proteins L1 to L34) of *E. coli* ribosomes[39]. *Figure 6.6* shows a two-dimensional electrophoretogram of all 54 proteins of the *E. coli* 70S ribosome.

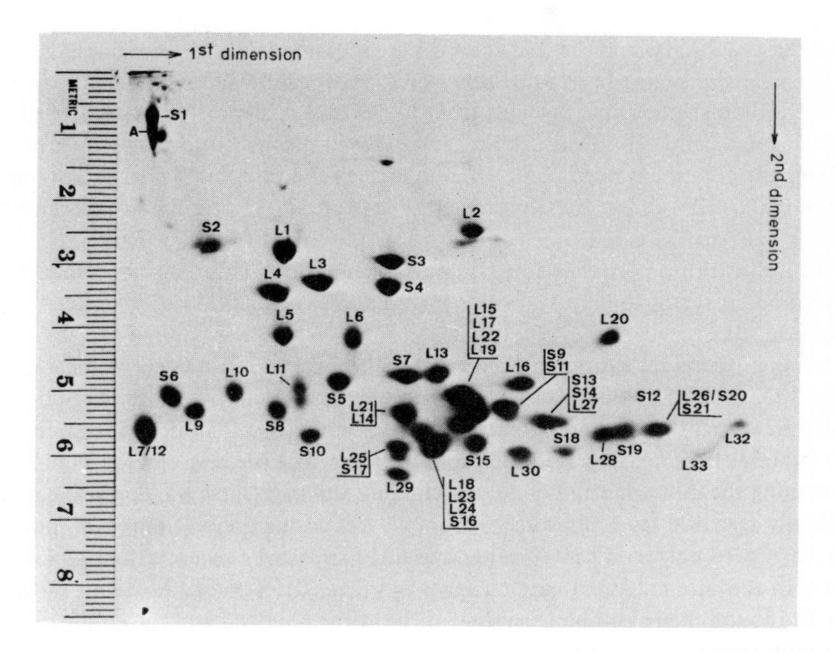

Figure 6.6 Separation of the proteins of *E. coli* ribosomes by two-dimensional gel electrophoresis. By Kyriakopoulos and Subramanian[74]. (Photograph by courtesy of Dr A. R. Subramanian.)

There has been some controversy about the molar ratios of the proteins in the ribosomal particles. Results of early analyses indicated that most but not all proteins were present in one copy per particle. A few proteins, which were called fractional proteins, were detected in less than 1:1 molar ratio. It was first assumed that the population of ribosomal particles was heterogeneous, the fractional proteins being present on some and absent from other particles, probably depending on the functional state of the ribosome. Lately, however, serious doubts have arisen as to the validity of these early data. Although attachment and release of proteins at different functional stages during translation cannot be *a priori* excluded (extra-ribosomal protein factors are known to bind temporarily to ribosomes, the same might apply also to ribosomal proteins), nevertheless, the lower yield of fractional proteins can more probably be ascribed to losses during isolation and analysis. Results with improved extraction and gel electrophoretic techniques suggest that each protein is present in at least one copy in each ribosome[40]. Two proteins of the large subunit, L7 and L12, which contain the same polypeptide chain and differ only in that L7 is acetylated in the N-terminal position, are present in four copies, the proportion of the acetylated to the non-acetylated form varying in the different particles[41]. As mentioned above, one protein has been detected on both the 30S and the 50S subunit. The 70S ribosome contains only one molecule of this protein but as it is located at the interface of the two subunits, it may attach to one or the other when the two subunits are separated and the proteins extracted.

Ribosomal proteins from some other prokaryotes show great similarities to the *E. coli* proteins: ribosomal proteins from *E. coli* and *B. stearothermophilus* can even be mixed and can substitute for each other in reconstitution experiments. A striking difference has been observed, however, in the protein pattern of *B. stearothermophilus* ribosomes: these particles do not contain S1 protein, nor has another protein yet been detected which would substitute for S1 (ref. 42). This is the more surprising, as this protein has a very important function in binding ribosomes to mRNA and it seems to play a role in controlling the recognition of specific initiation sites.

Some proteins are easily removed from the ribosomes. S1 protein is released from 30S particles by dialysis against Mg-free buffer. Other proteins are removed when ribosomal subunits are subjected to CsCl density gradient centrifugation. The proteins thus released are called split proteins, and the particles left behind, containing the more tightly bound proteins, are the core particles. Proteins can be gradually removed from ribosomal subunits by exposing them to high salt concentration, e.g. to buffer–salt mixtures containing increasing concentrations of LiCl. The split proteins are able to attach again to the particles and to re-form a physically and functionally complete subunit if the salt is removed and the core particles and split proteins are incubated together at 40°C in a phosphate buffer–$MgCl_2$–KCl mixture, the so-called reconstitution buffer. Removal of some proteins and reconstitution of active ribosomes proved a helpful technique in studying the function of individual ribosomal proteins.

Reconstitution of ribosomal particles

In 1968, Nomura and coworkers achieved complete reconstitution of 30S particles from pure 16S RNA and the 21 ribosomal proteins[43]. The RNA and the proteins were mixed in reconstitution buffer (see above) and incubated at 40°C. The 22 components assembled into a ribonucleoprotein particle which proved to be identical to the native 30S subunit, not only with respect to physicochemical characteristics, but also in their ability to carry out protein synthesis. Reconstitution of the 50S particle proved a more difficult task. It was first achieved with *B. stearothermophilus* ribosomes by Nomura and Erdmann[44] and later also with *E. coli* ribosomes by Dohme and Nierhaus[45], who used a more complex two-step procedure to produce active particles. These reconstitution experiments were important not only because they revealed that in such a complex structure each component has an unambiguously defined position which it can take up spontaneously, but they were also of great importance for the further elucidation of the topography of the ribosome and the function of individual proteins.

The function of ribosomal proteins

Many different approaches have been used to study the function of individual ribosomal proteins. The difficulties encountered in these studies arise not so much from the great number of proteins as from their manifold interactions. In each ribosomal function the participation of a number of proteins could be demonstrated. Some of these proteins may have an essential role, others a stimulating effect, some proteins may act indirectly by stabilising the conformation of RNA and proteins. The problem is closely connected with that of the topography of the ribosome: in an intertwined structure of proteins and RNAs, several molecules contribute to the formation of active sites and consequently several molecules (RNAs and proteins) cooperate in every ribosomal function. This has led to the concept of ribosomal 'domains', areas comprising a number of ribosomal components which are together responsible for a definite function. Instead of attempting to ascribe an exact function to each ribosomal protein, the aim of recent investigations has been the elucidation of the structural and functional organisation of the ribosome, by locating the areas, and establishing the groups of proteins and RNA stretches which, in a functionally cooperative way, bring about the different translational events.

We are still far from building an exact functional or structural model of the ribosome. In the following, the different methods which shed some light on different aspects of these problems will be described. For a valid interpretation of the data, results obtained by different methods, and information on the structural as well as on the functional aspects have to be considered.

(1) Early experiments were based on the reconstitution technique, which seemed to offer promising possibilities for detecting in which processes a ribosomal protein is involved. 30S particles were reconstituted in incomplete systems, lacking one or another of the S-proteins, under the assumption that the particles so produced would be deficient in one or another ribosomal function which then could be ascribed to the missing protein. Although such incomplete ribosomal particles could indeed be produced, no far-reaching conclusions could be drawn from the results, indicating that the proteins did not act independently. A strong interaction existed between a number of ribosomal proteins, with the consequence that leaving out one affected the structure of the whole particle and thus also the function of other proteins.

These ambiguities could be eliminated by using partial rather than complete reconstitution systems. An increasing number of proteins can be removed from the ribosomes by washing with salt solutions of increasing concentrations. The loss of proteins is accompanied by the loss of different functional activities. In some cases these functions can be restored by adding back one or a few of the missing proteins. This provides good evidence that these particular proteins are essential for a given ribosomal function. This method was used to study the peptidyl transferase centre of the 50S ribosomal subunit. Moore *et al.*[46] and Nierhaus and Montejo[47] studied the peptidyl transferase activity of core particles lacking 10 or 18 ribosomal proteins, respectively. The core particles were inactive, but activity could be restored by supplementing them with proteins L16 (ref. 46) or L11 (ref. 47). It was concluded that these proteins must have an important role in forming the active peptidyl transferase centre.

(2) An altogether different approach, affinity labelling, aims at defining the ribosomal proteins and RNAs which interact with extra-ribosomal components of the translation system. Radioactively labelled derivatives of aminoacyl-tRNA, peptidyl-tRNA or mRNA are cross-linked to the ribosome by UV irradiation or by a chemical reaction which fixes these molecules in the position where interaction occurred. A covalent bond is thus formed between these RNAs and one or a few components of the ribosome. After disruption of the ribosome, the radioactive label remains attached to these components which can then be isolated and identified[8]. Pongs *et al.*[48] found that an analogue of the initiation codon, AUG, was cross-linked to protein S1. Experiments with other codon analogues suggested the participation of further proteins of the 30S subunit in mRNA binding: S4, S11, S12, S13, S18 and S21. Peptidyl-tRNA labelled in the peptide moiety could be cross-linked to proteins L2 and L27 and to a lesser extent also to five other proteins of the 50S subunit, whereas aminoacyl-tRNA derivatives were linked to L16.

These results were interpreted as suggesting that proteins L2 and L27 form part of the P-site, whilst L16 forms part of the A-site on the ribosome. These proteins must also be close to the peptidyl transferase site, as the cross-linked parts of the molecules were near the CCA ends of the tRNAs which is the site where peptide transfer occurs. Applying the same technique to peptidyl-tRNA derivatives with longer peptide chains revealed that the growing peptide chain also attaches to the ribosome, and the order of ribosomal proteins along the peptide chain was established as L2, L27, L32 (or L33) and L24 (ref. 49). Contact has also been detected between the peptide chain and 23S RNA, all along from the CCA end of tRNA to the N-terminus of the peptide[20]. This led to the conclusion that 23S RNA is present at or close to the peptide donor site of the peptidyl transferase centre and also along the path where the growing peptide chain is attached.

(3) Further characterisation of the peptidyl transferase centre has been achieved by Nierhaus, who succeeded in reconstituting an active ribonucleoprotein particle, comprised of 23S RNA and 18 proteins, which in itself could carry out this reaction. Some of these proteins seemed to play an indirect role, in forming the proper ribonucleoprotein structure in which the proteins directly involved in the function can occupy the positions which are optimal for carrying out peptide transfer. Proteins L3, L16 and L20 were found to be essential for reconstitution

and are therefore assumed to be directly involved in the reaction.

(4) Chemical modification of some proteins or reaction with specific antibodies inhibit individual ribosomal functions, but such data are often difficult to interpret. In the case of the peptidyl transferase centre, the results were in good agreement with those from reconstitution experiments: antibodies to L11 and L16 caused inactivation, confirming the importance of these proteins in the peptidyl transfer reaction[50].

(5) Specific properties of some ribosomal proteins made it possible to investigate their function in special ways.

Studies on the function of protein S1 were facilitated by the easy removal of this protein from the 30S subunit. Comparison of the translational activities of complete ribosomes and ribosomes lacking S1 revealed that loss of this protein impairs the capacity of ribosomes to form initiation complexes. Adding S1 protein to the S1-deficient ribosomes restores their activity[51]. Optimal stimulation is obtained with one copy of S1 protein per ribosomal particle[52]. An excess of S1 protein inhibits translation, probably because under these conditions the protein binds to mRNA and interferes with the binding and/or the movement of the ribosome.

S1 protein is also connected with the selection of specific initiation sites on mRNA. Steitz *et al.*[24] observed that in addition to the three specific initiation sites on R17 RNA, S1-deficient ribosomes also attach to other nucleotide sequences. *B. stearothermophilus* ribosomes which do not contain S1 protein can still initiate translation (whether another protein of these ribosomes takes over the function of S1 is not known), but *in vitro* studies with R17 and MS2 viral messengers showed that on a polycistronic mRNA they do not select the same initiation sites as *E. coli* ribosomes. While the latter attach preferentially to the initiation site of the coat protein cistron, *B. stearothermophilus* ribosomes initiate most efficiently at the A-protein cistron, and only to a very low extent at the other two initiation sites. Addition of protein S1 to *B. stearothermophilus* protein synthesising systems results in increased initiation at the coat protein cistron[53]. S1 protein may be involved in initiation at the step of annealing of 16S rRNA to mRNA. This is indicated by its localisation in the ribosome near to the 3′ terminus of 16S RNA[54] and by its ability to bind to the 3′-terminal sequence of this RNA[25] and alter the conformation of this RNA stretch[26]. The latter effect may only be a reflection of another property of S1 protein detected simultaneously by Bear *et al.* and Szer *et al.*, viz. that it can unwind pyrimidine-rich double-stranded parts in RNA molecules[55]. With respect to this activity, S1 protein behaves similarly to DNA unwinding protein (*see* Chapter 2): it acts stoichiometrically rather than catalytically, about one protein molecule being needed for every 40 nucleotides in the RNA to disrupt the secondary structure. It has been shown recently that the unwinding activity of S1 protein is connected with its role in initiation complex formation: the two functions disappear together upon chemical modification[56].

The various effects of S1 protein may all be related to each other: it may stimulate initiation complex formation by disrupting the secondary structure of rRNA or by locally unwinding hairpin loop structures in mRNA near the initiation site. Its similarity to DNA unwinding protein may be due to the fact that it binds nonspecifically to various sites in mRNA, which in turn may be the cause of the inhibitory effect high concentrations of S1 protein have on translation.

S1 protein has another function which, however, seems quite unrelated to those discussed above. It forms the α subunits of $Q\beta$ RNA polymerase, the enzyme responsible for transcription of the RNA of $Q\beta$ phage[57]. It is interesting in this respect that while S1 protein binds nonspecifically to a number of pyrimidine-rich stretches in R17 RNA and also in other RNA molecules, in $Q\beta$ RNA it binds to a specific site near the 3′ terminus where transcription starts[58]. It has been suggested that its unwinding activity may be important for the transcription of $Q\beta$ RNA, which has a rather strong secondary structure, but there is still no evidence to connect its role in transcription with that in translation.

Special interest attaches to protein S12 the function of which has been studied in relation to the effect of streptomycin on bacterial ribosomes. This antibiotic causes mis-translation in *in vitro* protein synthesising systems and it was shown to exert its effect through the 30S ribosomal subunit. Studies on streptomycin-resistant and streptomycin-dependent mutants revealed that the protein affected by these mutations, S12, had an important role in controlling the codon-dependent interaction of aminoacyl-tRNA with the ribosome[59]. Another type of mutation which affects the fidelity of translation occurs in the *ram* locus (for 'ribosomal ambiguity mutation'), in the structural gene of protein S4. These two proteins are thus involved in the control of the accuracy of translation.

THE TOPOGRAPHY OF THE RIBOSOME

Ribosomal subunits

The spatial arrangement of proteins and nucleic acids within the ribosomal particles has been the subject of intensive investigations for many years. Various experimental approaches have been applied to obtain information on the sites occupied by different proteins, on the interactions between different ribosomal proteins and on the interactions of proteins with rRNA.

The first attempts were based on an oversimplified image of the ribosomal particles: it was expected that globular proteins attach to each other and to well defined stretches of rRNA. Most of the analytical techniques applied to ribosomes were developed with this idea in mind: they aimed at determining the binding sites of ribosomal proteins on rRNA, distinguishing between 'accessible' proteins at the surface of the particle and 'inaccessible' proteins on the inside, and defining which proteins are in close proximity to each other. Only recently has it been recognised that the structural organisation of the ribosome is very different indeed.

On the one hand, a number of ribosomal proteins are highly asymmetrical, elongated molecules; on the other, rRNAs are present in strongly folded tertiary structure. Interaction between these components can occur at multiple sites, resulting in an intertwined structure of proteins and RNA which leads to ambiguous results with many of the above analytical techniques[30]. Even the most sophisticated modern methods have been unable to resolve this complex structure, in which one protein can extend from one site on the surface through the inside of the particle to a second, quite distant, site on the surface, and in which different stretches of the RNA, far apart in the linear sequence, can be joined to form a binding site for one protein. It is understandable that in spite of the great progress made in the last few years our knowledge of the topography of ribosomes is still limited and that the large amount of data available can be interpreted only if the results obtained by different experimental methods are compared and combined. In the following, some approaches to this problem will be described which contributed to our present ideas on the topography of the *E. coli* ribosome[8,30,65].

(1) One of the early approaches was based on studying the mechanism of reconstitution, i.e. on following step-wise the attachment of the ribosomal proteins in the *in vitro* assembly of the ribosomal particle. Nomura's group established the order in which the different proteins join to form a 30S subunit (assembly map)[60]. They found that some proteins were able to bind directly to 16S RNA (primary binding proteins) while others required the presence of several other ribosomal proteins. This seemed to indicate that the first group of proteins, comprising S4, S7, S8, S15, S17 and S20, were in direct contact with rRNA in the ribosomal particle, whereas the proteins of the latter group were attached through interaction with other proteins. Although some of the conclusions drawn from these experiments have been questioned since, they helped to form the first idea of the topography of a ribosomal particle.

Nierhaus and coworkers studied the assembly of the 50S subunit of *E. coli* ribosomes[61]. They defined two groups of primary binding proteins: 11 proteins interact directly with 23S RNA, 3 (L18, L25, L5) with 5S RNA. The complete process of reconstitution of the large subparticle is more complex and comprises more steps than that of the small subunit. Three intermediate particles with different sedimentation coefficients were isolated, but their analysis showed that two of them contained the same 17 or 18 proteins bound to 23S RNA. The change in their sedimentation behaviour reflects a change in conformation which is essential for the formation of an active particle[62]. Conformational changes also occur in the course of assembly of the 30S particle[31].

The order in which proteins assemble to form an active 30S or 50S subunit reflects to some extent their relative positions within the particle. Interpretation of these data is, however, ambiguous. Recent investigations called attention to the importance of RNA conformation for interaction with proteins and also to the changes such interactions can induce in RNA conformation. In view of such results, the distinction between primary binding proteins and proteins which do not react directly with rRNA seems somewhat arbitrary. It may depend more on the conformation of rRNA as to which proteins can directly attach to it. Hochkeppel and coworkers found that both 16S RNA[31,63] and 23S RNA[76] take up different conformations if extracted by phenol or by acetic acid and urea. rRNA preparations obtained by the latter technique interact with more proteins: an additional six or seven ribosomal proteins bind directly to 16S RNA and two more proteins to 23S RNA.

Both direct and indirect evidence is also available that interaction with proteins may affect the conformation of RNA. Bear *et al.*[64] have shown that binding of L18 and L25 alters the conformation of 5S RNA; Hochkeppel and Craven[31] found that at least three of the primary binding proteins of the 30S subunit, S4, S7 and S8, alter the conformation of 16S RNA. Moreover, this conformational change enables 16S RNA to interact also with protein S9. Considering that conformational changes are known to occur in the course of ribosome assembly, it seems probable that early binding of some proteins in the assembly process may be required merely for producing a suitable three-dimensional structure which can accommodate further proteins. The ribosomal proteins which join late may interact with rRNA indirectly, through proteins already present, or they may merely require the specific RNA conformation which has been formed by attachment of the primary binding proteins.

(2) Direct interaction of some ribosomal proteins with rRNA provides an opportunity to determine the binding sites of these proteins on 16S, 23S and 5S RNA. The classical method of isolating and sequencing the nucleic acid fragment protected by a protein (*see* p. 85 and p. 156) was applied to rRNAs with known nucleotide sequences, whereas in the case of 23S RNA, the nucleotide sequence of which is only partially known, approximate location of protein binding sites was also attempted by reacting the proteins with different fragments of the RNA molecule and identifying the fragment to which a protein could bind[65]. The results of these studies reflected the complex structural organisation of the ribosome. The binding sites of a number of proteins on 16S RNA were not confined to small areas of the rRNA, corresponding in size to that of the protein, nor were they built of a simple linear stretch of the RNA. Quite large regions containing several hairpin loops were protected by some ribosomal proteins, as shown in *figure 6.7* for the binding sites of proteins S8, S15 and S20. It can be seen in the figure that the protected RNA regions also contain excisions; they consist of several shorter RNA stretches held together by base pairing or by the protein itself. Proteins S4 and L24 protect exceptionally large areas, about 500 nucleotides long, in 16S and 23S RNA, respectively.

It is strongly indicated, however, that only minor parts of these long stretches are in direct contact with protein, the rest being held together mainly by secondary and tertiary RNA-RNA interactions[67]. Sloof *et al.* investigated protein binding sites under the electron microscope and provided direct evidence that at these sites the RNA was present in a strongly folded secondary and tertiary structure[68]. 16S and 23S RNA were reacted with different ribosomal proteins and were denatured so as to abolish secondary and tertiary interactions completely. The structure was

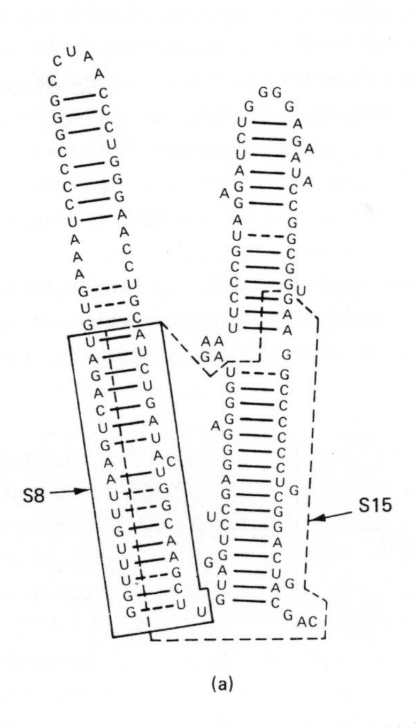

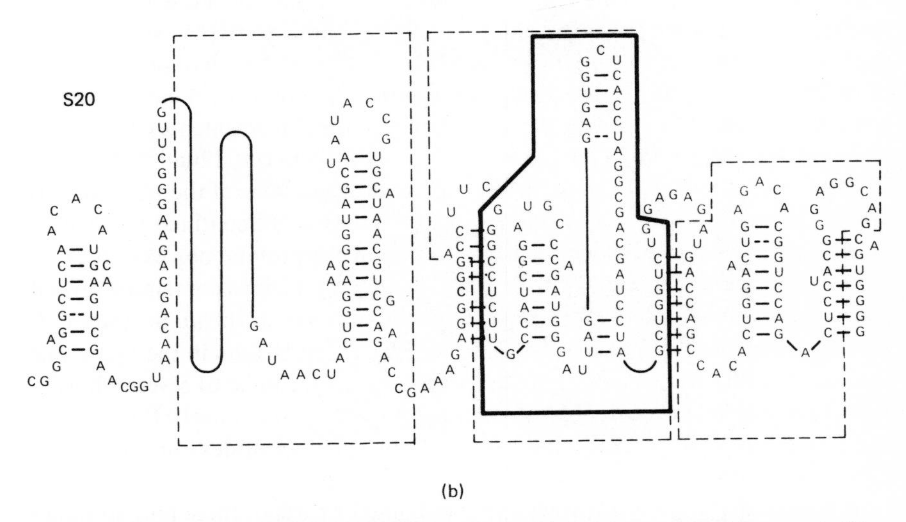

Figure 6.7 Protein binding sites on 16S rRNA. (a): RNA stretches protected by protein S8 (solid line) and S15 (dotted line) (Ungewickel *et al.*[66]). (b): RNA stretch protected by protein S20 (dotted line). Only the sequence in the region marked by the heavy line was shown to be directly linked to protein. The rest of the protected stretch is probably involved in RNA–RNA interactions rather than in RNA–protein interactions (Ehresmann *et al.*[67]). The complete structure of 16S RNA was not known at the time the binding site was located. The missing sequences are shown as solid lines.

stabilised, however, at the sites where the proteins were bound, and in the electron microphotographs these sites showed up as small dense coils, while the free RNA stretches appeared as linear threads (*figure 6.8*). From the length of these linear stretches, the position of the protein binding site in the rRNA could also be established. Using a somewhat similar approach, Nomura and coworkers found that the binding of different proteins preserved a different and characteristic configuration of the RNA, so that the binding sites of different proteins could be recognised under the electron microscope[77].

These results are in good agreement with our present ideas on the structural organisation of ribosomal particles, according to which the rRNA is present in an elaborate three-dimensional structure in which relatively distant stretches of the

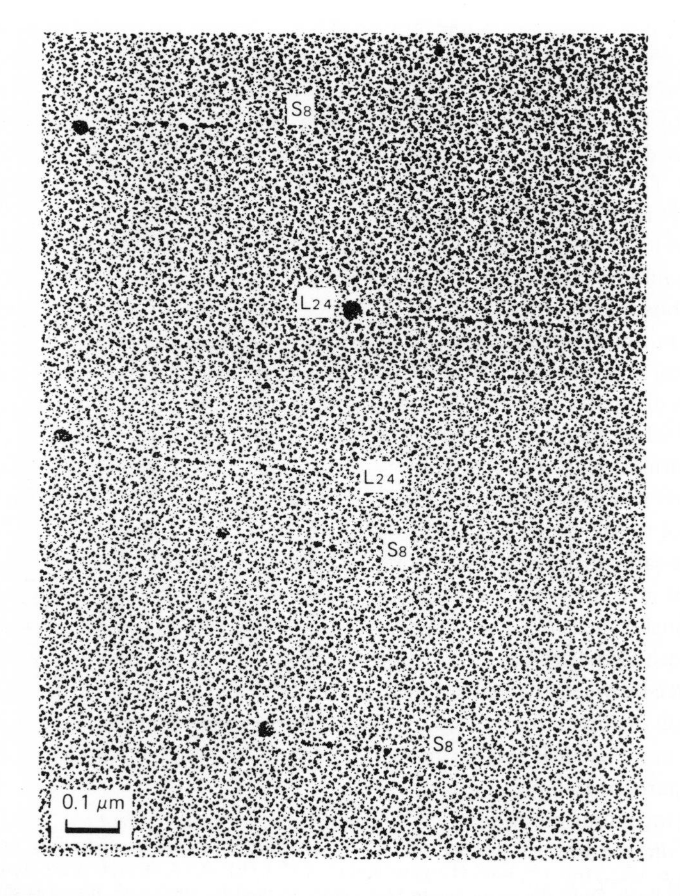

Figure 6.8 Electron micrograph of rRNA associated with different ribosomal proteins. The RNA has been denatured after binding the proteins. The protein-protected part remains folded, the unprotected RNA can be seen as a linear, denatured stretch. (By Sloof *et al.*[68]. Reproduced with permission from *Mol. Gen. Genet.*, **147**, 129 (1976). Photograph by courtesy of Dr P. Sloof.)

polynucleotide chain can be brought into close proximity, and a number of protein molecules have the shape of a long thin thread which can interact with more than one site in the RNA. The intertwined protein–RNA structure results in multiple, three-dimensional binding sites rather than single linear sites of interaction.

The same overall picture also emerges from studies on ribonucleoprotein fragments removed from ribosomes by mild digestion. Such ribonucleoprotein fragments, comprising a few proteins and a part of the rRNA, were analysed to determine which proteins and which RNA stretches are in close proximity in the ribosome[65]. The results showed here, too, that the RNA fragments present covered a rather large area in the rRNA molecules and had some stretches excised and others strongly folded.

(3) Another approach to determine topographical neighbourhoods is based on chemical cross-linking of proteins[65]. Chemical reagents or UV irradiation can form covalent bonds between adjacent proteins or between proteins and adjacent stretches of RNA. In this way a number of neighbouring protein pairs have been characterised, and the proteins near the 3′ terminus of 16S RNA have also been identified. These latter data have particular importance, as the 3′ end of this RNA participates in the initiation of translation (p. 194), and it can be assumed that the proteins in its close proximity may also be involved in this process. Proteins S1 and S21 could be cross-linked to the 3′ terminus of 16S RNA[54]. The role of S1 in binding ribosomes to the specific initiation sites in mRNA has already been determined from functional studies.

Chemical cross-linking is a very useful tool in mapping the inside of the ribosome. The difficulty is, however, that elongated protein molecules may extend across the whole particle and may have a number of different neighbours[8,30]. Exact information on the topography will be obtained only when the particular amino acids which are in close proximity in two proteins have been identified and located in the amino acid sequence.

(4) The same difficulty is encountered with the most modern technique introduced into the study of ribosome structure. The method of singlet–singlet energy transfer measures the distance between two protein molecules which are labelled with two different fluorescent dyes[8,30]. In the case of elongated proteins, the distance between the centres of mass does not yield sufficient information on the position occupied by these proteins inside the ribosomal particle. The method will be of more use if it can be further developed to determine distances between specified parts of two proteins.

(5) An ingenious method which brought great progress in the mapping of the ribosome surface and in locating functional sites has been developed in Wittmann's laboratory[8]. Specific antibodies were prepared to all ribosomal proteins and were allowed to interact with 30S, 50S and 70S ribosomes. The idea behind this line of investigation was that only the proteins on the surface of the particle will react with antibody, thus revealing even at this early stage of the experiment, some details of the arrangement of ribosomal proteins.

A further possibility for exact location of these protein–antibody complexes was provided by an electron microscopic technique which showed up directly to which part of the particle the antibody was attached. These investigations brought the first clear-cut evidence that ribosomal proteins have an elongated shape. Antibodies reacted with every ribosomal protein, and some of the antibodies attached to two or more sites on the surface of the particle, often at a considerable distance from each other. This implied, on one hand, that all ribosomal proteins, or at least part of their polypeptide chains, were at the surface, and on the other, that several proteins took up an elongated conformation in the ribosome, the two ends of the polypeptide chain pointing to two distant sites on the ribosomal surface. The sites where the antibodies attach to the ribosome could be mapped, and a model of the 30S particle constructed on this basis is shown in *figure 6.9*. Not all proteins on

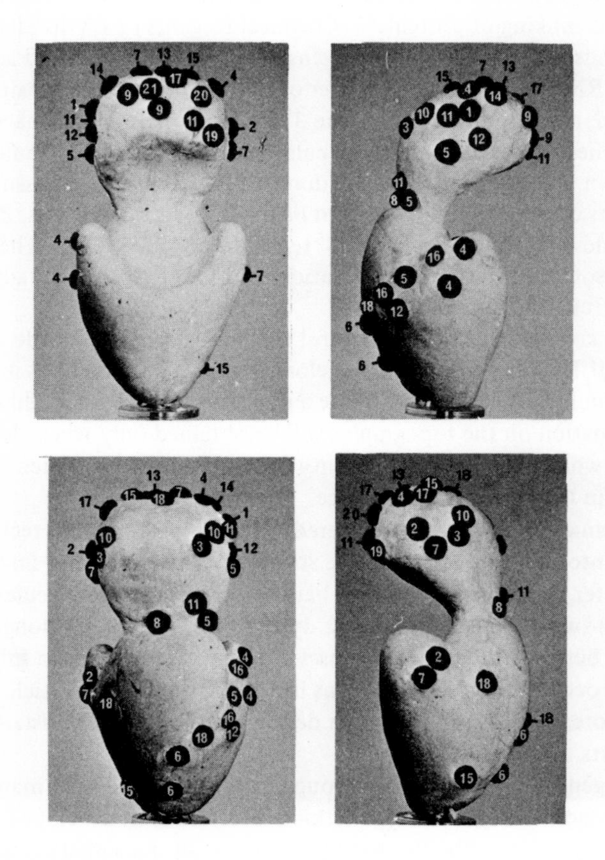

Figure 6.9 Model of the 30S ribosomal subunit, showing the positions of the different ribosomal proteins. The four photographs are related to each other by successive rotations of the model by 90°. (From Stöffler and Wittmann[8]. Reproduced with permission from *Molecular Mechanisms of Protein Biosynthesis*, Academic Press (1977). Photograph by courtesy of Professor H. G. Wittmann and Dr G. Stöffler.)

the surface of the 50S particle have yet been mapped, but the position of 19 proteins can be seen in *figure 6.10.* For some ribosomal proteins it has been established in which ribosomal function they participate, and so mapping of the ribosome surface also gives an idea of the localisation of functional sites. The area on the 30S subunit, where proteins participating in initiation and mRNA binding are present, is located at the anterior and right side of the 'head' (*figure 6.9*). The proteins mapped on the 50S subunit in *figure 6.10* respectively form the peptidyl transferase centre (L11, L16), accommodate the CCA end of peptidyl-tRNA (L2, L27) and bind 5S RNA (L5, L18, L25).

Figure 6.10 Model of the 50S ribosomal subunit, showing the positions of proteins of the peptidyl transferase centre (L11, L16), and of those associated with 5S RNA (L5, L18, L25) and with peptidyl-tRNA (L2, L27). (From Stöffler and Wittmann[8]. Reproduced with permission from *Molecular Mechanisms of Protein Biosynthesis*, Academic Press (1977). Photographs by courtesy of Professor H. G. Wittmann and Dr G. Stöffler.)

The model of the 70S ribosome

In *figure 6.1*, the model of the 70S ribosome is shown as it has been constructed from electron microscopic studies. Mapping of the two subunits revealed the location of some functional sites; it is thus possible to locate the sites in the 70S ribosome at which the different components of the protein synthesising system attach[8]. This will be discussed in Chapter 8.

Upon association of the two subunits, some proteins and RNA stretches in the two particles come into contact with each other. It has been noted that the reactivity of some proteins and the accessibility of some sites in 16S and 23S RNA are altered. This does not necessarily mean that all these proteins and RNA stretches are localised at the interface. At most ribosomal events, association and dissociation are accompanied by conformational changes which may result in increased as well as decreased accessibility of some components. Some interface proteins could be unambiguously identified: proteins S11 and S12 of the 30S subunit bind to 23S RNA, protein S16 could be cross-linked to proteins of the 50S subunit. The identity of S20 and L26 locates this protein at the interface also. The two rRNAs, 16S RNA and 23S RNA, anneal to each other with their complementary stretches near their 3′ termini (p. 195).

As can be seen in *figure 6.1*, a tunnel-like cavity is formed between the two subunits in the 70S ribosome. Evidence has been obtained by different experimental techniques that the mRNA is accommodated in this tunnel[8]. The 'mRNA binding domain' has been located at a part of the 30S subunit which is in contact with the 50S subunit; these areas surround the tunnel. Electron micrographs also suggest that the mRNA may pass through a tunnel between the two subunits (*see* also Chapter 8, *figure 8.4*). The length of the tunnel as measured by electron microscopy is about 100 Å, which agrees well with the length of the mRNA fragment protected by ribosomes.

BIOSYNTHESIS OF RIBOSOMES

The synthesis of different ribosomal components is very exactly coordinated in both prokaryotic and eukaryotic cells: neither excess proteins nor excess RNAs accumulate. Much of our knowledge of the expression of ribosomal genes in *E. coli* comes from the investigations of Nomura's group, who studied the genetic loci of the rRNA and r-protein genes, the transcription of rRNAs and the transcription and translation of the genes which code for r-proteins[33,69,70].

There are seven rRNA transcription units in the *E. coli* genome. Each comprises the three rRNA genes and also tRNA genes in the spacer regions[33,71]. The organisation of these transcription units explains the coordinated synthesis of the three rRNA species (*see* also Chapter 3). The coordinated production of more than 50 proteins and the balanced synthesis of r-proteins and rRNAs must be under a more complex regulatory mechanism. Studies on the synthesis of ribosomal proteins under a variety of conditions and in a number of mutants indicated that control is exercised mainly at the stage of transcription[69]. The genes for ribosomal proteins are organised into several transcription units which function like large operons, responding coordinately to signals affecting one or a few promoters.

Two promoter regions have been characterised and their nucleotide sequences determined by Nomura's group[78]. 35 to 39 r-protein genes are clustered near the *str* locus; 5 others have been located at a distant part of the chromosome, and the genes for S18 and S20 proteins are in a third, separate part of the *E. coli* genome. DNA stretches containing a number of genes from the clustered region were incorporated into transducing phages. Expression of these genes could be studied *in vitro* as well as *in vivo*. The organisation of several transcription units could thus be established[70,78]. In some of these, r-proteins are co-transcribed with other protein genes, e.g. with subunits of RNA polymerase. The synthesis of the proteins is controlled by the amount of mRNA available for translation, the primary control being exercised at the initiation of transcription of these mRNAs. This still leaves a possibility of further regulation through the stability of mRNA, and of an additional mechanism for more subtle regulation at the translational level. It is not known how the expression of genes in unlinked transcriptional units is coordinated, e.g. of r-protein genes at different chromosomal loci, or of r-protein and rRNA genes.

The situation is even more complicated in eukaryotic organisms, where rRNAs are synthesised in the nucleolus and r-proteins in the cytoplasm and where even the synthesis of all rRNAs is not under a common control. Three of the eukaryotic rRNA species, 28S, 18S and 5.8S RNA, are synthesised together in the nucleolus. They are transcribed from one transcription unit by class I RNA polymerase[34,73]. The transcript then undergoes processing to yield the three mature rRNA molecules (p. 114). 5S RNA is synthesised separately (p. 197), by a different RNA polymerase, the class III enzyme[73] (*see* also p. 105).

Gorenstein and Warner[72] studied the synthesis of ribosomal proteins in yeast. Although it is probable that the sixty or more genes for ribosomal proteins do not form polycistronic transcription units in yeast (they are probably not even adjacent), the results nevertheless suggested that transcription of these genes is coordinately controlled. Mutations in two genes, which may be regulatory genes rather than structural genes coding for proteins, caused coordinate inhibition of the synthesis of at least 40 r-proteins. Moreover, studies on the *in vitro* translation of mRNAs for ribosomal proteins strongly indicated that the synthesis of r-proteins is controlled at the level of transcription. Warner and Gorenstein proposed that a control mechanism not unlike that functioning in *E. coli* may exist also in eukaryotic organisms.

Assembly of the ribosomal particles

In vivo assembly of ribosomal particles from their RNA and protein components does not occur along the same pathway as that observed in *in vitro* reconstitution experiments. Precursor ribosomal particles have been isolated from *E. coli* cells which contained precursor RNA, i.e. larger RNA molecules than the mature rRNA, with a number of ribosomal proteins attached to it. The order in which r-proteins attach to RNA could thus be compared with the *in vitro* assembly map, and indicated that after an initial similarity in the attachment of the first few proteins, the two processes diverge. It is clear, however, from the presence of precursor RNA in the pre-ribosomal particles, that *in vivo* attachment of proteins starts before post-transcriptional processing of the rRNAs. The finding of Spillman *et al.*[62] that the proteins which attach first to 23S RNA are all bound to the 5′-terminal part of the molecule suggests that the first proteins may bind to the rRNA while the latter is still being transcribed.

REFERENCES

1 Kurland, C. G. in *Molecular Mechanisms of Protein Biosynthesis* (eds Weissbach, H. and Pestka, S.) Academic Press, New York, 81 (1977).
2 Roberts, B. E. and Paterson, B. M. *PNAS,* **70**, 2330 (1973).
3 Gross, K., Probst, E., Schaffner, W. and Birnstiel, M. *Cell,* **8**, 455 (1973).
4 Strohman, R. C., Moss, P. S., Micou-Eastwood, J., Spector, D., Przybyla, A. and Paterson, B. *Cell,* **10**, 265 (1977).
Heywood, S. M., Kennedy, D. S. and Bester, A. J. *PNAS,* **71**, 2428 (1974).
5 Lane, C. D., Gregory, C. M. and Morel, C. *Eur. J. Biochem.,* **34**, 219 (1973).
6 Lodish, H. F. *J. Mol. Biol.,* **50**, 689 (1970); *Nature,* **226**, 705 (1970).
7 Hall, N. D. and Arnstein, H. R. V. *Biochem. Biophys. Res. Comm.,* **54**, 1489 (1973).
8 Stöffler, G. and Wittmann, H. G. in *Molecular Mechanisms of Protein Biosynthesis* (eds Weissbach, H. and Pestka, S.) Academic Press, New York, 117, (1977).
9 Boublik, M. and Hellmann, W. *PNAS,* **75**, 2829 (1978).
10 Brownlee, G. G., Sanger, F. and Barrell, B. G. *Nature,* **215**, 735 (1967).
10a Erdmann, V. A. *Nucl. Acids Res.,* **6**, r29 (1979).
11 Ofengand, J. and Henes, C. *J. Biol. Chem.,* **244**, 6241 (1969).
12 Herr, W. and Noller, H. F. *FEBS Lett.,* **53**, 248 (1975).
13 Weidner, H., Yuan, R. and Crothers, D. M. *Nature,* **266**, 193 (1977).
14 Erdmann, V. A. *Progr. Nucl. Acid Res. Mol. Biol.,* **18**, 45 (1976).
15 Fox, G. E. and Woese, C. R. *Nature,* **256**, 505 (1975).
16 Nishikawa, K. and Takemura, S. *FEBS Lett.,* **40**, 106 (1974).
17 Bear, D. G., Schleich, T., Noller, H. F. and Garrett, R. A. *Nucl. Acid Res.,* **4**, 2511 (1977).
18 Branlant, C., Sri Widada, J., Krol, A. and Ebel, J. P. *Eur. J. Biochem.,* **74**, 155 (1977).
19 Van Duin, J., Kurland, C. G., Dondon, J., Grunberg-Manago, M., Branlant, C. and Ebel, J. P. *FEBS Lett.,* **62**, 111 (1976).
20 Sonenberg, N., Wilchek, M. and Zamir, A. *Eur. J. Biochem.,* **77**, 217 (1977).
21 Carbon, P., Ehresmann, C. and Ebel, J. P. *FEBS Lett.,* **94**, 152 (1978).
22 Bowman, C. M., Dahlberg, J. E., Ikemura, T., Konisky, J. and Nomura, M. *PNAS,* **68**, 964 (1971).
23 Shine, J. and Dalgarno, L. *Nature,* **254**, 34 (1975).
24 Sprague, K. U., Steitz, J. A., Greney, R. M. and Stocking, C. E. *Nature,* **267**, 462 (1977).
Steitz, J. A., Wahba, A. J., Laughrea, M. and Moore, P. B. *Nucl. Acids Res.,* **4**, 1 (1977).
25 Dahlberg, A. E. and Dahlberg, J. E. *PNAS,* **72**, 2940 (1975).

26 Steitz, J. A. personal communication.
27 Baan, R. A., Hilbers, C. W., Van Charldorp, R., Van Leerdam, E., Van Knippenberg, P. H. and Bosch, L. *PNAS,* **74**, 1028 (1977).
28 Cox, R. A. *Progr. Biophys. Molec. Biol.,* **32**, 193 (1977).
29 Kurland, C. G. in *Ribosomes* (eds Nomura, M., Tissieres, S. and Lengyel, P.) Cold Spring Harbor Laboratories, New York, 309 (1974).
30 Brimacombe, R. in *Relations between Structure and Function in the Prokaryotic Cell* (eds Stanier, R. Y., Rogers, H. J. and Ward, J. B.) Cambridge University Press, 1 (1978).
31 Hochkeppel, H. K. and Craven, G. R. *J. Mol. Biol.,* **113**, 623 (1977).
31a Luoma, G. A. and Marshall, A. G. *J. Mol. Biol.,* **125**, 95 (1978).
32 Wrede, P. and Erdmann, V. A. *PNAS,* **74**, 2706 (1977).
33 Jaskunas, S. R., Nomura, M. and Davies, J. in *Ribosomes* (eds Nomura, M., Tissieres, S. and Lengyel, P.) Cold Spring Harbor Laboratories, New York, 333 (1974).
34 Reeder, R. H. in *Ribosomes* (eds Nomura, M., Tissieres, S. and Lengyel, P.) Cold Spring Harbor Laboratories, New York, 489 (1974).
35 Walker, T. A. and Pace, N. R. *Nucl. Acids Res.,* **4**, 595 (1977).
36 Hagenbüchle, O., Santer, M. and Steitz, J. A. *Cell,* **13**, 551 (1978).
37 Subramanian, A. R., Haase, C. and Giesen, M. *Eur. J. Biochem.,* **67**, 591 (1976).
38 Martini, O. H. W. and Gould, H. J. *Mol. Gen. Genet.,* **142**, 317 (1975).
39 Kaltschmidt, E. and Wittmann, H. G. *PNAS,* **67**, 1276 (1970).
40 Hardy, S. J. S. *Mol. Gen. Genet.,* **140**, 253 (1975).
41 Subramanian, A. R. *J. Mol. Biol.,* **95**, 1 (1975).
42 Isono, K. and Isono, S. *PNAS,* **73**, 767 (1976).
43 Traub, P. and Nomura, M. *PNAS,* **59**, 777 (1968).
44 Nomura, M. and Erdmann, V. A. *Nature,* **228**, 744 (1970).
45 Dohme, F. and Nierhaus, K. H. *J. Mol. Biol.,* **107**, 585 (1976).
46 Moore, V. G., Atchison, R. E., Thomas, G., Moran, M. and Noller, H. F. *PNAS,* **72**, 844 (1975).
47 Nierhaus, K. H. and Montejo, V. *PNAS,* **70**, 1931 (1973).
48 Pongs, O., Stöffler, G. and Bald, R. W. *Nucl. Acids Res.,* **3**, 635 (1976).
49 Eliat, D., Pellegrini, M., Oen, H., Lapidot, Y. and Cantor, C. R. *J. Mol. Biol.,* **88**, 831 (1974).
50 Tate, W. P., Caskey, C. T. and Stöffler, G. *J. Mol. Biol.,* **93**, 375 (1975).
51 Van Dieijn, G., Van der Laken, C. J., Van Knippenberg, P. H. and Van Duin, J. *J. Mol. Biol.,* **93**, 351 (1975).
Szer, W., Hermoso, J. M. and Leffler, S. *PNAS,* **72**, 2325 (1975).
52 Bosch, L. in *Proceedings of the 10th FEBS Meeting* (eds Chapeville, F. and Grunberg-Manago, M.) North Holland/American Elsevier, New York, 275 (1975).
53 Isono, S. and Isono, K. *Eur. J. Biochem.,* **56**, 15 (1975).
54 Czernilofsky, A. P., Kurland, C. G. and Stöffler, G. *FEBS Lett.,* **58**, 281 (1975).
55 Bear, D. G., Ng, Ray, Van Derveer, D., Johnson, N. P., Thomas, G., Schleich, T. and Noller, H. F. *PNAS,* **73**, 1824 (1976).
Szer, W., Hermoso, J. M. and Boublik, M. *Biochem. Biophys. Res. Comm.,* **70**, 957 (1976).
Thomas, J. O., Kolb, A. and Szer, W. *J. Mol. Biol.,* **123**, 163 (1978).
56 Kolb, A., Hermoso, J. M., Thomas, J. O. and Szer, W. *PNAS,* **74**, 2379 (1977).
57 Carmichael, G. G., Weber, K., Niveleau, A. and Wahba, A. J. *J. Biol. Chem.,* **250**, 3607 (1975).
58 Senear, A. W. and Steitz, J. A. *J. Biol. Chem.,* **251**, 1902 (1975).
59 Gorini, L. *Nature NB,* **234**, 261 (1971).
60 Mizushima, S. and Nomura, M. *Nature,* **226**, 1214 (1970).

61 Nierhaus, K. H. and Dohme, F. *PNAS*, **71**, 4713 (1974).
62 Spillman, S., Dohme, F. and Nierhaus, K. H. *J. Mol. Biol.*, **115**, 513 (1977).
63 Hochkeppel, H.-K., Spicer, E. and Craven, G. R. *J. Mol. Biol.*, **101**, 155 (1976).
64 Bear, D. G., Schleich, T., Noller, H. F. and Garrett, R. A. *Nucl. Acids Res.*, **4**, 2511 (1977).
65 Brimacombe, R., Nierhaus, K. H., Erdmann, V. A. and Wittmann, H. G. *Progr. Nucl. Acid Res. Mol. Biol.*, **18**, 1 (1976).
66 Ungewickel, E., Garrett, R., Ehresmann, C., Stiegler, P. and Fellner, P. *Eur. J. Biochem.*, **51**, 165 (1975).
67 Ehresmann, B., Backendorf, C., Ehresmann, C. and Ebel, J. P. *FEBS Lett.*, **78**, 261 (1977).
68 Sloof, P., Garrett, R. A. and Nanninga, N. *Mol. Gen. Genet.*, **147**, 129 (1976).
69 Dennis, P. P. and Nomura, M. *J. Mol. Biol.*, **97**, 61 (1975).
70 Jaskunas, S. R., Lindahl, L. and Nomura, M. *Nature*, **256**, 183 (1975).
Nomura, M. *Cell*, **9**, 633 (1977).
Yamamoto, M. and Nomura, M. *J. Bacteriol.*, **137**, 584 (1979); *PNAS*, **75**, 3891 (1978).
71 Lund, E., Dahlberg, J. E., Jaskunas, S. R., Dennis, P. P. and Nomura, M. *Cell.*, **7**, 165 (1976).
Morgan, E. A., Ikemura, T., Lindahl, L., Fallon, A. M. and Nomura, M. *Cell.*, **13**, 335 (1978).
72 Gorenstein, C. and Warner, J. R. *PNAS*, **73**, 1547 (1976).
Warner, J. R. and Gorenstein, C. *Cell*, **11**, 201 (1977).
73 Blatti, S. P., Ingles, C. J., Lindell, T. J., Morris, P. W., Weaver, R. F., Weinberg, F. and Rutter, W. J. *Cold Spring Harbor Symp.*, **35**, 649 (1970).
Weinmann, R. and Roeder, R. G. *PNAS*, **71**, 1790 (1974).
74 Kyriakopoulos, A. and Subramanian, A. R. *Biochim. Biophys. Acta*, **474**, 308 (1977).
75 Dahlberg, J. E., Kintner, C. and Lund, E. *PNAS*, **75**, 1071 (1978).
76 Hochkeppel, H. K. and Gordon, J. *Nature*, **273**, 560 (1978).
77 Cole, M. D., Beer, M., Koller, T., Strycharz, W. A. and Nomura, M. *PNAS*, **75**, 270 (1978).
78 Post, L. E., Arfsten, A. E., Reusser, F. and Nomura, M. *Cell*, **15**, 215 (1978).

7 Transfer RNA: Its Role in Decoding the Message

ADAPTOR FUNCTION OF TRANSFER RNA

In replication and transcription, genetic information is transferred from one nucleic acid molecule to another, to an identical molecule or to a molecular species at least very similar in its chemical structure. Information transfer in these steps is basically a copying process. At the stage of translation a more complex form of information transfer is taking place: not only is the chemical structure of the mRNA which is carrying the message basically different from that of the protein molecule in which the genetic information is eventually expressed, but also in the former, the message is encoded in a triplet code which has to be translated into the sequence of single amino acids. Instead of a copying mechanism, a decoding mechanism is at work, and this requires specific adaptor molecules which can recognise and interact with both the trinucleotide codons and the corresponding amino acids. Early studies on the mechanism of protein synthesis had already recognised that the adaptor molecules fulfilling this function were the transfer RNAs (tRNAs). These small RNA molecules have specific sites for interaction with nucleotide triplets and also sites for binding amino acids. They specifically select amino acids which they can carry to the site of protein synthesis where, in the presence of the appropriate codon, they can bind to the ribosome and thus direct the amino acid into the right position in the polypeptide chain. This function is carried out in a series of consecutive steps, involving interactions with a number of specific proteins and also with other RNAs. These specific interactions ensure that translation of the genetic message occurs with very high accuracy: the frequency of errors in translation is less than 1 to 3000 per codon[1].

It follows from the adaptor function of tRNAs that they show a strong specificity for a given amino acid and also for the codon(s) corresponding to this amino acid. There are 40 to 60 different tRNA species in the cell, probably fewer in bacteria, more in higher organisms. Because of the degeneracy of the code, multiple tRNA species, recognising different codewords, may be specific for the same amino acid. In *E. coli*, there are, for example, six tRNA species specific for leucine, five for arginine, and three for glycine. $tRNA^{gly}$I interacts with the codon GGG, $tRNA^{gly}$II with GGG and GGA, $tRNA^{gly}$III with GGU and GGC. Still, the total number of different tRNAs does not equal the total number of codewords which implies that the same tRNA may recognise more than one codon.

The explanation was provided by Crick's 'wobble hypothesis' (ref. 2). The interaction of tRNAs with the codons in mRNA occurs via base pairing: each tRNA contains a trinucleotide sequence, the anticodon, which is complementary and can thus anneal to the trinucleotide codon. If strict complementarity were required for this codon–anticodon interaction, a specific tRNA would be needed for each codeword. Hydrogen bonds can also be formed, however, between bases other than G-C and A-T, even though the stability of such base pairs may be lower than that of the correct Watson-Crick base pairs. According to the wobble hypothesis, only the first two letters of the codon must form correct Watson-Crick base pairs, the third letter may enter into other types of base pairing (U-G, I-A, I-C, I-U), with the result that the same anticodon sequence may anneal to two or three codons differing only in their third letter. This explains how one $tRNA^{tyr}$ species (containing the anticodon G*UA) can recognise both codewords UAU and UAC of this amino acid, and how one species of $tRNA^{ala}$ (with the anticodon IGC) is sufficient to recognise three alanine codons: GCU, GCC and GCA. As, in the degenerate code, codons for the same amino acid often differ only in their last letter, wobbling markedly reduces the required number of specific tRNA species. No tRNA can, however, recognise more than three codewords and no wobbling is allowed in the first two letters of the codon. These restrictions still leave a requirement for quite a few 'iso-accepting tRNAs', i.e. different tRNA species with a specificity for the same amino acid but for different codewords.

Interaction with amino acid and with the codon occur in different steps at different stages of translation. Selection of the cognate amino acid depends on the specificity of the enzyme, aminoacyl-tRNA synthetase, which catalyses the binding of amino acid to tRNA. Codon-anticodon interaction takes place between the aminoacylated tRNA and the nucleotide triplet bound to ribosome. As a consequence of this interaction, the aminoacyl-tRNA binds to the ribosomal particle itself in a position facilitating the reaction between the amino acid it carries and the C-terminus of the growing polypeptide chain. Fidelity of decoding is thus controlled at two stages: at the selection of the right amino acid to bind to the tRNA and at the selection of the right aminoacyl-tRNA to bind to the ribosome.

STRUCTURE OF tRNA

tRNAs are among the smallest RNA molecules in the cell. They are built of a polynucleotide chain comprising 75 to 90 nucleotides, and have MWs of around 25 000. It is characteristic of tRNAs that they contain a very high percentage of minor bases, mainly methylated bases, dihydrouridine (D), pseudouridine (ψ) and occasionally substituted bases carrying more complex substituents. A list of the minor bases most commonly occurring in tRNA is given in *figure 7.1*.

Partly because of their small size, partly because of their high content of minor bases, tRNAs are particularly well suited for nucleotide sequence studies, and were, in fact, the first RNA molecules for which the primary structure was extensively studied. Today, the nucleotide sequences of over 75 tRNAs are known[2a] The expectation of these studies was that knowledge of the exact structure would also reveal the sites of interaction of tRNAs with the different enzymes, protein factors and RNAs, a question of great interest, as these interactions form the basis of the high fidelity of the translation process.

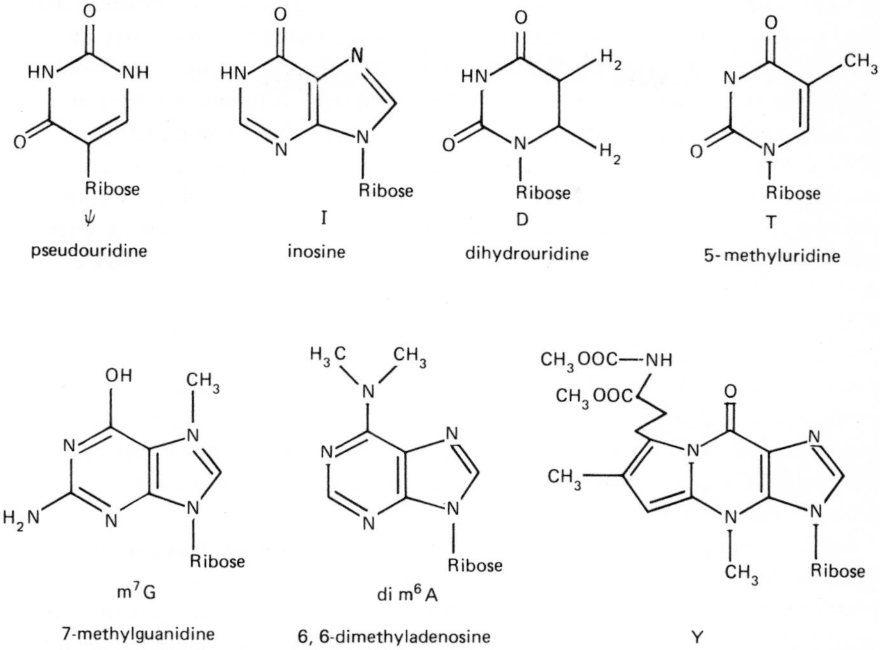

Ribose ψ pseudouridine | Ribose I inosine | Ribose D dihydrouridine | Ribose T 5-methyluridine

m^7G 7-methylguanidine | di m^6A 6, 6-dimethyladenosine | Y

Figure 7.1 Some minor nucleosides occurring frequently in tRNAs.

While much information has been collected on the primary structure of different tRNA species, of iso-accepting tRNAs, and of mutant tRNAs, the above problem has not been solved so far. Although quite extensive homologies between different tRNA species and a uniformity of the secondary structure of all tRNAs have been established, neither the similarities nor the variations in the primary structure have allowed any conclusions as to the exact sites of interaction with different protein molecules. Our knowledge is still restricted to the sites of interaction with RNAs and with the specific amino acid, the specific sites already detected in the earliest studies on tRNA structure. Years of research have been devoted to the characterisation of the sites of interaction with aminoacyl-tRNA synthetases, elongation factors and ribosomal proteins. The failure of such studies may be due to the very nature of these protein–RNA interactions: the specific sites in both molecules are probably part of a well defined three-dimensional structure; as in proteins, the active sites in tRNA may be built of several small stretches or even single nucleotides brought into close proximity by the specific folding of the molecule. However, recently, the tertiary structure of tRNA has been established, and further progress may also be expected in the elucidation of the tRNA–protein interactions.

If the structure of several tRNA molecules is compared, the most striking similarity is found in the secondary structure. All tRNA species can be fitted into a 'cloverleaf' structure, built of three or four 'arms' and a 'stem'. The primary and

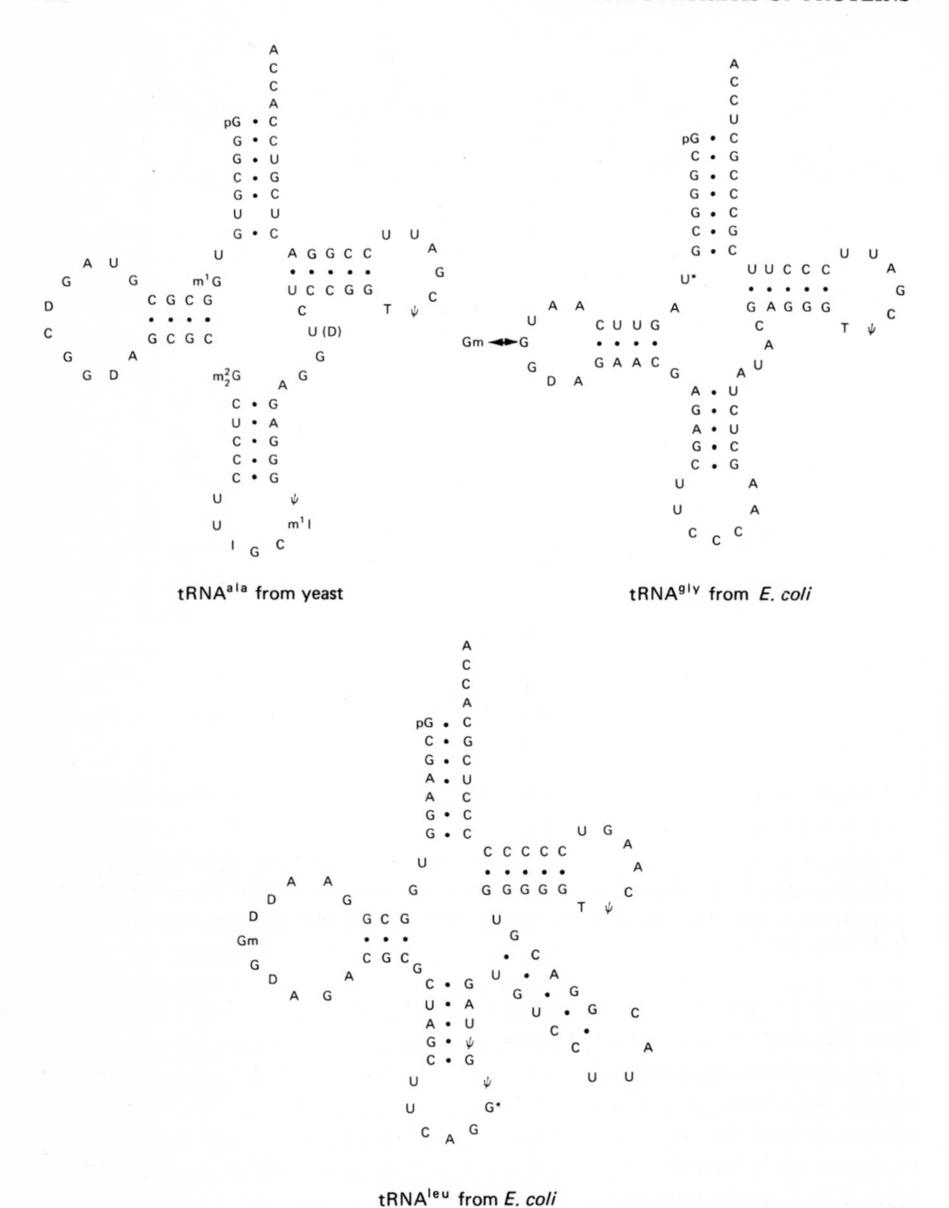

Figure 7.2 The cloverleaf structure of some tRNAs.

secondary structure of a few tRNAs is shown in *figure 7.2*. As far as the position of some individual nucleotides within this cloverleaf structure is concerned, a high degree of homology can be found between different tRNA species, although no long homologous sequences are present in any group of tRNAs. 21 out of about 80 nucleotides are invariant or semi-invariant in all tRNAs. The size of the arms,

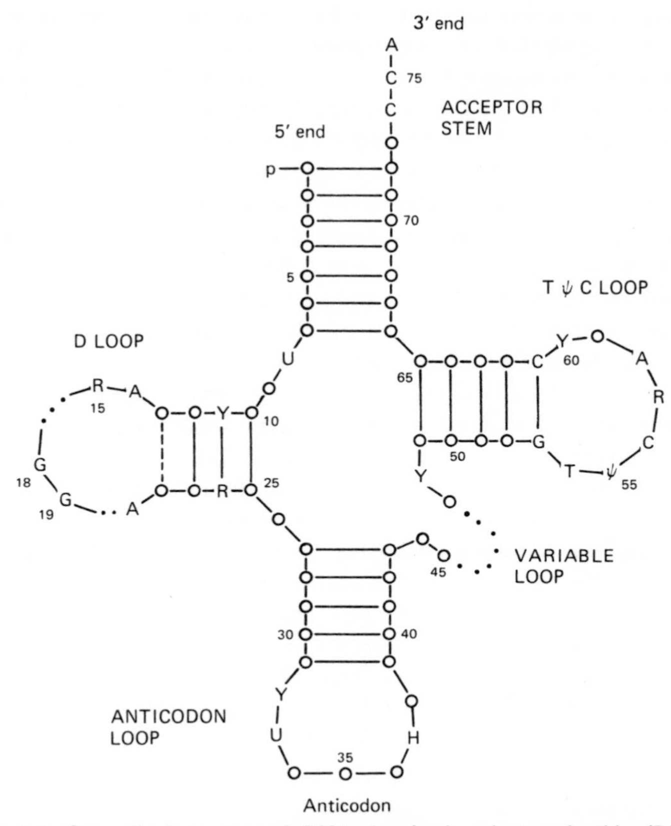

Figure 7.3 Generalised structure of tRNA, showing invariant nucleotides (Rich and RajBhandary[3]). Y: pyrimidine; R: purine; H: modified purine. (Reproduced, with permission, from the *Annual Review of Biochemistry*, **45**, 805. © 1976 by Annual Reviews Inc.)

the number of unpaired bases in the loops, and the number of base pairs in the stem, also show little variability among the different tRNA molecules.

The common characteristics in the cloverleaf structure of all tRNAs can be seen on the derived general structure in *figure 7.3*, which also shows the conserved nucleotides and the variations in the sizes of the loops[3]. The 5′ end and the 3′ end of the molecule form the amino acid acceptor stem, held together by seven base pairs. The 5′-terminal nucleotide is almost always G (except for the initiator tRNAs, see later), thus also defining a complementary nucleotide C in the 5th position from the 3′ end. The 3′-terminal four nucleotides do not take part in base pairing, the nucleotide sequence of the last three is CCA in all tRNAs. This CCA_{OH} end shows a low turnover in the cell; it is removed and can be replaced with the aid of the enzyme nucleotidyl transferase. The presence of the CCA terminus is an absolute requirement of the function of tRNA, as it is at the 2′ or 3′ hydroxyl of the end-standing adenosine that binding of the amino acid occurs.

Following the sequence further from the 5′ terminus, after the base-paired region of the acceptor stem two unpaired nucleotides follow, one of which is either U or D (dihydrouracil). There are one or more D's present also in 'arm' I, which is therefore called the D arm. Its total length varies from 15 to 18 nucleotides, comprising a stem of 3 or 4 base pairs and 7 to 11 nucleotides in the loop. No specific function has been ascribed so far to this part of the molecule. One nucleotide connects this arm to arm II, the structure of which has been strongly conserved: it comprises a 5-base pair stem and a 7-nucleotide loop, the anticodon loop. This loop contains a pyrimidine base, followed by a U, the three nucleotides of the anticodon, a purine nucleotide and a nonconserved nucleotide. The anticodon sequence varies, of course, from one tRNA to another. According to the wobble hypothesis, its first letter may form a less stable wobble base pair with the codon. This may be a G•U pair, or minor bases may be present in the first position of the anticodon, like I, which can enter into hydrogen bonded structure with U, C or A (*see* structure of $tRNA^{ala}$ in *figure 7.2*).

It may be of great importance that the size of the anticodon loop (7 nucleotides) is highly conserved. The structure of tRNA may be responsible for keeping up the right reading frame when decoding the message. A suppressor tRNA which cancels out the effect of a frameshift mutation has been isolated from a strain of *Salmonella typhimurium* carrying the *sufD* frameshift suppressor mutation[4]. The frameshift mutation occurred by insertion of an extra nucleotide and resulted in a change in the reading frame of the message. The suppressor tRNA could correct the reading frame. This mutant $tRNA^{gly}$ differed from $tRNA^{gly}$ (*figure 7.2*) by having CCCC as anticodon instead of CCC. This shows that lengthening the anticodon loop by one nucleotide has the effect of changing the reading frame from nucleotide triplets to nucleotide quadruplets.

The anticodon arm is followed by a variable region of tRNA, the extra arm (arm III), which may be short, comprising 4 or 5 bases in the loop with no stem (e.g. $tRNA^{ala}$, $tRNA^{gly}$I in *figure 7.2*), or long, built of 13 to 31 nucleotides with a variable degree of base pairing ($tRNA^{leu}$I in *figure 7.2*). Arm IV, again shows a high degree of conservation: the stem is formed of 5 base pairs the last of which is G–C and the nucleotide sequence of the 7-residue long loop shows a high degree of homology between different tRNA species: it always contains the trinucleotide sequence TψC, often preceded and followed by G and then by an A residue: GTψC(G)A. The stem of this arm continues directly into the acceptor stem with no connecting nucleotides in between. Arm IV is called the TψC arm because of the presence of this trinucleotide in the loop. The importance of this sequence became apparent when the sequence of 5S ribosomal RNA was determined. 5S RNA contains a complementary tetranucleotide sequence, CGAA, which suggested that the binding of tRNAs to ribosomes may occur through or be stabilised by annealing of the tRNA to 5S RNA by way of these complementary sequences (*see* Chapter 6). There is also a possibility of annealing between a GAAC sequence in 5S RNA and GTψC in tRNA.

The total number of invariant or semi-invariant residues is 21, as shown in *figure 7.3*. To seven of these, specific functions have been ascribed: the CCA end

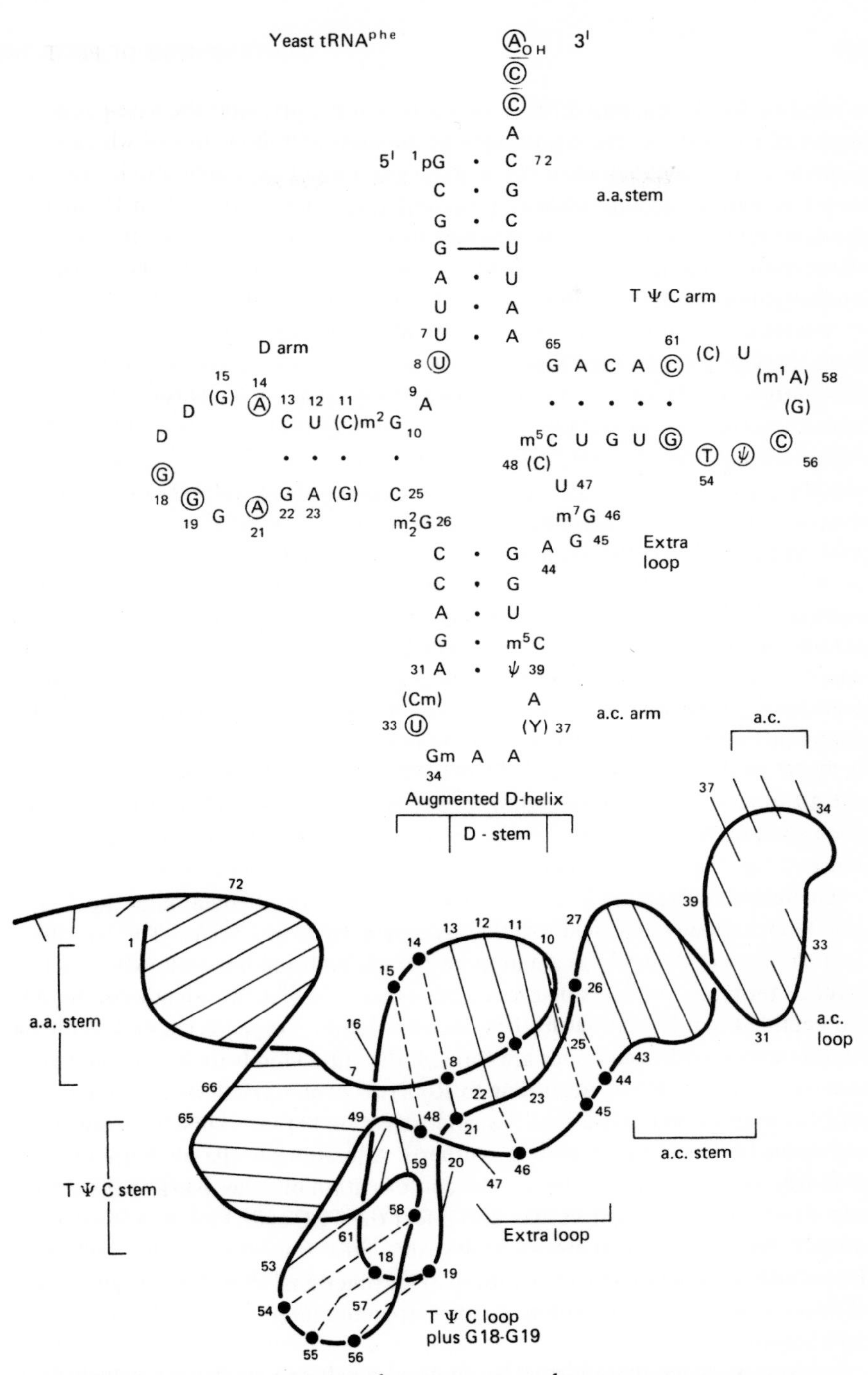

Figure 7.4 Tertiary structure of tRNA^{phe}. Above: tRNA^{phe} sequence arranged in the cloverleaf form. Invariant bases are circled. Below: schematic diagram of the tertiary structure. The RNA chain is represented by a continuous line; base pairs in the double-helical stems are connected by long thin lines. Shorter lines represent unpaired bases. Dotted lines represent base pairs in the tertiary structure, additional to those in the cloverleaf formula. (By courtesy of Dr A. Klug. Reproduced from Ladner *et al. PNAS*, 72, 4414 (1975) (ref. 6).)

is required for the binding of the amino acid, the GTψC (or TψCG) sequence for annealing to 5S RNA. The significance of the invariability of the remaining 14 nucleotides was revealed when the tertiary structure of $tRNA^{phe}$ was established: they take part in secondary interactions within the tRNA molecule which stabilise the three-dimensional structure. Whether they have a direct role in any of the tRNA–protein interactions or if they are needed only to provide and stabilise the conformation required for these interactions, is not known at present.

The three-dimensional structure of $tRNA^{phe}$ from yeast, the first to be derived from X-ray diffraction data, was established in 1974, simultaneously and independently at the Massachusetts Institute of Technology, Cambridge, Massachusetts[5], and at the Laboratory of Molecular Biology, Cambridge, UK[6]. *Figure 7.4* shows a schematic diagram of the folding of this tRNA molecule, indicating also the nucleotide–nucleotide interactions which stabilise the tertiary structure. The largest contribution to the stability of the molecule is made by stacking interactions: the orientation of most bases allows extensive stacking. In addition to these hydrophobic interactions, nucleotides in different loops are held together by a number of hydrogen bonds. These are often different from the type of hydrogen bonds encountered in the DNA double helix. Hydrogen bonding also occurs between bases which are not correct complementary pairs, e.g. between G in position 22 (D arm) and m^7G at position 46 (extra arm), or which are not in antiparallel orientation, e.g. between G at position 15 (D arm) and C at position 48 (TψC arm). Some bases form triples instead of pairs: for example, U12, A23 and A9 show such triple interaction. The sugar–phosphate backbone may also be involved in hydrogen bonding: it has been suggested that A21 is linked to the ribose of U8. Some of these unusual interactions are shown in *figure 7.5*.

The resulting structure is an L-shaped molecule with a bend around the TψC and D loops, these being held together by several hydrogen bonds. The acceptor stem and the anticodon loop are in more flexible end-standing positions. The CCA at the 3′ terminus remains single stranded, easily accessible for amino acid binding and peptide bond formation. In the anticodon loop there is considerable stacking but no tertiary hydrogen bonds are formed; the anticodon is thus available for base pairing with mRNA. According to Robertus *et al.*[6], the anticodon loop may exist in two different conformations, and transition between these conformations may ensue in the course of protein synthesis. Alterations in the structure of one loop may, of course, also influence the conformation of other parts of the molecule. Conformational changes may also affect the TψC and D loops, as tertiary interactions between them are not very strong. Such transitions in the conformation of tRNA, closely connected with distinct functional steps, have been suggested to occur upon aminoacylation, codon–anticodon interaction, and binding to the ribosome[3].

It seems probable that most tRNA molecules will fit a similar three-dimensional pattern. It is apparent from a comparison of *figures 7.3* and *7.4* that the nucleotides responsible for the stabilisation of the molecular structure are mainly invariant nucleotides. It can therefore be expected that the same nucleotides can enter

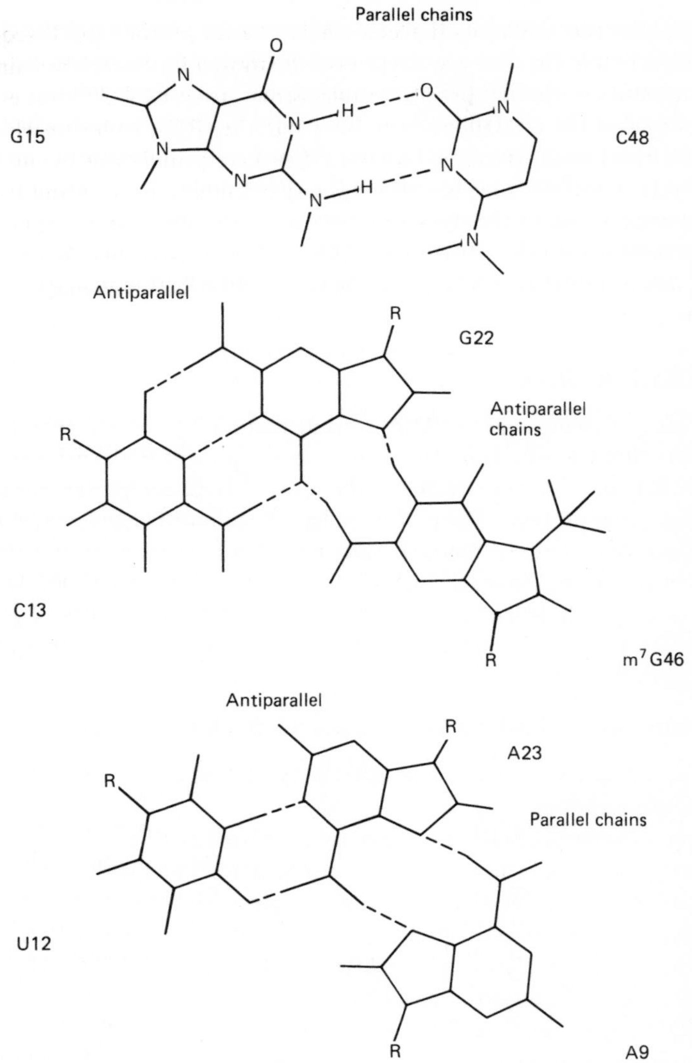

Figure 7.5 Some examples of non-Watson–Crick type interactions between nucleotides which stabilise the tertiary structure of tRNA (Klug *et al.*[31]). (Reproduced, with permission, from *J. Mol. Biol.*, **89**, 511 (1974). Copyright by Academic Press Inc. (London) Ltd.)

into the same interactions and cause similar folding in other tRNA molecules also. The above structure can accommodate variations in the numbers of nucleotides in the D loop and to some extent also in the extra loop, although the folding of tRNAs with long extra loops may be different from the above model. Indeed, it has been found that the crystal structure of yeast initiator tRNA (*see* p. 232) is almost identical to that of $tRNA^{phe}$, except for a small difference in the position of the anticodon arm[6a].

Tertiary structure determinations are carried out on crystals and the question to be asked before the data are interpreted in relation to the *in vivo* function of these molecules is whether the three-dimensional structure in solution is identical to that found in the crystalline form. Studies by indirect methods which did not reveal the exact secondary structure but yielded ample imformation on the accessibility of different nucleotides in the tRNA molecule to chemical reagents, complementary oligonucleotides, enzymes, etc. have given results which are in good agreement with the structure derived from X-ray crystallography[3]. It can thus be assumed that tRNA takes up the same conformation in crystals and in solution.

CHARGING OF tRNA

The binding of amino acid to tRNA is brought about by specific enzymes called aminoacyl-tRNA synthetases. These enzymes are present in the cell sap and also in organelles, and can be extracted in the form of large complexes containing also tRNA and other proteins. Many of them have been purified and, surprisingly, considering their identical function, have been found to show great differences in protein structure: some are single-chain proteins of about 100 000 MW, another group consists of dimers of similarly sized polypeptide chains, a third group shows dimer or tetramer structure built of different size subunits. They all catalyse the overall reaction

$$\text{amino acid} + \text{tRNA} + \text{ATP} \rightarrow \text{aminoacyl-tRNA} + \text{AMP} + \text{PP}_i$$

and show a strong but not absolute specificity for one amino acid and the corresponding tRNAs.

As not all aminoacyl-tRNA synthetases have been purified, the total number of different enzymes in any one cell cannot be determined with certainty. It seems probable, however, that their number agrees with that of the amino acids rather than with the number of tRNAs, i.e. the same aminoacyl-tRNA synthetase catalyses the aminoacylation of the different iso-accepting tRNAs. This presents a puzzling and so far unsolved problem with regard to the specific recognition of these tRNAs by the enzyme. Nucleotide sequence determination of iso-accepting tRNAs has not revealed any striking resemblance in their structure which might explain their recognition by the same enzyme, nor does it give a clue to the structural features which may enable the enzyme to distinguish between the cognate and the structurally very similar non-cognate tRNAs.

The charging of tRNA by the aminoacyl-tRNA synthetases occurs in a two-step reaction. The first step, the activation of amino acids (abbreviated as 'aa' in the equations) leads to the formation of an enzyme-aminoacyladenylate complex:

$$\text{aa} + \text{ATP} + \text{enzyme} \rightleftarrows \text{E}\cdot\text{aa-AMP} + \text{PP}_i$$

the β and γ phosphates of ATP are split off in the form of inorganic pyrophosphate. The energy of the split pyrophosphate bond is used in the formation of the ester linkage between the carboxyl group of the amino acid and the ribose-hydroxyl of adenosine.

The next step is the transfer of the aminoacyl group to the 3′-terminal adenosine residue of the tRNA:

$$\mathrm{E{\cdot}aa\text{-}AMP + tRNA \leftrightarrows aa\text{-}tRNA + E + AMP}$$

The aminoacyl group may attach in ester linkage to either the 2′ or the 3′ hydroxyl of the terminal adenosine. The synthetases show definite specificity in acylating one hydroxyl group or the other, but this specificity differs with the different enzyme–tRNA systems: Sprinzl and Cramer[7] defined three groups of tRNAs: the majority accepts the amino acid at the 2′ hydroxyl, the tRNAs for gly, his, lys, ser are aminoacylated at the 3′ hydroxyl, and those for tyr and cys can be aminoacylated at either hydroxyl. The specificity for the accepting hydroxyl group may not reside in the structure of the tRNA but may be the result of the interaction between enzyme and tRNA whereby one of the terminal hydroxyl groups comes into direct contact with the active site of the aminoacyl-tRNA synthetase. The aminoacyl group, once attached, can easily undergo an exchange between the 2′ and 3′ position.

Aminoacyl-tRNA synthetases also show a further enzyme activity: in addition to the activation and transfer reaction, these enzymes catalyse the deacylation of aminoacyl-tRNA

$$\mathrm{aa\text{-}tRNA \rightarrow aa + tRNA}$$

This hydrolytic activity has been detected in several (though not all) aminoacyl-tRNA synthetases. According to Fersht and Kaethner[9], two different active sites of the enzyme are responsible for the synthetic and hydrolytic activities. Van der Haar and Cramer[8] suggest that the non-accepting hydroxyl group may be involved in the deacylation reaction.

The deacylating activity, although seemingly counteracting the main function of aminoacyl-tRNA synthetases, in fact fulfils a very important control function over the specificity of the charging of tRNAs. As mentioned earlier, the fidelity of translation is very high indeed, the error rate is below 1 to 3000 for each codon, a fact which is hard to reconcile with the somewhat limited specificity observed in the interaction of aminoacyl-tRNA synthetases with their substrates. Errors in the activation of amino acids may occur with much higher frequency if the 'wrong' amino acid is of similar structure to the cognate amino acid of the synthetase. A much studied process of mis-activation is that of valine by the isoleucine tRNA synthetase. Although the enzyme is a hundred times more active with its specific substrate than with valine, even an error rate of 1 in 100 is much too high in comparison with the 30 times lower error rate in the overall process. In 1966, Baldwin and Berg[10] found that this error in the first step of the reaction is corrected in the second: formation of val-tRNA^{ile} cannot be observed, the mis-activated amino acid is instead released by a hydrolytic process.

$$\mathrm{val + ATP + E^{ile} \rightarrow E^{ile} \circ val\text{-}AMP + PP_i}$$
$$\mathrm{E^{ile} \cdot val\text{-}AMP + tRNA^{ile} \rightarrow val + AMP + E^{ile} + tRNA^{ile}}$$

The mechanism of decomposition of the enzyme-valyl-AMP complex has been studied by von der Haar and Cramer[8], who found that transiently, a transfer of the non-cognate amino acid to the tRNA occurs, but the val-tRNA^{ile} is rapidly hydrolysed by the ile-tRNA synthetase. Similar results were obtained by Fersht and Kaethner[9] in the case of threonine mis-activated by val-tRNA synthetase.

The hydrolytic activity of aminoacyl-tRNA synthetases can thus be interpreted as a control function, which serves to correct possible mistakes in the activation of amino acids. The rate of deacylation is much higher in the case of mischarged tRNAs than of cognate aminoacylated tRNAs. If the cognate amino acid is activated, its transfer to the cognate tRNA occurs with a velocity which exceeds the rate of hydrolysis by several orders of magnitude, while in the case of non-cognate amino acid, the rates of the two processes are comparable, preventing any net formation of aminoacyl-tRNA. Fersht and Kaethner determined the rate constants for the transfer of activated threonine to tRNA^{val} and for the decomposition of thr-tRNA^{val} as 36 s^{-1} and 40 s^{-1}, respectively, while the rate constants for the same processes with the cognate amino acid, valine, were 12 s^{-1} and 0.015 s^{-1}, respectively. Thus, thr-tRNA^{val} is deacylated faster than it is synthesised, while the formation of val-tRNA^{val} is hardly interfered with by the slow process of decomposition.

Rejection of mischarged tRNAs is a kind of 'proofreading' process which greatly increases the specificity of the charging reaction: to the specificity of the selection of amino acids a correction mechanism is added which very markedly reduces the chance of the 'wrong' amino acid being inserted into the polypeptide chain. At different stages of information transfer we often find such double safeguards built into the system. The very high accuracy of information transfer is achieved with the aid of proofreading mechanisms which can correct occasional mistakes. Hopfield[11] worked out a theory of such correction mechanisms, based on enzyme kinetics, and found that this 'kinetic proofreading' theory applies to DNA replication as well as to the translation process. We will see later that a further step of proofreading occurs at the stage of aminoacyl-tRNA binding to ribosomes.

CODON–ANTICODON INTERACTION

Selection of the specific amino acid to attach to the tRNA is the first step in the decoding process. The second step is the interaction of aminoacyl-tRNA with the cognate codon in mRNA, leading eventually to the insertion of the amino acid specified in the message into the protein molecule. It has long been known that at this stage selection of the specific aminoacyl-tRNA depends only on the structure of the RNA moiety, the nature of the aminoacyl group is not relevant. This has been unequivocally demonstrated in early studies on the function of tRNA by Lipmann and co-workers[12], who chemically reduced the cysteinyl group in cys-tRNA^{cys} to an alanyl group, and found that this ala-tRNA^{cys} inserted alanine residues into protein in response to codons specifying cysteine. Studies with mutant tRNAs also confirm that it is the RNA structure alone which determines the selection of aminoacyl-tRNA in response to a given codon. tRNAs which have suffered a point mutation in a base of their anticodon still retain their specificity for the original amino acid, but recognise a different codon. This leads to misreading of the code, i.e. to the introduction of the 'wrong' amino acid in response to a codon. $tRNA^{gly}$II recognises the codon GGA. A mutant $tRNA^{gly}$, however, interacts with the triplet AGA, a codon normally specifying arginine. This mutant $tRNA^{gly}$ introduces glycine in place of arginine into the newly synthesised polypeptide chain, because the selection of the aminacyl-tRNA depends on the codon–anti-codon interaction and not on the nature of the amino acid attached to the tRNA molecule.

Complementarity of a trinucleotide sequence is sufficient for the annealing of the two RNA molecules, but neither the stability of the annealed structure nor the specificity of interaction are high enough to meet the requirements for exact decoding. If direct annealing of different trinucleotides to the anticodon of a tRNA is studied, the association constant observed is rather low and the affinity of the tRNA for its cognate codon is found to be only ten times greater than that for some similar, non-cognate triplets. In the course of translation, however, this interaction does not take place between free mRNA and free aminoacyl-tRNA; the codon is attached to the ribosome and as a consequence of the codon–anti-codon interaction, the aminoacyl-tRNA also becomes attached to the ribosome. According to present theory, it is the coupling of this interaction with the binding of tRNA to ribosome which increases the specificity to the high levels needed to achieve accurate translation.

BINDING OF AMINOACYL-tRNA TO RIBOSOME

The binding of aminoacyl-t RNA to the programmed ribosome is a rather complex process which also requires the participation of elongation factors and GTP. In the course of translation, the aminoacyl-tRNA enters into a complex with elongation factor EF-Tu and GTP and it is in the form of this ternary complex that it is delivered to the ribosome and attached to the specific site where it eventually inserts its amino acid into the growing polypeptide chain (*see* Chapter 8). The formation of the ternary complex stabilises the codon–anticodon interaction: Schwarz *et al.*[13] found that the association constant of phe-tRNA and U_8 (UUU is the codon for phenylalanine) was greatly increased if phe-tRNA was present in the form of this ternary complex.

Binding of the aminoacyl-tRNA to the ribosome exercises a further stabilising effect, apart from introducing a control step in the selection of the specific aminoacyl-tRNA. As described in Chapter 6, the binding of aminoacyl-tRNA to the A-site of the ribosomal particle is brought about by RNA–protein and RNA–RNA interactions, with the participation of some ribosomal proteins and of 5S ribosomal RNA (5.8S RNA in eukaryotic ribosomes). Complementarity between, on the one hand, the CGAA sequence in prokaryotic 5S RNAs and eukaryotic 5.8S RNAs and, on the other, the (G)TψC(G) sequence in tRNAs, suggests that binding occurs through annealing between these molecules. More direct evidence was obtained by Ofengand and Henes[14], who found that the tetranucleotide TψCG inhibited aminoacyl-tRNA binding to ribosomes, presumably by competing with 5S RNA for the CGAA site. Annealing of the anticodon loop of tRNA to the codon in the A-site is thus stabilised by the annealing of the TψC arm to 5S RNA and by the binding of the CCA end to the peptidyl transferase centre of the ribosome, through ribosomal protein L16.

The most important aspect of aminoacyl-tRNA binding to ribosomes is that it is strictly codon dependent. This does not logically follow from the above considerations, as the sites in tRNA which interact with ribosomal proteins and rRNA are invariant sequences. All tRNAs contain the TψC arm as well as the CCA end, yet only the codon-specific aminoacyl-t RNA can attach to the ribosome. The idea has been put forward by Schwarz and coworkers[13] and has been extended by Kurland *et al.*[15], that codon–anticodon interaction is a prerequisite of tRNA–ribosome binding, because the former step triggers a conformational change in the tRNA molecule which enables it to anneal to 5S RNA. This hypothesis obtained strong support when the tertiary structure of $tRNA^{phe}$ was established, making it clear that the TψCG sequence is not freely accessible in the three-dimensional structure, but is hydrogen bonded to the dihydrouracil loop. A conformational change, however, can expose the TψC loop.

Previous model experiments had indicated that oligonucleotides complementary to TψCG bind only poorly to free tRNA. The binding is greatly increased, however, in the presence of oligo(U) (ref. 13). This is because the interaction between the codon UUU and the anticodon of $tRNA^{phe}$ alters the folding of the anticodon loop which in turn influences the conformation of the TψC loop, exposing the latter sequence so that it should easily anneal to 5S RNA. Kurland[15] considers the codon as an allosteric effector which, through a series of conformational changes, enhances the stability of interaction between the cognate tRNA and the ribosome. Assumption of such an allosteric effect implies that a selection step is introduced at the stage when aminoacyl-tRNA binds to ribosome. Although the trinucleotides of the codon and anticodon are not themselves highly specific in their interaction, nevertheless, only the specific aminoacyl-tRNA will attach to the ribosome, because only the perfect annealing of the two trinucleotide sequences can evoke the conformational change required for this attachment. Aminoacyl-tRNAs which do not match the codon at the A-site will be rejected. This theory, which is based on much supporting evidence although on no conclusive proof, gives a logical explanation of how a series of different interactions can control the precision of translation at the stage of decoding.

Objections have been raised against the assumption that the codon–anticodon interaction itself could bring about a conformational change which exposes the TψC loop. An alternative mechanism has also been proposed, according to which ribosomal proteins would be responsible for altering the conformation of tRNA.

Lake and coworkers recently proposed a molecular model which describes the events on the ribosome from the recognition of the specific aminoacyl-tRNA up to its stable binding at the site where peptide bond formation can occur[16]. Although they approach the problem from a different angle, they arrive at a not

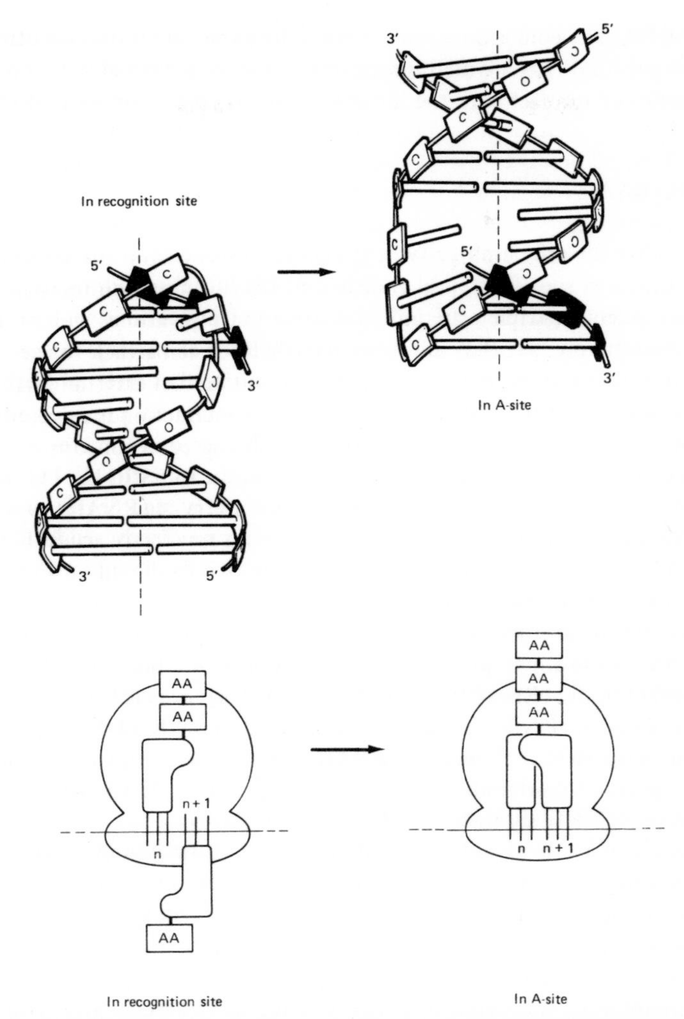

Figure 7.6 Above: the switch in the conformation of the anticodon loop, suggested by Lake[16]. Black areas: nucleotides of the codon forming base pairs with the anticodon. Below: schematic presentation of the two positions occupied by aminoacyl-tRNA in the above two conformations. (Courtesy of J. A. Lake[16].)

dissimilar conclusion. Their model is also based on the assumption that the conformation of the anticodon loop of tRNA is altered upon interaction with the cognate codon: they suggest a switch in the stacking of the bases in the anticodon loop. The tertiary structure (*figure 7.4*) allows such a change[6]. The consequence is, however, that the aminoacyl-tRNA can take up two different positions on the ribosome. While at the stage of codon recognition it is attached to a site on the small ribosomal subunit (recognition site), distal from the peptidyl transferase centre: the conformational change in the anticodon loop moves the CAA end to the A-site on the large subunit, as shown diagrammatically in *figure 7.6*. The aminoacyl-tRNA thus comes into contact with the peptidyl transferase centre and with the peptidyl-tRNA which is attached to the adjacent P-site; it is now suitably positioned to react with the latter and to form a new peptide bond.

INITIATOR tRNAs

It is amazing that the only ambiguity in the genetic code affects the codons which function in the most specific step of translation: the initiator codons AUG and GUG. These nucleotide triplets are indistinguishable from codons for methionine and valine, respectively, yet they are never mistaken for an internal codon by the protein synthesising system, neither is translation initiated at internal methionine or valine codons. The structural features of mRNA which may provide additional information to specify initiator codons have been discussed in Chapter 5. Here, we are concerned with the specific tRNA molecules, the initiator tRNAs, which recognise these codons and interact with them with a very high degree of accuracy.

Two tRNA species responding to the AUG codon were discovered in *E. coli* by Clark and Marker in 1966 (ref. 17). One of these tRNAs, designated $tRNA_F^{met}$, after being charged with methionine, undergoes a formylation reaction by the bacterial enzyme transformylase which, using formyltetrahydrofolate as formyl donor, substitutes the amino group of the tRNA-bound methionine, thus producing a formyl-methionyl-$tRNA_F^{met}$ (fmet-$tRNA_F^{met}$). The enzyme does not react with other aminoacyl-tRNAs, nor does it formylate the methionyl group if this is bound to the other $tRNA^{met}$ species, $tRNA_M^{met}$. Ability to accept formyl group is a specific property of methionine bound to $tRNA_F^{met}$. This tRNA functions only in the initiation of translation, and is called initiator tRNA.

The presence of a substituent on the amino group of methionine prevents the formation of a peptide bond between this amino group and the carboxyl group of any other amino acid. As protein synthesis proceeds from the amino to the carboxyl terminus, it follows that this methionine must be the N-terminal amino acid in a newly synthesised polypeptide chain.

Formyl-methionine was detected at the N-terminus of nascent viral proteins synthesised in *E. coli* extracts[18]. A very high proportion of methionine was found in the N-terminal residues of total *E. coli* protein[19]. All this supported the key role of fmet-$tRNA_F^{met}$ in the initiation of protein synthesis. The formyl-methionine residue could be detected only in nascent polypeptide chains because the bacterial cell contains enzymes which remove this group immediately after or even before translation is terminated: formylase splits off the formyl group; its action is often, but not always, followed by proteolytic cleavage of the methionine.

Initiator tRNAs also differ from the tRNAs which take part in elongation with respect to their interactions with protein factors. While other aminoacyl-tRNAs form complexes with elongation factor EF-Tu, fmet-$tRNA_F^{met}$ does not interact with this factor[20]. As formation of the ternary complex aminoacyl- tRNA•EF-Tu•GTP is required for binding aminoacyl-tRNA to a ribosome engaged in elongation (*see* Chapter 8), the lack of reaction between fmet-$tRNA_F^{met}$ and EF-Tu provides a further (and probably more important) safeguard to ensure that the methionine carried by the initiator tRNA will not be inserted at an internal position. Internal methionine codons are recognised by met-$tRNA_M^{met}$ only, by methionine bound to the non-initiator tRNA species. However, initiator tRNAs do interact with other protein factors involved in translation, with initiation factors. Of the three prokaryotic initiation factors IF1, IF2 and IF3, IF2 is directly involved in binding fmet-$tRNA_F^{met}$ to the ribosome. The interaction between IF2 and initiator tRNA is highly specific, no other aminoacyl-tRNA can substitute for fmet-$tRNA_F^{met}$ (ref. 21).

Specific interaction between initiation factors and initiator tRNA plays an essential role in eukaryotic systems also. Eukaryotes lack the enzyme transformylase; no formyl-methionyl-tRNA is thus produced in these organisms. Still, the two different $tRNA^{met}$ species also exist in eukaryotic cells, fulfilling the same functions as the two methionine tRNAs of prokaryotes: met-$tRNA_F^{met}$ donates a methionine residue at the N-terminus of the polypeptide chain, while met-$tRNA_M^{met}$ introduces methionine residues at internal positions. This shows clearly that a formylated amino group is not an indispensable requisite for the selection of initiation sites: the unformylated met-$tRNA_F^{met}$ of eukaryotic organisms, which, like its prokaryotic counterpart, interacts with initiation factors and not with elongation factors, shows the same specificity for the initiator codon.

It is interesting that although the enzyme which carries out formylation is not present in eukaryotes, the methionine bound to eukaryotic initiator tRNA is, nevertheless, able to undergo formylation. Eukaryotic met-$tRNA_F^{met}$ can be formylated by the bacterial enzyme while eukaryotic met-$tRNA_M^{met}$ cannot. This suggests that some structural feature common to all initiator tRNAs is required for this reaction. It might be assumed that the same structural feature is also responsible for other specific reactions of initiator tRNAs.

It is understandable that studies on the nucleotide sequences of initiator tRNAs aroused great interest, for it was expected that elucidation of the primary structure would reveal some specific structural features in initiator tRNAs which could explain how these molecules fulfil their specific function. The first results[22] caused some disappointment, as no dramatic differences could be detected between the nucleotide sequence of $tRNA_F^{met}$ and other tRNAs of *E. coli* (*figure 7.7*). As, however, the structure of more initiator tRNAs from different organisms became known, the importance of some common structural characteristics was revealed. The primary structure of initiator tRNAs is highly conserved, their nucleotide sequence is practically identical in organisms as far apart in the evolutionary scale as yeast, salmon, rabbit liver and human placenta[23]; there are differences, though, between prokaryotic and eukaryotic initiator tRNAs. Bacterial initiator tRNAs have only six base pairs in the acceptor stem, the 5′-terminal base is unpaired. The functional significance of this is, however, uncertain. Eukaryotic initiator tRNAs all share the sequence GAUCG, which replaces the GTψCG sequence in arm IV in these tRNA molecules. Since in other tRNAs this sequence is involved in ribosome binding, this consistent difference must be significant for the function of initiator

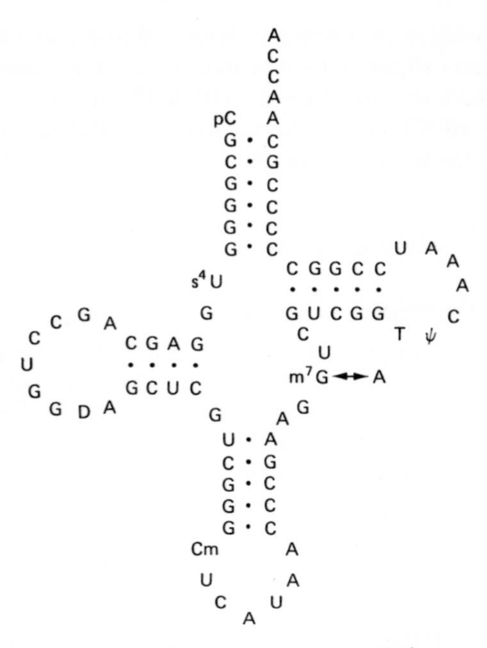

Figure 7.7 The nucleotide sequence of $tRNA_F^{met}$ from *E. coli.*

tRNAs. Eukaryotic initiator tRNAs, unlike other eukaryotic tRNAs, cannot anneal to 5.8S RNA, which is probably present at the A-site of eukaryotic ribosomes. As far as complementarity is concerned, these tRNAs could anneal to GAUC or CGAU sequences in 5S RNA, which is supposed to occupy the P-site in eukaryotic ribosomes[24] (*see* also Chapter 6). This would be in line with the function of initiator tRNAs in translation; they do indeed bind to the P-site when forming an initiation complex (p. 247–248 and 253). Although such interaction between eukaryotic initiator tRNA and 5S rRNA has been repeatedly suggested, it has not been confirmed so far by direct evidence.

Mammalian initiator tRNAs also lack the invariant U residue in position 33, adjacent to the anticodon. Lake proposed that in tRNAs engaged in elongation, this nucleotide may serve a tRNA–tRNA interaction between the aminoacyl-tRNA and the peptidyl-tRNA bound to the ribosome and may therefore have a role in peptide bond formation[16]. One may speculate that initiator tRNAs which do not engage in peptide bond formation on the amino group may not enter into the same kind of tRNA–tRNA interaction.

The anticodon of initiator tRNAs, CAU, shows an unusual capacity for forming wobble-base pairs: it recognises both initiation codons AUG and GUG which implies that U–A and U–G pairs can both be formed between the first letter of the codon and the last letter of the anticodon. In all other codon–anticodon interactions, correct Watson–Crick base pairs are formed in this position.

Taniguchi and Weissmann[35] suggested another unusual characteristic for this anticodon loop. When comparing initiation complex formation at initiation codons followed by either an A or a G residue, they found much stronger interaction with the former structure than with the latter. They interpreted this effect by assuming that the U residue adjacent to the anticodon may also interact with an A residue in the mRNA and can thus strengthen the interaction between the initiator tRNA and the initiation site.

TRANSFER RNA IN THE CONTROL OF PROTEIN SYNTHESIS

tRNA patterns in different cells

Efficiency of protein synthesis may be controlled by the availability of the appropriate tRNA species. In most cells, because of the variety of proteins synthesised, there is a roughly equal requirement for every tRNA species. In some cells, which synthesise one protein in very large amounts, it is, however, possible to compare the pattern of tRNAs with the special requirements for translation of this particular message. A parallelism could be found in several tissues between the tRNA profile and the amino acid composition of the major protein synthesised[25]. The best example is the silk gland of the silkworm *Bombix mori*, which specialises in the production of silk fibroin, a protein with a rather unusual amino acid composition, being built mainly of three amino acids: 44% glycine, 29% alanine, 12% serine. Garel *et al.*[26] studied the tRNA profile of the silk gland and found that the relative abundance of $tRNA^{gly}$, $tRNA^{ala}$ and $tRNA^{ser}$ corresponded well to the above proportions of amino acids. Sprague *et al.*[27] determined the nucleotide sequence of the two major alanine tRNA species in the silk gland and showed that both had IGC anticodons which again corresponds to the requirement for decoding GCU, the codon which is almost exclusively used for alanine in fibroin mRNA. The high demands for $tRNA^{ala}$ are met by the appearance of a novel tRNA species, specific for this tissue. These data suggest that the tRNA pattern of the cell is adapted to the needs of the protein synthesising system.

Dramatic changes can occur in the tRNA pattern upon virus infection. In prokaryotes, a number of new, phage-coded tRNA species are synthesised: in T4 phage-infected *E. coli*, new tRNAs are produced for leu, pro, gly, arg, ile and ser; in T5 phage-infected cells the number of phage-specific tRNAs is at least 14 (ref. 28). These new tRNAs may promote the efficient translation of the viral message. Although the evidence in this respect is not quite conclusive, such a role of phage-specific tRNAs is indicated by the fate of $tRNA^{leu}$ in T4 phage-infected *E. coli*. One of the iso-accepting leucine tRNAs, $tRNA^{leu}I$, which recognises the codon CUG, is inactivated by a phage-specific nuclease which splits the molecule into two (*see* p. 154). The phage directs the synthesis of another $tRNA^{leu}$ species; this, however, responds to the codon UUA. This seems to reflect the difference in the use of codewords in messages of the host cell and the viral message. Unfortunately, the majority of data on tRNA patterns in different cells were collected years ago, when little was known about the primary structure of mRNAs or genes. The role of the tRNA patterns in the control of translation should be reassessed now that a considerable amount of information has accumulated on the use of codewords in different messages.

Suppressor tRNAs

A very important form of control is exercised by suppressor tRNAs. As mentioned earlier (p. 229), mutation in a tRNA may lead to mis-translation, to the insertion of an amino acid which does not correspond to the codeword present in the message. Mis-translation can, however, work as a positive control; it can correct an otherwise damaging mutation which has occurred in the structural gene of an enzyme. A well studied missense mutation occurs, for example, in the tryptophan synthetase gene of *E. coli*: a G → A substitution transforms the GGA codon (glycine) into AGA (arginine) in the mutant strain trpA36. However, substitution of arginine for glycine in the active site of the enzyme leads to the production of an inactive protein. The effect of this mutation can be cancelled by a suppressor mutation which affects the $tRNA^{gly}$II gene of *E. coli*[29]. A single base substitution in the anticodon of this tRNA causes a change in codon recognition: the mutant tRNA will recognise AGA instead of CGA but will still insert glycine in response to this codeword. The result is suppression of the original mutation; active tryptophan synthetase, containing glycine in the active centre, will be produced. This type of tRNA mutant is called missense suppressor.

Another important group of mutant tRNAs are the nonsense suppressors, which, by a similar mechanism, can correct nonsense mutations in the structural gene of a protein. An amber mutant of *E. coli*, for example, in which mutation has produced a UAG nonsense triplet in an internal position of the alkaline phosphatase gene, cannot synthesise this enzyme because of early termination of translation. Goodman and coworkers[30] studied the suppression of this mutation and identified a suppressor tyrosine tRNA, which differed from the wild-type tRNA by a change in the anticodon from G*UA to CUA, a triplet complementary to UAG. This suppressor $tRNA^{tyr}$ misreads the termination codon for a tyrosine codon and thus enables the cell to produce alkaline phosphatase.

Suppressor mutation may occur at another part of the tRNA molecule and may still lead to altered codon recognition. A $tRNA^{tyr}$ UGA suppressor carries a mutation in the stem of arm I. Its anticodon, CCA, is unchanged but it still inserts tryptophan in response to the nonsense codon UGA. It is not well understood how the CCA triplet can anneal to UGA instead of UGG. Possibly, the mutation distorts the conformation of the tRNA so that it affects the codon–anticodon interaction.

A frameshift suppressor mutant has been shown by Riddle and Carbon[4] to produce a $tRNA^{gly}$ species which has a four-base anticodon. This suppressed mutation by correcting the shift in the reading frame of the message.

tRNAs are also involved in more complex regulatory mechanisms which control the expression of genes. The extent of charging of $tRNA^{trp}$ controls transcription of the trp operon[32] and phe-tRNA as well as his-tRNA may have a similar regulatory effect on the expression of the phe and his operons[33]. Other tRNAs also play a role in the regulation of enzyme synthesis[34]. tRNAs can act as primers for DNA synthesis by reverse transcriptase (p. 72). Several viral RNAs contain at their 3′ end tRNA-like structures, the function of which is not yet clear. All these data emphasise the fact that tRNAs are multifunctional molecules and that, in addition to their basic function of decoding the message, they play an important part in a number of different processes connected with gene expression. A discussion of these different roles of tRNA is, however, beyond the scope of the present chapter.

REFERENCES

1 Loftfield, R. B. and Vanderjagt, D. *Biochem. J.*, **128**, 1353 (1972).
2 Crick, F. H. C. *J. Mol. Biol.*, **19**, 548 (1966).
2a Gauss, D. H., Gruter, F. and Sprinzl, M. *Nucl. Acids Res.*, **6**, r1 (1979).
3 Rich, A. and RajBhandary, U. L. *Ann. Rev. Biochem.*, **45**, 805 (1976).
4 Riddle, D. L. and Carbon, J. *Nature NB*, **242**, 230 (1973).
5 Kim, S. H., Suddath, F. L., Quigley, G. J., McPherson, A., Sussman, J. L., Wang, A. H. J., Seeman, N. C. and Rich, A. *Science*, **185**, 435 (1974).
6 Ladner, J. E., Jack, A., Robertus, J. D., Brown, R. S., Rhodes, D., Clark, B. F. C. and Klug, A. *PNAS*, **72**, 4414 (1975).
Jack, A., Ladner, J. E. and Klug, A. *J. Mol. Biol.*, **108**, 619 (1976).
Robertus, J. D., Ladner, J. E., Finch, J. T., Rhodes, D., Brown, R. S., Clark, B. F. C. and Klug, A. *Nature*, **250**, 546 (1974).
6a Schevitz, R. W., Podgarny, A. D., Krishnamachary, N., Hughes, J. J. and Sigler, P. B. *Nature*, **278**, 188 (1979).
7 Sprinzl, M. and Cramer, F. *PNAS*, **72**, 3049 (1975).
8 Van der Haar, F. and Cramer, F. *Biochemistry*, **15**, 4131 (1976).
9 Fersht, A. R. and Kaethner, M. M. *Biochemistry*, **15**, 3342 (1976).
10 Baldwin, A. N. and Berg, P. *J. Biol. Chem.*, **241**, 839 (1966).
11 Hopfield, J. J. *PNAS*, **71**, 4135 (1974).
Yamane, T. and Hopfield, J. J. *PNAS*, **74**, 2246 (1977).
12 Chapeville, F., Lipmann, F., von Ehrenstein, G., Weisblum, R., Ray, W. J. Jr and Benzer, S. *PNAS*, **48**, 1086 (1962).
13 Schwarz, U., Menzel, H. M. and Gassen, H. G. *Biochemistry*, **15**, 2484 (1976).
14 Ofengand, J. and Henes, C. *J. Biol. Chem.*, **244**, 6241 (1969).
15 Kurland, C. G., Rigler, R., Ehrenberg, M. and Blomberg, C. *PNAS*, **72**, 4248 (1975).
Kurland, C. G. in *Molecular Mechanisms of Protein Biosynthesis* (eds Weissbach, H. and Pestka, S.) Academic Press, New York, 81 (1977).
16 Lake, J. A. *PNAS*, **74**, 1903 (1977); in *Gene Expression, Proc. 11th FEBS Meeting*, **43**, 121 (1978).
17 Clark, B. F. C. and Marcker, K. A. *Nature*, **211**, 378 (1966); *J. Mol. Biol.*, **17**, 394 (1966).
18 Adams, J. M. and Capecchi, M. R. *PNAS*, **55**, 147 (1966).
Webster, R. E., Engelhardt, D. L. and Zinder, N. D. *PNAS*, **55**, 155 (1966).
19 Waller, J. P. *J. Mol. Biol.*, **7**, 483 (1963).
20 Ono, Y., Skoultchi, A., Klein, A. and Lengyel, P. *Nature*, **220**, 1304 (1968).
Schulman, L. H., Pelka, H. and Sundari, R. M. *J. Biol. Chem.*, **249**, 7102 (1974).
21 Rudland, P. S., Whybrow, W. A., Marcker, K. A. and Clark, B. F. C. *Nature*, **222**, 750 (1969).
22 Dube, S. K., Marcker, K. A., Clark, B. F. C. and Cory, S. *Nature*, **218**, 232 (1968).

23 Gillum, A. M., Urquhart, N., Smith, M. and RajBhandary, U. L. *Cell,* **6**, 395 (1975).
Piper, P. W. and Clark, B. F. C. *Nature,* **247**, 516 (1974).
Simsek, M., RajBhandary, U. L., Boisnard, M. and Petrissant, G. *Nature,* **247**, 518 (1974).
24 Erdmann, V. A. *Progr. Nucl. Acid Res. Mol. Biol.,* **18**, 45 (1976).
25 Garel, J. P. *J. Theor. Biol.,* **43**, 225 (1974).
26 Garel, J. P., Hentzen, D. and Daillie, J. *FEBS Lett.,* **39**, 359 (1974).
27 Sprague, K. U., Hagenbuchle, O. and Zuniga, M. C. *Cell,* **11**, 561 (1977).
28 Scherberg, N. H. and Weiss, S. B. *PNAS,* **67**, 1164; **69**, 1114 (1970).
29 Hill, C. W., Squires, C. and Carbon, J. *J. Molec. Biol.,* **52**, 557 (1970).
30 Goodman, H. M., Abelson, J., Landy, A., Brenner, S. and Smith, J. D. *Nature,* **217**, 1019 (1968).
31 Klug, A., Ladner, J. and Robertus, J. D. *J. Mol. Biol.,* **89**, 511 (1974).
32 Lee, F. and Yanofsky, C. *PNAS,* **74**, 4365 (1977).
33 Zurawski, G., Brown, K., Killingly, D. and Yanofsky, C. *PNAS,* **75**, 4271 (1978).
Di Nocera, P. P., Blasi, F., Di Lauro, R., Frunzio, R. and Bruni, C. B. *PNAS,* **75**, 4276 (1978).
Barnes, W. M. *PNAS,* **75**, 4281 (1978).
34 Brenchley, J. E. and Williams, L. S. *Ann. Rev. Microbiol.,* **29**, 251 (1975).
35 Taniguchi, T. and Weissmann, C. *J. Mol. Biol.,* **118**, 533 (1978).

8 The Mechanism of Translation

Synthesis of a protein molecule requires a technical machinery of high complexity. As compared to information transfer between nucleic acid molecules where direct copying occurs on the basis of complementarity of nucleotides, the translation of a nucleotide sequence into an amino acid sequence will necessarily involve a greater number of chemical reactions and the participation of additional specific nucleic acid and protein components. The overall process of translation has been briefly described in the Introduction, and the properties of the major components of the translational system—the mRNA which carries the coded message, the tRNAs which select the amino acids specified by each codon in the mRNA, the ribosomes which bring together these components and actually form the peptide bonds—have been discussed in detail in the previous chapters. Here we will describe the mechanism of the individual steps, the RNA-RNA and RNA-protein interactions, which bring about the faithful translation of the message. An attempt will also be made to characterise the structural features of the RNAs and proteins involved and the interactions between them which are responsible for the specificity of the process and which therefore may participate in the control of protein synthesis at the translational level.

There are three distinct steps in the synthesis of a polypeptide chain. The first, *initiation*, results in the formation of an initiation complex in which the ribosome is bound to the specific initiation site on the mRNA while the initiator tRNA is annealed to the initiator codon and bound to the ribosome. This step is probably the most complex one, and it seems to be the determining step for the efficiency and rate of protein synthesis. Translational controls usually act by influencing the initiation process.

The second stage, *elongation*, consists of joining amino acids to the carboxy terminus of the growing polypeptide chain according to the sequence specified by the coded message. Incorporation of each amino acid occurs by the same mechanism, the same steps are repeated over and over again until the termination codon is reached in the message. Elongation is less susceptible to regulation, as under normal conditions this is not the rate-limiting step. The lack of amino acids, tRNAs or elongation factors may, however, under special conditions (e.g. starvation) reduce the rate of elongation or prevent the synthesis of a polypeptide chain.

The termination codon gives the signal for the third and last stage of protein synthesis, the *termination* and *release* of the polypeptide chain. Some control may also be exercised at this stage, and readthrough may sometimes occur at the termination site, leading to the synthesis of larger molecules.

In addition to the main components of the translational system listed above, each stage requires the participation of a number of protein factors which catalyse the binding of the RNAs to ribosomes, stabilise the complexes formed, bring about the release of the final product, etc. Most of these factors show a specificity and selectivity in their interactions with the RNAs concerned which greatly contribute to the fidelity of translation. Different protein factors are involved in the three stages of translation; accordingly, they are called initiation, elongation and release factors. It has been claimed that the initiation of protein synthesis in mammalian systems also involves a group of specific small RNA molecules, the translational control RNAs (tcRNAs).

A further component required in both initiation and elongation is the nucleoside triphosphate, GTP. The presence of Mg^{2+} ions is necessary throughout the translation process.

Different stages in translation can be studied separately in *in vitro* systems by different techniques. Initiation complex formation can be detected by separating ribosome-bound initiator tRNA and mRNA from the free RNAs by sucrose density gradient centrifugation or by membrane filtration. Another assay for initiation is based on the fact that in the absence of one of the elongation factors, EF-G, translation will stop after the formation of the first peptide bond. Different antibiotics inhibit different steps in protein synthesis and can therefore also be used for studying the initiation or elongation steps separately. The overall process is usually followed by the incorporation of labelled amino acids into protein, which can be assayed either by the radioactivity of the total protein fraction or by the labelling of a purified protein or of some characteristic peptide fragments of this protein.

INITIATION OF PROTEIN SYNTHESIS

Whenever genetic information is transferred from one molecule to another, it is essential for the accuracy of the transfer that a specific site should be recognised where the process is initiated. In translation, a definite startpoint is of even greater importance than in transcription, as a shift in the site where decoding of the mRNA starts may also cause a shift in the reading frame. Because of the 'commaless' nature of the genetic code, only the position of the initiation codon determines the reading frame, and faulty initiation might result in complete distortion of the message. In some viruses the same nucleotide sequence has been found to code for different proteins when read in different reading frames; this makes it obvious that recognition of the genuine initiation site is a basic requirement not only in order to synthesise proteins starting from their genuine N-terminus but also in order to synthesise proteins the amino acid sequence of which actually corresponds to the message encoded in their specific genes.

Initation is therefore the key step in precise translation. The mechanism of this step, and especially the nature of the initiation signal and its recognition by the translation system, have been the subjects of extensive studies ever since information on the nucleotide sequences of initiation sites became available. Elucidation of the mechanism of initiation complex formation, as far as it is known today, has also shed light on some structural features of the RNAs concerned and on the nature of various protein factors which may all contribute to the specific selection of initiation sites. In spite of the large amount of data which have accumulated, however, we still cannot explain unequivocally how the genuine initiator codon is recognised in an RNA sequence which may contain dozens of the same nucleotide triplets. The formation of an initiation complex at a specific initiation site is probably the result of a series of well balanced interactions between RNA and RNA, RNA and protein, or protein and protein molecules. Each of these interactions may influence the association of ribosomes with mRNA and it may be because of this interdependent action of a number of different factors that we cannot define exactly which component or which reaction is responsible for directing the ribosome to the specific initiation site.

Apart from their importance for the fidelity of translation, the selection of specific initiation sites and the conditions which may influence this selection, are of great interest also from another point of view. As initiation is the rate-limiting step in translation and is subject to control mechanisms which regulate protein synthesis at the translational level, the efficiency of the synthesis of a protein molecule will depend on and will be controlled through the efficiency of initiation complex formation at the corresponding cistron. Although protein synthesis is mainly controlled at the transcriptional level, which means that the availability of mRNAs will primarily determine whether and in what quantity different proteins will be synthesised, there is still considerable variability in the efficiency with which different mRNAs are translated in the same cell, and this efficiency may be altered under various physiological conditions or in response to environmental effects. Synthesis of some proteins may become partially or even completely inhibited, in spite of the presence of the appropriate mRNA, if ribosomes do not attach to the initiation sites.

Such phenomena are known both in prokaryotic and eukaryotic systems, but the importance of translational controls is much greater in the latter. At different stages of differentiation or even in the same cell at different stages of the cell cycle, the synthesis of some proteins may be triggered off or slowed down, according to the needs of the cell, or in response to hormonal effects. The conditions which allow or prevent attachment of ribosomes to mRNA, or which determine how efficient this interaction will be, are therefore of great significance for these control mechanisms. We have to look at two aspects of the problem of how specific initiation sites are selected: (a) how a specific sequence is recognised in one RNA molecule, (b) how ribosomes interact with initiation sites of different messages; whether they show a preference for one, or an inability to attach to the other.

Later in this chapter we will discuss how ribosomal RNA, ribosomal proteins, initiation factors and also other proteins may influence the selection of initiation sites. Although our present knowledge of the function of these components does not enable us to suggest a mechanism of general validity, the study of these interactions can give us some insight into the possible ways in which the translation of different messages is regulated. Before we can discuss these problems in greater detail, however, we have to get a clear picture of the different steps which lead to the formation of the initiation complex and of the different RNAs and proteins involved in these steps.

The structure of initiation complexes and the main pathway of their assembly are analogous in many respects in prokaryotic and eukaryotic organisms. There is, however, considerable difference between the number and the nature of initiation factors in the two systems and also in the sequence of their interactions with different components and in the stability of such interactions. We know much more about the events in prokaryotic cells, especially in *E. coli*, where we have a fairly good knowledge of the structure of the ribosome as well as of initiation factors, and where the possible interactions between all components have been thoroughly investigated. In eukaryotes a greater variability seems to exist between different organisms, and although a number of studies have been conducted in organisms as different as wheat germ, *Xenopus* oocytes and mammals, a reasonably exact scheme of the individual steps in initiation complex formation has been obtained only for the last group. Results obtained in different laboratories with various tissues, under slightly different conditions or with initiation factors purified in different ways, are sometimes difficult to compare; also, controversial data exist which at present cannot be resolved. The most clearly analysed system is the translation of globin mRNA by different mammalian ribosomes. In the following, initiation in prokaryotes and eukaryotes will therefore be described separately, concentrating in the former case on results obtained in *E. coli*, in the latter on mammalian systems using globin messenger.

The mechanism of initiation in *E. coli*

(a) The 30S initiation complex

Although the synthesis of proteins is carried out by 70S ribosomes, the process can be initiated only if the particles are in the dissociated state. The small subunit starts the process by forming a 30S initiation complex. (This mechanism is of general validity: in other prokaryotes, as well as in eukaryotes, it is only the small subunit of the ribosome which is involved in the first stage of initiation.) The large subunit, which is directly involved mainly in reactions connected with the elongation of polypeptide chains, joins the complex in the second stage of initiation, forming the 70S initiation complex.

The first requirement for initiation is therefore that free ribosomal subunits should be available. Some free subunits are always present in the cell, in equilibrium with the 70S ribosomes. With the aid of a dissociation factor (initiation factor IF3 possesses this activity), this equilibrium can be shifted towards the dissociated state, making free 30S and 50S particles available for initiation complex formation.

The 30S initiation complex comprises the 30S subunit, mRNA and fmet-$tRNA_F^{met}$, attached to each other at specific sites: the 'initiation domain' of the

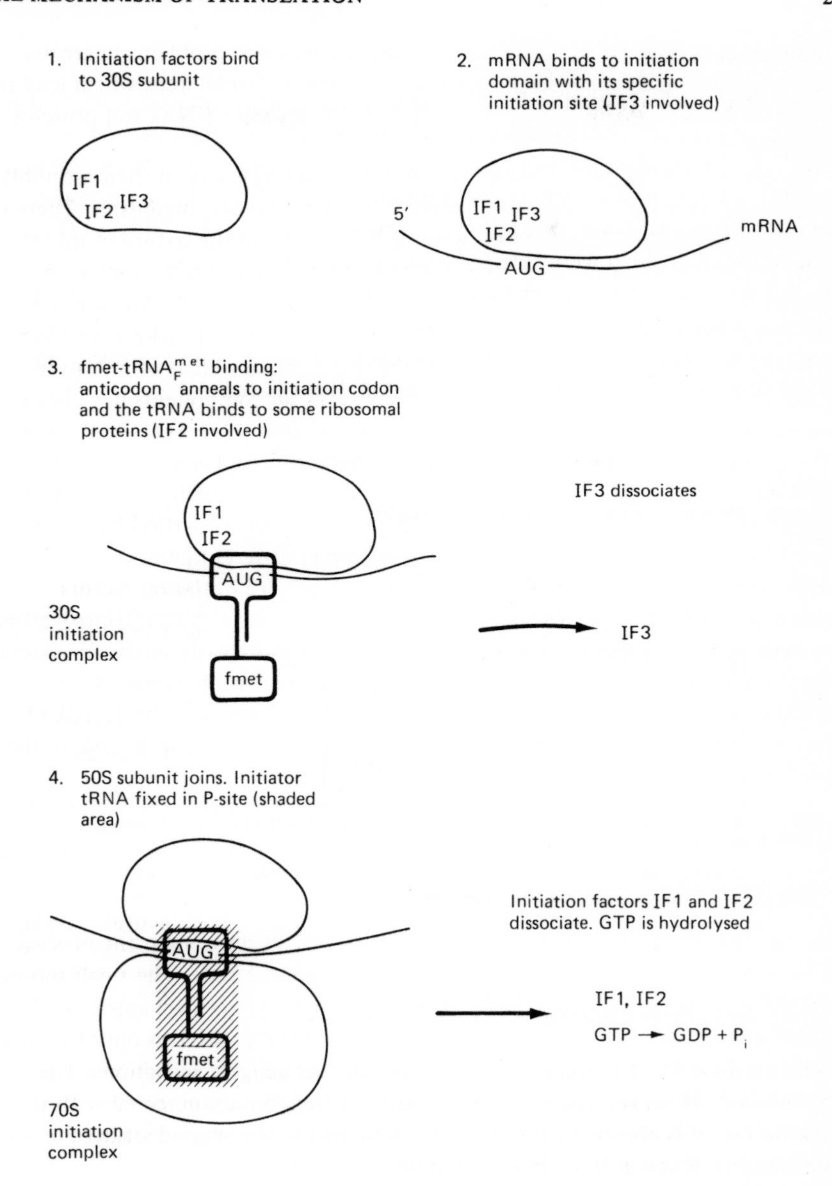

Figure 8.1 Schematic representation of consecutive steps in initiation complex formation.

ribosome (*see* Chapter 7) binds to the initiation site on the mRNA, the initiator tRNA is annealed with its $^{3'}$UAC$^{5'}$ anticodon to the AUG (or GUG) initiation codon on the mRNA, and is also attached to some ribosomal proteins (*figure 8.1*). In addition, three initiation factors and GTP are needed for the formation of this complex and these molecules are also present on the ribosome at the early stage of initiation.

(b) Initiation factors

The requirement for initiation factors was discovered when ribosomes washed with solutions of high salt concentration proved inactive in forming initiation complexes[1]. In the presence of high concentrations of salt (e.g. 1 to 2 M ammonium chloride), a fraction of loosely bound proteins is removed from the ribosomes. From this ribosomal wash, three proteins, initiation factors IF1, IF2 and IF3, could be isolated which were able to restore the activity of the salt-washed particles. All three initiation factors have been purified to homogeneity[2,3,4]; their molecular weights, physicochemical characteristics and, in the case of IF3, also the amino acid sequence, have been determined[5]. *Table 8.1* shows some properties and the functions of the three initiation factors from *E. coli*.

Table 8.1 Initiation factors in *E. coli*

Initiation factor	Multiple forms	Molecular weight	Shape	Role in initiation
IF1		9500		No specific function; enhances the activity of both IF2 and IF3
IF2	$IF2_a$	95 000–117 000	Elongated	Selective binding of fmet-$tRNA_F^{met}$ to the 30S ribosomal subunit
	$IF2_b$	85 000 (probably produced by proteolytic cleavage of $IF2_a$ (ref. 3))	Elongated	
IF3	At least two forms, possibly more	20 668 (181 amino acids)[5] 19 997 (lacking the first 6 N-terminal amino acids of the larger variant)	Probably not globular	(1) Binding 30S ribosomes to the initiation site of mRNA (2) Dissociation factor activity; keeping 30S and 50S ribosomal subunits in the dissociated form

The need for IF3 to bind ribosomes to natural messengers was shown in early experiments[1]. However, there has been some controversy about whether this initiation factor is also required when synthetic oligo- or polynucleotides are used as messengers. Some authors have found little or no effect of IF3 under such conditions[6], while others claim that the translation of artificial messengers, like poly(U), is also greatly enhanced by IF3 (ref. 7). The question is of interest because artificial polynucleotides do not contain specific nucleotide sequences which are characteristic of the initiation sites of natural mRNAs (*see* Chapter 5). If IF3 were required only under conditions of specific initiation to bind ribosomes to specific initiation sites, this would imply that this initiation factor may be involved in the selection of the specific sites. Such a role has indeed been attributed to IF3, as will be seen later. Looking at the contradictory evidence, it still seems probable that the requirement for IF3 is at least more pronounced if initiation occurs under specific conditions. The situation is further complicated by the fact that the need for IF3 also varies in specific initiation on different natural messengers or at different cistrons of the same mRNA. Initiation at the coat protein cistron of R17 RNA is, for example, much more dependent on IF3 than initiation at the A-protein cistron[8].

IF3 is a multi-functional protein, as in addition to its role in binding the ribosomes to mRNA it also exhibits dissociation factor activity[9]. Kinetic analysis of the association and dissociation of ribosomal subunits showed that IF3 alters the equilibrium constant by decreasing the rate of association about fourfold without affecting the rate constant for dissociation[3,14]. IF3 acts by preventing association rather than by dissociating 70S ribosomes. It combines with the 30S particles and attaches to them near to the ribosomal interface, where it binds to ribosomal proteins S7 and S12 and to 16S ribosomal RNA[10]. According to Van Duin *et al.*[11], the binding site of IF3 covers a sequence in 16S RNA which can anneal to 23S RNA. The dissociation factor activity may thus be based on prevention of the base pairing between the RNAs of the small and the large subunit.

The function of IF2 is to bind the initiator tRNA to the 30S ribosome. IF2, in the form of a free protein, can associate with fmet-$tRNA_F^{met}$ and GTP. The association constants of both IF2·fmet-$tRNA_F^{met}$ and IF2·GTP·fmet-$tRNA_F^{met}$ are, however, very low. It has been assumed, therefore, that the complexes with free IF2 are not genuine intermediates in the initiation process and that more stable complexes are formed *in vivo* with ribosome-bound initiation factor[12]. The specificity of interaction between IF2 and fmet-$tRNA_F^{met}$ is very strict: no other aminoacyl-tRNA is bound by IF2, and the initiation factor discriminates even between the formylated and unformylated form of met-$tRNA_F^{met}$, in favour of the former[13].

IF1 has no well defined function; it stimulates the activity of the other two factors, and also contributes to the dissociating effect of IF3. In some prokaryotes no initiation factor corresponding to IF1 has been detected; it seems that in some organisms initiation can be carried out by IF2 and IF3 alone.

The three initiation factors in *E. coli* probably act in a cooperative way. This fits in well with their location on the ribosome: all three are attached to the 30S subunit in roughly the same area. Cross-linking experiments show that IF2 and IF3 are in close proximity to an overlapping set of ribosomal proteins[10].

(c) The sequence of events in initiation complex formation

The first step in initiation is the binding of initiation factors to a free 30S ribosomal subparticle. The direct interaction of ribosomes with initiation factors has been studied by Fakunding *et al.*[15], who found that IF1 and IF3 stabilised the binding of IF2 to the ribosome. Although both free IF2 and free IF3 can interact specifically with the initiator tRNA and the mRNA, respectively, the efficiency and stability of these interactions are greatly increased if the factors are attached to ribosomes which suggests that the binding of factors occurs prior to the binding of the RNAs. This is also supported by the fact that initiation factors are found in the cell attached to 30S ribosomes.

There is some uncertainty about whether the next event is the binding of mRNA or of fmet-$tRNA_F^{met}$; controversial results have been obtained in direct binding assays. Vermeer *et al.*[16] observed mRNA binding independently of initiator tRNA, while Jay and Kaempfer[17] claim that fmet-$tRNA_F^{met}$ binds to ribosomes in the absence of mRNA. This latter complex is very unstable, and may have escaped detection by conventional assay techniques. It cannot be decided at present whether this labile association might represent a genuine intermediate in *in vivo* initiation. It seems certain, however, that for stable binding of fmet-$tRNA_F^{met}$, 'programmed' ribosomes are required, i.e. ribosomes which carry a nucleotide stretch containing an initiation codon. This supports the view that mRNA binding occurs prior to initiator tRNA binding.

In *in vitro* assays, synthetic oligonucleotides or even an AUG triplet are sufficient to cause stable binding of fmet-$tRNA_F^{met}$. On natural mRNAs, whether *in vitro* or *in vivo*, the ribosomes bind to specific initiation sites so that a genuine initiation codon is present in the proper position for the fmet-$tRNA_F^{met}$ to anneal to it. Initiation factor IF3 and ribosomal protein S1 are directly involved in the binding of 30S ribosomes to mRNA, and 16S RNA also takes part in this process by annealing to a specific site near the initiation codon. Different aspects of the specificity of this process will be discussed later.

The programmed ribosome binds fmet-$tRNA_F^{met}$ with the aid of IF2. A ribosome-bound ternary complex, fmet-$tRNA_F^{met}$·IF2·GTP, is formed. GTP is bound to the ribosome and will be hydrolysed at a later stage. The high selectivity of the IF2–initiator tRNA interaction contributes to the specificity of the initiation process: no faulty interactions can occur, only formylmethionine can be inserted into the polypeptide chain in response to an initiator codon.

The initiation factors are present only temporarily on the ribosome, while the initiation complex is being formed. IF3 is probably the first to be released: Vermeer *et al.*[16] found that it dissociates from the 30S complex as soon as the initiator tRNA has been bound. This also supports the idea that tRNA binding takes place after the binding of mRNA. IF3 must certainly be removed at an early stage, as the next step, the joining of the 50S ribosomal subunit to form the 70S initiation complex, would be prevented in the presence of IF3.

Concomitantly with the association of the two subunits, IF1 is released from the ribosome. Removal of IF2 occurs after the 70S ribosome has been formed and it requires the hydrolysis of the ribosome-bound GTP. This reaction is catalysed by an IF2-activated ribosomal GTPase which produces GDP + P_i; these are also removed from the initiation complex. Studies with a GTP analogue, GMPP(CH_2)P, which cannot be hydrolysed by GTPase, showed that this nucleotide can replace GTP in the step of fmet-$tRNA_F^{met}$ binding, but prevents the release of IF2 from the 70S initiation complex[15]. As a consequence of this, the next aminoacyl-tRNA does not bind to the ribosome, and elongation is not possible[18]. Dissociation of the initiation factors is thus necessary in order to obtain a functional initiation complex. The factors, once released, can attach to another 30S particle; they are constantly recycled and thus function in a catalytic way.

(d) The structure of the 70S initiation complex

Upon association of the two subunits, the other major components of the initiation complex, mRNA and initiator tRNA, take up their proper positions on the ribosome (*figure 8.1*).

(1) The mRNA is accommodated in the tunnel formed between the two ribosomal subunits. With its purine tract it anneals to the 3′ end of 16S RNA, with the initiator codon to the initiator tRNA. It is in close proximity to nine proteins of the 30S subunit, including protein S1, which are all situated at the 'head' of the particle (*see* Chapter 6, *figure 6.9.*)

(2) fmet-$tRNA_F^{met}$ is still annealed to the initiator codon and is in touch with the corresponding proteins of the 30S subunit, but it is also bound to proteins of the 50S subunit and thus occupies the P-site on the particle (*see* p. 202 and p. 234). According to recent data of Dahlberg *et al.*[19], the D arm of the initiator tRNA may also anneal to a 17-nucleotide-long stretch in the middle of the 23S ribosomal RNA molecule. The formylmethionyl group is in contact with the peptidyl transferase area of the 50S particle (*see* Chapter 6).

Although all incoming aa-tRNAs bind to the A-site of the ribosome (by definition, A-site means the aminoacyl-tRNA site), several different observations prove that initiator tRNA behaves differently, and is located at the peptidyl-tRNA site

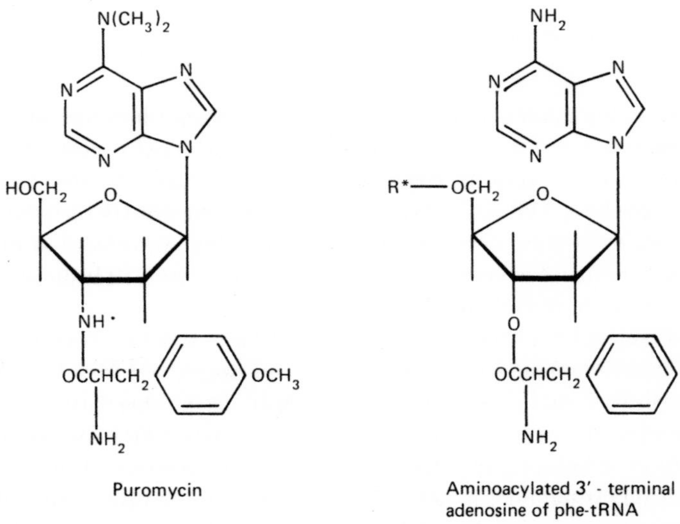

Puromycin

Aminoacylated 3′ - terminal adenosine of phe-tRNA

Figure 8.2 Comparison of the structure of puromycin and of the aminoacylated terminal adenosine of a tRNA R*: rest of the $tRNA^{phe}$ molecule.

(P-site) in the initiation complex. By affinity labelling, Collatz *et al.*[20] found fmet-$tRNA_F^{met}$ attached to protein L27 which had previously been shown to be part of the P-site[10]. At the same time, no binding of the initiator tRNA to L16, a characteristic protein of the A-site, has been observed.

Further evidence for the location of the initiator tRNA at the P-site of the ribosome has been provided by the effect of puromycin on the 70S initiation complex. The structure of this antibiotic shows some similarity to the aminoacylated 3′ end of aa-tRNAs (*see figure 8.2*). It can therefore compete for the A-site, and can form a peptide-like bond with the peptidyl-tRNA present in the P-site, but as it does not contain the rest of the tRNA molecule which binds to the ribosome and anneals to the mRNA, further reactions cannot occur, the puromycin–peptide is released from the ribosome. It follows from this mechanism that puromycin can react only with tRNA derivatives present in the P-site. If, after formation of the initiation complex, puromycin is added to an *in vitro* system, fmet-puromycin is produced, proving that fmet-$tRNA_F^{met}$ was present in the P-site on the ribosome. This does not necessarily mean that fmet-$tRNA_F^{met}$ directly enters the P-site when it is first bound to the particle. It has been suggested that it attaches temporarily to the A-site and is transferred to the P-site when initiation is completed. This, however, seems not to be the case. There might be a specific site for initial binding of the initiator tRNA on the ribosome, different from both the A- and P-sites, from which, at a later stage, fmet-$tRNA_F^{met}$ is transferred to its final position in the P-site.

At completion of the initiation step, the ribosome is ready to start elongation, i.e. the fmet-$tRNA_F^{met}$ is in a suitable position to form a peptide bond with the first incoming aa-tRNA.

Initiation of translation in eukaryotes

The main stages in initiation complex formation are analogous in eukaryotic and prokaryotic organisms, but in the former the process shows greater complexity both with respect to the individual steps and to the number and structures of the protein factors involved. This complexity, together with the greater variations between different eukaryotic organisms and our rather poor knowledge of the structure of eukaryotic ribosomes, makes it difficult to obtain a clear picture of each individual event in the initiation of translation in eukaryotic systems. However, initiation factors have been isolated and initiation of specific proteins has been assayed by different authors in various cells and tissues[21,22,23,23a]. Comparison seems straightforward in some cases but is ambiguous in others; often it is not possible to identify the various protein factors obtained from different tissues by different isolation techniques, and this makes a proper comparison of the various protein synthesising systems extremely difficult. However, the main events in initiation seem to be similar in these different tissues. Instead of attempting to describe a generalised mechanism of initiation, which would necessarily be ambiguous, we will therefore discuss the steps leading to initiation complex formation in one specific system, that in which globin mRNA is translated by mammalian ribosomes.

Translation of globin mRNA has been thoroughly studied in reticulocytes as well as in heterologous systems, and also in highly purified, mixed protein synthesising systems containing initiation factors from reticulocytes and ribosomes from other tissues. The work of Staehelin's group[21] and of Benne and Hershey[23a] on this process has produced the most complete model so far obtained for the sequence of events which take place on the mammalian ribosome in the course of initiation.

Formation of the initiation complex occurs in two stages with eukaryotic ribosomes also. First, the small ribosomal subunit binds the initiator tRNA and the mRNA to form a 40S initiation complex; this is followed by the attachment of the large subunit (60S), resulting in the 80S initiation complex which is ready to start elongation of the polypeptide chain. A protein with dissociation factor activity[24] is needed to produce free subunits, and initiation factors are required for the binding of each component to the ribosome. GTP binding and hydrolysis occur in the same way as on prokaryotic ribosomes. However, instead of three initiation factors, at least seven proteins participate in the process, and there is an additional requirement for ATP, which is hydrolysed when the 40S complex is formed. Staehelin's group purified seven, Benne and Hershey eight initiation factors from a crude KCl wash of reticulocyte ribosomes and studied their protein structure as well as their role in the binding of the different compounds to the ribosomal particle*.

As can be seen in *table 8.2*, these factors differ widely in size and complexity. The smallest (eIF-1) has a MW of 15 000 and consists of a single polypeptide chain, while the largest (eIF-3) constitutes a protein complex in itself, comprising 9 to 10 subunits of different sizes and in different amounts. Its MW is around 300 000. As we will see later, the structure of this initiation factor may be the most variable in different organisms or tissues. Its structural heterogeneity may therefore reflect a functional heterogeneity, suggesting that it may play a role in controlling the specificity of initiation.

The individual steps which lead to initiation complex formation in this mammalian system differ in some respects from those observed in prokaryotes. In eukaryotic systems it could be unequivocally shown that the binding of initiator tRNA precedes the binding of mRNA. Staehelin and coworkers, using ^{3}H-labelled met-$tRNA_F$ and globin mRNA labelled by iodination with ^{125}I, followed the binding of these molecules to 40S ribosomal subunits in the presence of initiation factors. They incubated 40S particles from mouse liver with the two RNAs, adding the latter separately or together. The mixtures were then fractionated by sucrose density gradient centrifugation, which separated the ribosomes from the free

*In a purified fraction of the ribosomal wash which contains the initiation factors, Sonenberg *et al.*[80] detected a specific polypeptide which binds to the capped 5′ termini of reovirus mRNA. As it has been assumed that the cap structures form part of the recognition site or of the binding site of ribosomes, this cap-binding protein may also be involved in initiation.

Table 8.2 Mammalian initiation factors*

Initiation factor	Molecular weight	Subunit structure	Main role in initiation of globin synthesis
eIF-1†	15 000	Single chain	
eIF-2	32 000 + 47 000 + 50 000	3 Subunits	Binding of initiator tRNA followed by ribosome binding
eIF-3	sedimentation coefficient 17S	9 to 10 subunits, some in non-stoichiometric amounts	Binding of mRNA to ribosome
eIF-4A	50 000	Single chain	Cooperates in mRNA binding
eIF-4B	80 000	Single chain	Cooperates in mRNA binding; ATPase activity
eIF-4C	17 000	Single chain	Joining of ribosomal subunits
eIF-5	170 000	Single chain	Joining of ribosomal subunits

*From rabbit reticulocytes. (Data according to Staehelin *et al.*[21].)

†It causes some confusion in the literature that different authors use different symbols for the initiation factors they isolated. The symbols shown here are those defined at the Fogarty Meeting (1976) as the generally accepted abbreviations for eukaryotic initiation factors.

RNAs; the RNAs bound could thus be detected by the radioactivity sedimenting with the ribosome fraction. It was found that met-tRNA$_F^{met}$ could bind equally well to the ribosomal particles whether globin mRNA was present or not, while binding of mRNA could be achieved only in the presence of initiator tRNA. These experiments also showed an absolute requirement for both initiation factors eIF-2 and eIF-3 for the binding of globin mRNA. Further studies on the role of the initiation factors made it clear, however, that only eIF-3 was directly involved in mRNA binding. The function of eIF-2 is to bind met-tRNA$_F^{met}$ to the ribosome. This process is required, however, to enable the ribosome to bind mRNA. The lack of eIF-2 therefore prevents the attachment of mRNA as well as of initiator tRNA to the 40S particle.

Another difference, as compared with prokaryotic systems, was found in the stability of the complex formed between initiator tRNA and free initiation factor. The lability of such complexes in prokaryotes ruled out their role as intermediates in initiation. In mammalian systems, however, the interaction of met-tRNA$_F$ and GTP with free eIF-2 produces a stable ternary complex[25], and this is considered to be the first intermediate in the process of initiation. The met-tRNA$_F$·eIF-2·GTP complex binds as such to the small ribosomal subunit. The binding of mRNA follows, mediated by initiation factor eIF-3. The reaction is absolutely dependent on this factor but is also greatly stimulated by factors eIF-4A and eIF-4B.

In this system, too, we see a cooperative action of the different initiation factors: eIF-1 stimulates both initiator tRNA binding and mRNA binding, eIF-3, while primarily involved in the binding of mRNA to the ribosome also stabilises the 40S·met-tRNA$_F$·GTP complex. Hydrolysis of one molecule of ATP takes place when mRNA is bound to the ribosome. Splitting into ADP and inorganic phosphate probably occurs by the catalytic action of eIF-4B which has been

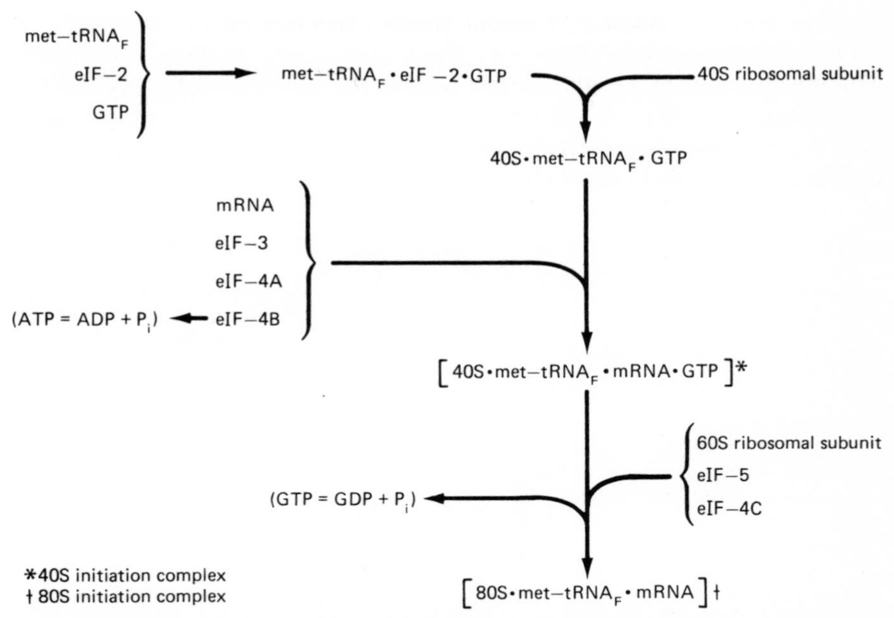

Figure 8.3 The main events in initiation of translation in mammalian systems. (According to Staehelin and coworkers[21].) Symbols for initiation factors as in *table 8.2*.

found to possess ATPase activity. As a result of all these reactions, as represented in *figure 8.3*, a 40S initiation complex is formed which still contains elongation factors and an intact molecule of GTP. The next step is the attachment of the 60S subunit and thus formation of the 80S initiation complex. This is catalysed by factors eIF-4C and eIF-5 and is accompanied by release of eIF-2 and GTP and by hydrolysis of the latter to GDP + P_i. Factor eIF-5 and the 60S subunit together may be responsible for the GTPase activity[21,25a]. eIF-3 also may be released at this stage[23a].

GTP plays a role in two steps of initiation: it is needed for binding initiator tRNA to eIF-2 and to ribosomes, and its hydrolysis is required to allow association of the two ribosomal subunits. If GMPP(CH_2)P is substituted for GTP, met-RNA_F binding proceeds unhindered, but, as no hydrolysis can occur, only 40S initiation complex is formed, and the large subunit does not join. Hydrolysis of ATP is necessary for the binding of mRNA; if ATP is omitted or replaced by the non-hydrolysable nucleotide AMPP(CH_2)P, globin mRNA does not attach to the 40S ribosome, and no 40S initiation complex is formed.

Very little is yet known about the topography of eukaryotic ribosomes, and it is therefore not possible to define the positions of initiator tRNA and mRNA in a eukaryotic initiation complex as exactly as can be done on the *E. coli* ribosome. Some information is available, however, and this points to a structure rather similar to the prokaryotic initiation complex. Electron microphotographs[26] of silk gland polysomes (*see* p.272) are in good agreement with the prokaryotic model in which the mRNA is 'threaded' through the tunnel in the ribosomes between the two subunits (*figure 8.4*). The stretch of mRNA which is in contact with the ribosomes in the 80S complex is about the same size as the ribosome binding sites in prokaryotic mRNAs[27,28]. At an earlier stage of initiation, however, the 40S ribosomes protect longer sequences in mRNAs which in most cases include the 5′-terminal capped structures[27]. In the course of initiation complex

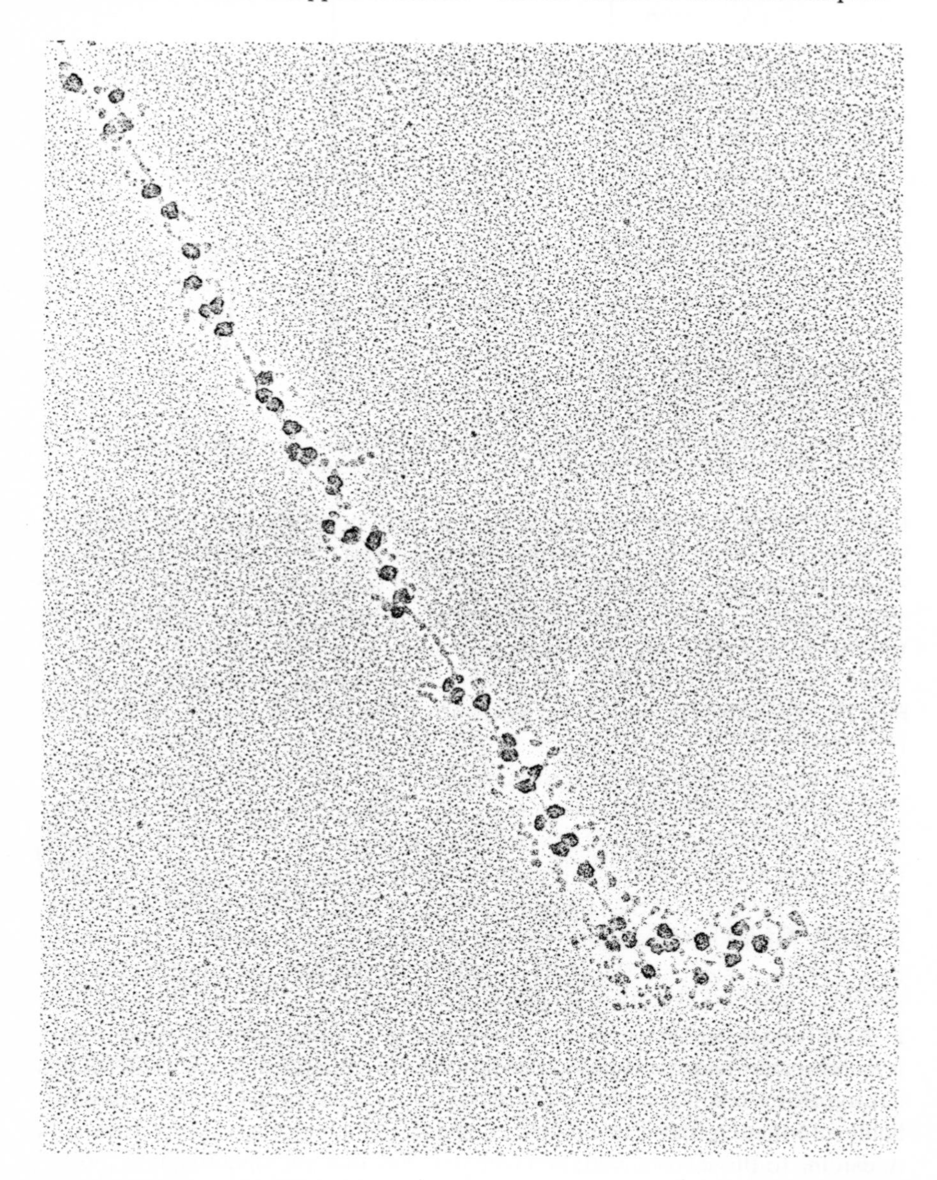

Figure 8.4 Electron micrograph of silk gland polysomes. (By courtesy of Dr S. McKnight.)

formation, the mRNA undergoes a conformational change. In the 40S initiation complex it probably has a folded structure which brings the 5′ end nearer to the initiation codon. Upon formation of the 80S complex, the mRNA becomes unfolded, and the capped 5′ terminus is released from the ribosome (*see* p. 178). It is in good agreement with such a mechanism that a protein with mRNA unwinding activity has been found associated with eIF-3 in the KCl wash from reticulocyte ribosomes[29]. (An analogous phenomenon also exists in prokaryotes: IF3 can associate with ribosomal protein S1 which also has RNA unwinding activity. The function of this protein is, however, different; it is thought to disrupt the secondary structure at the 3′ end of 16S RNA at this stage of initiation, to facilitate annealing of 16S RNA and mRNA (*see* p. 194).)

The reaction of met-$tRNA_F^{met}$ with puromycin shows that the initiator tRNA is situated in the P-site in eukaryotic initiation complexes also. This is further supported by nucleotide sequence studies on eukaryotic initiator tRNAs and on the two small ribosomal RNAs 5S RNA and 5.8S RNA. It was revealed that annealing can occur between complementary sequences in arm III of initiator tRNAs and in 5S RNA. All other aminoacyl-tRNAs anneal to 5.8S RNA, which is thought to be in the A-site (*see* Chapters 6 and 7). The initiator tRNA thus occupies a position different from that of aa-tRNAs on the eukaryotic ribosome.

The specificity of initiation

It has been discussed earlier in this Chapter that the problem of how the specificity of initiation is maintained in the cell can be approached from two different viewpoints. We may be interested in how a genuine initiation site is recognised in a long nucleotide sequence which contains a great number of AUG and GUG triplets; or we may investigate another aspect of the problem which is connected with the regulation of translation, viz. how the initiation system discriminates between genuine initiation sites of different messages. Whichever approach we choose, the basic questions to be answered are: which components of the initiation system take part in the selection of specific sites on the mRNA and what are the conditions which can influence this selection? In Chapter 5 the different structural features in mRNAs which play a part in initiation of translation have been discussed in detail. In order to see how the selection of initiation sites can be influenced by these features, we have to study the interactions of mRNA with different components of the initiation system.

(a) Annealing of ribosomal RNA to mRNA

In prokaryotic mRNAs, near the initiation sites, purine-rich sequences have been detected which are complementary to the 3′-terminal sequence of 16S ribosomal RNA (p. 157). Shine and Dalgarno suggested a mechanism for ribosome binding to these mRNAs based on the annealing of the complementary sequences[30] According to this mechanism, ribosomal particles would recognise the complementary sequences and would thus be guided to the specific initiation sites where their attachment to mRNA would be stabilised by the annealing of the two mRNAs. The idea of such a mechanism found support in the high degree of homology observed in the nucleotide sequences near the 3′ termini of different 16S RNA species[31] and in the widespread occurrence of complementary purine-rich stretches at the initiation sites of different prokaryotic mRNAs (*table 5.1*). Steitz and coworkers demonstrated that annealing could indeed occur between such complementary RNA fragments[32]. Further support was provided by recent studies of Taniguchi and Weissmann[77], who compared the roles played by this interaction and by the interaction between initiator tRNA and the initiation codon. They used synthetic oligonucleotides comprising sequences complementary to the 3′ terminus of 16S RNA. In the presence of an excess of such oligonucleotides, the 3′-terminal region of the ribosomal RNA was blocked and this resulted in a marked inhibition of initiation complex formation on Qβ RNA. Ribosome binding to the trinucleotide AUG was not affected.

The results showed the importance of both interactions, but they left no doubt that preventing the Shine–Dalgarno-type annealing very markedly reduced the ability of ribosomes to bind to genuine initiation sites. In view of the above data it can be accepted today that, at least in prokaryotes, the interaction of rRNA with mRNA plays an important role in initiation. We will see later, however, that this is only one of several events which are essential for specific initiation. In itself, this interaction cannot explain the specificity in the selection of initiation sites, or the stability of the initiation complexes formed.

The results discussed so far refer to prokaryotic systems only. In eukaryotes the situation is still ambiguous. Complementary sequences have been detected in eukaryotic rRNAs and many, but not all, eukaryotic mRNAs[33]. The variable positions of the latter sequences in relation to the initiation codon make it also doubtful whether they can play a similar role to the purine tracts in prokaryotic mRNAs (*see* p. 163). Kozak and Shatkin[78], as well as Baralle and Brownlee[79], emphasise the importance of other types of interactions in eukaryotic systems. Their data do not support the participation of a Shine-Dalgarno-type annealing in the initiation process in eukaryotes.

As mentioned above, even in prokaryotes there are some unanswered questions as regards the specificity and stability that can be achieved by this annealing. In a

UAGCGAGACGCUACCAUG	UGUCAUGGGAUCCGGAUG
CAAGAGUUUCUUCCUAUG	UCCAACGGUGCUCCUAUG
GUGGGAUGCUCCUACAUG	GGCACAAGUUGCAGGAUG
AGGGGUACAAUCCGUAUG	UGUAAGGAGCCUGAUAUG
ACCGGAGUUUGAAGCAUG*	GAGCCUGAUAUGAAUAUG
AAGGUCUCCUAAAAGAUG	AGCAGGGCAGCGUAGAUG
ACAUGAGGAUUACCCAUG*	UGGCGAGACGAUACGAUG
CCUUAGGUUCUGGUAAUG	UCCCAGGUGCCUUCGAUG

Figure 8.5 Some sequences in MS2 RNA which possess the characteristic features of initiation sites. The two sequences marked by asterisks are genuine initiation sites, the others are not.

large RNA molecule, a number of nucleotide sequences may occur which contain complementary bases to the relevant rRNA sequence and are in the neighbourhood of an AUG or GUG triplet; however, no initiation occurs at these sites. *Figure 8.5* lists a few nucleotide sequences from MS2 RNA which, as far as the primary structure is concerned, could all bind ribosomes, yet only two of which act as genuine initiation sites. In the ΦX174 genome, over 20 potential ribosome binding sites can be found, but only 10 function as initiation sites. In human β-globin mRNA, there are 4 tetranucleotide sequences complementary to 18S RNA, at distances from non-initiator AUG triplets which are within the limits of variation found around genuine initiation sites in other eukaryotic mRNAs. It is thus clearly indicated that some additional structural signal and/or some further interactions must participate in the discrimination between specific and nonspecific ribosome binding sites.

The genuine initiation sites show great variations in their efficiency. The same ribosomes may attach preferentially to one or another message, and different ribosomes may initiate the translation of the same message with different efficiency[22,34,35]. Shine and Dalgarno originally proposed that the stability of annealing between the initiation sites and the RNA of the small ribosomal subunit could determine the efficiency of initiation at these sites. More thorough investigation of nucleotide sequences in rRNAs and mRNAs revealed, however, that there was no correlation between efficiency of initiation and extent of base pairing in the two RNAs. Sprague *et al.*[31] found that 16S RNA from *B. stearothermo-*

Table 8.3 Annealing between 16S rRNA and the initiation sites of mRNAs

mRNA	Number of base pairs formed with *E. coli* 16S rRNA
R17 A protein	7
R17 coat protein	3 (5)
R17 replicase	4 (6)
Qβ coat protein	3 (4)
Qβ replicase	5
Qβ A-protein	4
ΦX174 G protein	5
ΦX174 D protein	4
ΦX174 B protein	5
ΦX174 H protein	9 (10)
trp leader	3 (4)
trp E	3
trp A	3 (4)
lac z	4
gal E	4
gal T	6

Numbers in parentheses include G•U base pairs.

philus ribosomes and from *E. coli* ribosomes contain an identical nucleotide sequence near their 3′ termini (although not in exactly the same position); still, the former particles initiate almost exclusively at the A-protein cistron of R17 RNA, while the latter attach preferentially to the coat protein cistron. When comparing the extent of possible base pairing at the three cistrons of R17 RNA (*table 8.3*), it seems obvious that the binding of *E. coli* ribosomes which occurs primarily to an initiation site which anneals poorly to 16S RNA must be governed by some other factors.

From the different components of the initiation system, initiation factor IF3 and ribosomal protein S1, two proteins shown to play a direct role in binding ribosomes to mRNA, are the most probable candidates for interaction at this stage, and could increase both the specificity and the stability of the attachment of ribosomes to a specific initiation site. Steitz *et al.*[8] compared the requirement for IF3 and protein S1 in initiation complex formation at the three cistrons of R17 RNA. They found that initiation complex formation at the coat protein cistron, which contains only three complementary bases, is absolutely dependent on the presence of these proteins, while at the A-protein cistron, which can form seven base pairs with 16S RNA, IF3 and protein S1 have very little effect on initiation.

There is thus an inverse relationship between the stability provided by base pairing and the requirement for IF3 and S1 protein. Steitz *et al.* came to the conclusion that one, or possibly both, of these proteins may have a stabilising effect on the base-paired structure formed between mRNA and 16S RNA. Although there is no direct evidence to support this assumption, it is in good agreement with some other observations. Both IF3 and S1 protein are localised on the ribosome near the 3′ end of 16S RNA[10] which would facilitate such a function. IF3 has been found to bind preferentially to base-paired hairpin loop structures in RNA[36]. S1 protein, on the other hand, has an unwinding effect on base-paired structures[37] and binds to single-stranded pyrimidine-rich RNA stretches. This points to a different kind of involvement in initiation complex formation. It is therefore more probable that IF3 may stabilise mRNA–rRNA annealing.

The need for some stabilising action is also apparent from calculations of the thermodynamic stability of the base-paired structures which can be formed between mRNA and rRNA. As can be seen in *tables 5.1* and *8.3*, some ribosome binding sites contain only three, and many only four, complementary bases. Although formation of three base pairs also occurs at other stages of translation (the obvious example is the annealing of the anticodon of tRNA to the codon in mRNA), it is nevertheless true that the thermodynamic stability of such short base-paired regions is extremely low. In codon–anticodon interaction, base pairing serves primarily the recognition of the cognate codon and induces conformational changes; the aminoacyl-tRNA is stabilised on the ribosome by additional interactions with ribosomal proteins and with RNA. It seems therefore that the weak annealing of some mRNAs to rRNA should also be stabilised by protein–RNA interactions.

(b) Specific role of initiation factors

In the course of purification of initiation factors from different cells, observations were made which suggested that initiation factors may have a more specific function in initiation than the one described above. Several authors claimed to have obtained different variants of IF3 which directed ribosomes to different mRNAs[4,38] or to different cistrons in a polycistronic messenger[39]; or to have isolated different eukaryotic initiation factors from various tissues (corresponding in their function to eIF-3) which enhanced translation of one specific message[22,40,41]. It soon emerged, however, that the interpretation of some of these results was ambiguous. The IF3 preparation which specifically stimulated initiation on T_4 phage mRNA but not on MS2 RNA, probably contained another protein, called interference factor (i-factor, later designated factor iα) which prevented attachment of ribosomes to the MS2 coat protein initiation site but did not affect the T_4 mRNA initiation site[12,42]. The same interference factor was found to stimulate ribosome binding to the synthetase cistron of MS2 RNA; in combination with IF3, it therefore behaved like a specific initiation factor which directed ribosomes to the synthetase rather than to the coat protein initiation site. Further purification of the different variants of IF3 pointed to the existence of a homogeneous IF3 species in combination with different interference factors which had a positive or negative influence on the binding of ribosomes to different initiation sites[43]. Recent results on the physicochemical properties and the amino acid sequence of highly purified IF3 revealed that more than one molecular species of this protein exist, but the difference found so far in their structure is restricted to the absence of the first six amino acids at the N-terminus of one species[5,44].

It thus seemed that specific selection of initiation sites, at least in prokaryotes, could be attributed to a series of interference factors rather than to specific initiation factors. However, this idea had to be revised, too, when interference factor iα was identified as ribosomal protein S1 by Wahba *et al.*[45]. This was a rather unexpected result and it seemed difficult to reconcile the facts that on the one hand this protein could inhibit initiation on MS2 RNA and on the other it was known to be needed for translation, specifically in the process of binding ribosomes to mRNA[46].

Bosch's group, who investigated the effect of S1 protein on MS2 RNA translation under different conditions[47], resolved the problem. They found that ribosomes depleted of S1 protein are unable to translate MS2 RNA. If S1 is added in low concentrations, the protein binds to the ribosomal particles and thus enables them to form initiation complexes with the mRNA. If a large excess of S1 protein is present, however, this will bind to the mRNA and will block some active sites, preventing the attachment of ribosomes to these sites. This explains the interference factor activity of protein S1, but at the same time, throws doubt on the relevance of such an activity for the *in vivo* translation process, where ribosomal proteins are probably not present in more than equimolar amounts.

This affects the interpretation of all studies on the different interference factors, as they were all carried out in *in vitro* protein synthesising systems. The other interference factors have not yet been identified, but the possibility exists that they also are ribosomal proteins: they have been isolated, together with initiation factors, from a crude ribosomal wash. If this were the case, their *in vivo* function should be considered as doubtful as that of factor iα. At present, there is still considerable controversy about the existence of interference factors and the possibility of selection of specific initiation sites by the combined action of IF3 and ribosomal proteins.

A different approach to elucidate whether IF3 may contribute to the recognition of specific initiation sites was attempted by isolating and locating the binding site of this protein on MS2 RNA[36]. This result was also ambiguous: IF3 could definitely recognise and protect a specific site in this RNA; however, this site was far from the initiator regions of the three cistrons. It is questionable whether free IF3 interacts with mRNA in the same way as IF3 bound to the ribosome.

The situation is also complicated in eukaryotic systems. Several authors claim that specific initiation factors are required for the synthesis of specific proteins. Few initiation factors have been obtained, however, in sufficiently pure form to make certain that the specific effects observed are indeed due to the specific functions of different protein species. This can also lead to equivocal results: an initiation factor, first believed to be specific for the translation of proteins of EMC virus, later proved to be identical with the eIF-4A of reticulocytes[21]. Still, a specific function of initiation factors seems more strongly established in these systems than in prokaryotes.

Translation in heterologous systems (see below) often depends on the presence of homologous initiation factors. The factor responsible for the specificity of mRNA selection is most probably eIF-3. This is indicated by its function as well as by its protein structure. The structural complexity of eIF-3, and the non-equimolar ratio of its subunits, point to the possibility of a functional heterogeneity: changes in the subunit structure or alteration of one of its subunits may easily lead to a modification in the specificity of the protein complex. Heywood found that globin mRNA, myosin mRNA and myoglobin mRNA could be translated by heterologous ribosomes, but the translation of each of these mRNAs required a specific initiation factor, an eIF-3 species isolated from reticulocytes, red muscle and white muscle, respectively[22,48]. Heywood also claims that the ribosomal wash of muscle ribosomes contains a group of small specific RNA molecules (translational control RNAs) which specifically stimulate translation of different mRNAs[49].

Another specific role of an initiation factor, which may be connected with the selection of mRNAs, was observed in poliovirus-infected HeLa cells. Poliovirus mRNA is actively translated in extracts of these cells, while other mRNAs are not. Rose *et al.*[81] found that this selective inhibition of translation is caused by inactivation of initiation factor eIF-4B in cells infected with poliovirus.

(c) Specific properties of the ribosomal particles

The question of whether ribosomes possess an inherent specificity has been the subject of long argument since the very first results with *in vitro* protein synthesising systems. According to one view, ribosomes merely provide an exactly functioning machinery which collects the amino acids and produces the polypeptide chain, depending for the specificity of the process entirely on the message with which they have been programmed. As opposed to this, it has also been claimed that the particles themselves possess some specificity and are able to select the messages which they translate. Opinions have changed over the years, according to results in different systems which seemed to support one or the other of these hypotheses. We can certainly accept today that an absolute specificity does not exist: studies on heterologous systems provided ample proof that ribosomes can translate 'foreign' mRNAs. The efficiency of translation may vary, however, if ribosomes are programmed with different mRNAs, and such quantitative differences may be sufficient to exercise some control of translation *in vivo*, where a preferential attachment of ribosomes to one rather than another message may determine how efficiently the different proteins are synthesised.

Some very active *in vitro* protein synthesising systems have been used for the translation of a number of mRNAs of different origin. Such an active system can be obtained from wheat germ, in which, among others, globin[50], histones[51] and proteins of reovirus[27], and polyoma virus[52] have been synthesised. In reticulocyte lysates, synthesis of myosin and myoglobin could be achieved provided that the specific initiation factors were present[22].

A more complicated situation can be expected if prokaryotic and eukaryotic components are mixed, as the two kinds of ribosomes and the two kinds of mRNAs differ exactly in the structural features which influence their interaction. Prokaryotic 16S rRNA and eukaryotic 18S rRNA contain complementary sequences to the corresponding group of mRNAs only. Furthermore, the capped 5′ termini of eukaryotic mRNAs can attach to eukaryotic ribosomes at the stage of recognition. It is difficult to envisage how components with such different structures can interact with each other; nevertheless, it has been found that some combined systems of this type can function quite efficiently.

Bacteriophage mRNAs have been translated in different mammalian systems and in wheat-germ extracts[53]. Although some nonspecific initiation seems to occur under these conditions, the main products of translation are genuine viral proteins. Legon *et al.*[54] compared the interaction both of eukaryotic ribosomes with prokaryotic message and of prokaryotic ribosomes with eukaryotic message. They found that rabbit reticulocyte ribosomes could attach to the specific initiation sites in bacteriophage f1 mRNA. The ribosomes also protected some nonspecific sites, and the heterogeneity of ribosome binding sites was even greater when 40S rather than 80S particles were used. In contrast to these results, *E. coli* ribosomes failed to protect any specific region in rabbit globin mRNA. The inactivity of some heterologous systems and the reduced specificity of others is not surprising considering the lack of important structural features which normally contribute to the specific interactions between homologous mRNAs and ribosomes. In these heterologous systems, recognition may occur by a different mechanism or may be based on additional structural signals and interactions which play a minor role in the homologous system.

Studies in *in vitro* protein synthesising systems have some disadvantages. On the one hand, nonspecific processes may occur in cell extracts which would not take place in the cell, and on the other, the efficiency of translation achieved is often much lower than in the *in vivo* process. Because of this, we cannot be certain that observations like those described above reflect exactly the situation in the living cell. It is therefore rather important that heterologous translation can also be studied *in vivo*. There are two different ways in which this can be accomplished.

It was discovered some years ago that the oocytes of *X. laevis* could translate faithfully almost any kind of eukaryotic mRNA which had been injected into the oocytes[55]. This system has since been used for assaying different mRNAs, and the oocyte ribosomes have proved very efficient and quite nonspecific in their translation. This points to a lack of specificity in ribosomes *in vivo* also. However, the objection may be made that oocytes, being at an early stage of differentiation, may be expected to be multi-functional and might contain ribosomes which are less specific than those of highly differentiated tissues.

The other approach is still at an early stage, as it makes use of the possibilities opened up by the methods of genetic engineering. It is based on the construction of organisms with new genetic material which can produce new proteins. With the fast progress in the techniques of cloning eukaryotic genes in bacterial cells, the question of whether these genes can actually be expressed in prokaryotic organisms acquired great importance (*see* also p. 105). Methods have been worked out for fusing eukaryotic genes to genes in *E. coli* DNA. It was found that such foreign proteins could be synthesised by the transcription–translation machinery of the bacterial cell[56]. So far production of two yeast enzymes, of an ovalbumin-like protein and of the A and B chains of human insulin, has been achieved in *E. coli.*

Apart from the significance of these results for future possibilities in genetic engineering, its relevance to our present problem lies in the fact that *E. coli* ribosomes can translate eukaryotic messages *in vivo*. These messages, however, have been co-translated with *E. coli* proteins. It has not yet been shown that *E. coli* ribosomes can specifically initiate translation at eukaryotic initiation sites. Further studies may provide more decisive evidence as to whether prokaryotic ribosomes can recognise eukaryotic initiation signals.

All the above studies emphasise the non-specific nature of ribosomes: the particles can indeed attach to a great variety of mRNA molecules. Still, we cannot go as far as to deny the possibility that ribosomal particles can influence the selection of initiation sites. As mentioned before, even though they do not possess an absolute specificity, the particles may interact with different efficiency with the different mRNAs, and this in itself may contribute to the regulation of protein synthesis in the cell. Furthermore, we know of at least a few cases in which selective initiation occurs at one cistron of a polycistronic mRNA, and this selectivity is definitely and entirely due to the structural characteristics of the ribosomal particles. The best example is the different translation of R17 (and MS2 as well as f2) RNA by *E. coli* and by *B. stearothermophilus* ribosomes.

The three cistrons of these RNAs are translated with very different efficiency by both ribosomes. However, while initiation by *E. coli* ribosomes at the synthetase, A-protein and coat protein cistrons occurs at a ratio of 1:5:20, *B. stearothermophilus* ribosomes initiate primarily and almost exclusively at the A-protein cistron (*see* Chapter 5). Comparison of mixed reconstituted systems, containing initiation factors, ribosomal proteins and ribosomal RNAs from *E. coli* or *B. stearothermophilus*, led to the conclusion that the component responsible for the different initiation properties was present in the ribosomal protein fraction. Eventually, Isono and Isono[57] detected the only difference between the two kinds of ribosomes: the *Bacillus* ribosomes did not contain protein S1. These workers also showed that addition of this protein to a *B. stearothermophilus* system altered the selection of initiation sites: it enhanced the translation of coat protein and synthetase with the result that the three proteins initiated in similar proportions as in *E. coli*.

We do not know how protein S1 influences the selection of initiation sites. It is known that this ribosomal protein is multi-functional (*see* Chapter 6), but none of its known functions explains this effect. Steitz *et al.*[8] made an observation which may be relevant to this phenomenon. *E. coli* ribosome depleted of S1 protein also attached to R17 RNA at nonspecific sites, while in the presence of S1 protein such nonspecific interactions seemed to be suppressed. It is thus possible that this ribosomal protein plays a role not only in the binding of the particles to mRNA but also in the recognition of the specific sites where this binding occurs.

(d) Interaction of ribosomes with the capped termini of eukaryotic mRNAs

The difference in the length of RNA protected in the 40S and the 80S initiation complexes suggests that the additional sequences which participate in the formation of the first complex are involved in an early event, probably in the recognition of the initiation site. These additional sequences include in many cases (but not always) the 5′-terminal caps (*table 5.3*) which would support the view that interaction of ribosomes with these structures forms part of the recognition process in eukaryotic systems. As this interaction does not occur with all eukaryotic mRNAs, it can be considered as one of several alternative mechanisms or as a process which enhances, but is not absolutely required for, initiation complex formation.

Several eukaryotic mRNAs contain a relatively short (up to about 50 nucleotides long) leader sequence which contains no AUG sequence. It is tempting to envisage here a recognition mechanism by which interaction with the capped 5′

termini helps the ribosome to find the first AUG triplet where it can bind and form an active initiation complex. Such a mechanism has been proposed by Kozak and Shatkin[78], who compared initiation complex formation on different length fragments from the 5′-terminal parts of reovirus mRNAs. An absolute requirement for the AUG initiation codon was observed. The length of the fragment did not influence the ribosome binding capacity, while removal of the cap structure reduced but did not abolish ribosome binding. These data are in agreement with a two-step model of initiation: the 40S ribosome may bind first to the 5′ terminus of the mRNA, and then move along until it encounters the first AUG triplet. At this stage and at this site, the 60S subunit may join; thus the active 80S complex is formed and may initiate translation. Experiments in the presence of edeine, an antibiotic which prevents joining of the 60S subunit, provided further support for this hypothesis: several 40S particles were bound under these conditions to multiple sites on the mRNA. The particles seemed to have a common entry site at the 5′ terminus of the mRNA. Kozak and Shatkin assumed that the particles moved along the mRNA from the 5′ terminus, a phenomenon which may be part of the normal initiation process. This model explains several observations about ribosome binding sites on eukaryotic mRNAs; it does not agree, however, with all the known data. It implies that the leader sequence does not contain non-initiator AUG triplets, a requisite which may not be fulfilled in every eukaryotic mRNA. There is in fact evidence for AUG codons in leader sequences and probably not all of these are used as initiation sites. This may mean that the above model has only limited validity or that the 40S ribosomes, sliding along the mRNA, recognise not merely the AUG codon but also an additional signal which is present at the genuine initiation sites.

Related observations have been made by Legon[82] who found 40S subunits at several sites in the leader region of polyoma mRNA as well as 80S ribosomes at the initiation site. Such a separation of a 40S binding site from 80S initiation site similarly implies some sort of movement of the ribosomal subunit from the leader sequence to the initiation codon.

A two-step model, which involves the interaction of the 40S subunit with the 5′ terminus and the formation of the 80S initiation complex at the nearest genuine initiation site, has further implications. It may explain the puzzling phenomenon that in eukaryotic mRNAs internal initiation sites are not recognised[59]. mRNAs from several eukaryotic viruses carry messages for more than one protein, but–in contrast to the translation of prokaryotic polycistronic mRNAs–here only one message is translated, that which starts nearest to the leader sequence. The other proteins are synthesised on smaller spliced, processed mRNAs in which a leader sequence is attached to the coding region near the initiation site. Translation of the late mRNAs of SV40 and polyoma viruses provide good examples of this mechanism.

The organisation of the genome is very similar in these viruses (*see figures 1.1* and *3.10*). In both viruses three spliced mRNA molecules are produced from the part of the genome which codes for the proteins VP1, VP2 and VP3. The largest mRNA molecule, 19S RNA, contains information for protein VP1 and VP3 as well as for protein VP2, but the VP1 and VP3 messages start at internal initiation sites. In an *in vitro* translation system, Smith *et al.*[52,59] found that only VP2 was translated from the 19S mRNA of polyomavirus; the internal initiation sites were not recognised. Translation of the VP1 and VP3 messages occurs from smaller, 16S and 18S spliced mRNAs, respectively. In these mRNAs the initiation site of the protein which is translated, is adjacent to the leader sequence.

It is too early to draw far-reaching conclusions, but these observations seem to be in good agreement with a translation mechanism which can initiate only at the initiation site nearest to the leader sequence.

(e) The influence of mRNA conformation on the accessibility of initiation sites

The secondary and tertiary structure of mRNA may have a positive and a negative effect on initiation. As to the positive effect, proteins interacting with mRNA may recognise a specific three-dimensional structure at the initiation site rather than merely a nucleotide sequence. We have no direct evidence for specific secondary structures being formed at these sites; the assumption is based on the general property of proteins that they do often recognise a definite spatial arrangement, and on an analogy with specific sites in DNA which interact with proteins: they often contain symmetrical sequences which may form an unusual looped-out secondary structure, such as the sequences around the origin of replication (p. 32) or sequences in the operator region (p. 95).

Much more evidence is available on the negative control function exercised by the secondary and tertiary structure of mRNA. Chapter 5 discussed in detail how initiation sites may be made inaccessible to ribosomes by being involved in a strong secondary structure or by being buried in the tertiary structure. The conformation of MS2 RNA is responsible for the different efficiencies with which *E. coli* ribosomes can attach to the three initiation sites: the coat protein cistron is easily accessible to ribosomes, while the elaborate tertiary structure hinders access to the other two initiation sites. This may be a general mechanism to control ribosome binding to different sites in mRNA. It may be valid for many mRNAs and may result not only in the determination of the efficiency of different initiation sites, but also the same mechanism may ensure that ribosomes do not attach to nonspecific sites on an mRNA. Discrimination between genuine initiation sequences and the numerous other sequences in the same molecule which could potentially bind ribosomes (*see figure 8.5*) may be brought about by specific strong folding of the RNA, resulting in a tertiary structure in which all nonspecific sequences are inaccessible. There is some evidence in favour of this hypothesis: if MS2 or R17 RNA are mildly denatured, initiation complex formation becomes more efficient but ribosome-protected fragments can be isolated in small amounts which do not correspond to any of the three specific initiation sites. The inactivity of some internal initiation sites, mentioned above, might also be explained in this way.

It is important to remember, however, that the conformation of RNA molecules, especially that of large molecules, is not rigid, but can undergo changes upon interaction with other RNAs or with protein molecules. Hairpin loops can be unwound by protein S1 at the site of ribosome binding; this reaction may even be a necessary requisite for stable attachment of ribosomes to mRNA. Conformational changes are also expected to occur at later stages of translation: in elongation, a folded RNA structure would prevent the ribosomes from moving along the molecule. A disruption of the secondary structure very probably occurs concomitantly with ribosome movement. Whether this is also brought about by protein S1 or by some other mechanism is still not known. How the conformation of R17 RNA is changed during translation of the coat protein cistron, and how this change enables ribosomes to initiate also at the synthetase cistron, are described on p. 174.

Another step in initiation where conformational changes are involved has also been described earlier. In different eukaryotic mRNAs, the 5′-terminal caps and the sequences complementary to 18S rRNA are present at varying distances from the initiation codon. At the earliest stage of initiation, all three sites may take part in the formation of the 40S initiation complex which implies that, depending on the structure of the leader sequence, the mRNA may be present in different folded structures. Attachment of the 60S subunit may be accompanied by the opening up of these structures (p. 175–178). However, too little is known today of the structure of eukaryotic mRNAs and eukaryotic initiation complexes to decide whether this folding and unfolding of mRNA during initiation has a role in controlling the specificity of the process.

An important structural characteristic of eukaryotic mRNAs which has also been assumed to be connected with the control of initiation, is their association with proteins. Both the mature mRNAs in the cytoplasm and their precursors in the nucleus bind specific proteins. In the rabbit globin messenger ribonucleoprotein complex, two main protein components can be seen under the electron microscope attached to the two ends of the RNA molecule. It has been shown in the case of globin mRNA as well as other mRNAs that the protein components present in the cytoplasmic complex are different from the nuclear proteins attached to the RNA precursor[60]. In spite of many attempts to identify the proteins associated with mRNA[61], so far we have no evidence as to what their function may be. Although it has been claimed that some of these proteins show a similarity to eukaryotic initiation factors and might function as such, and also that the ribonucleoprotein structure increases the stability of mRNA, or preserves a definite tertiary structure, the field has not yet been sufficiently explored to support any of these assumptions with direct evidence.

(f) Special ways of control

The interactions described above represent general mechanisms which, together, contribute in a well balanced way to the control of initiation of protein synthesis. With different mRNAs, in different protein synthesising systems, one or the other of these controlling factors may have a predominant role but neither of them could be responsible, alone, for the specificity of initiation. It seems probable that it is the result of a well balanced effect of all these interactions that: (a) no faulty initiations occur at nonspecific initiation sites and (b) the efficiency of initiation at genuine initiation sites is under a flexible translational control.

There is a possibility that in bacterial protein synthesis, where transcription and translation occur in a coupled process, a different control mechanism may be functioning. As ribosomes attach to the mRNA when only a short stretch of it has been synthesised, the selection of the genuine initiator codon may present a simpler problem and may be controlled differently from a system where the ribosomes have to recognise the initiation sites within a complete large RNA molecule. Too little is known, however, about the coupled transcription–translation process to allow even speculation as to the control mechanisms which may be at work.

In addition to the general interactions described above which control the translation of many mRNAs, there are also different special mechanisms in the different cells which regulate the synthesis of one specific protein. Without going into the details of the many different regulatory mechanisms, we will briefly discuss one principle of regulation which shows some similarity to the processes already described: the blocking of an initiation site by a specific protein. This same principle is also used in the control of the synthesis of other proteins.

In the reproduction of RNA bacteriophages, the synthesis of replicase is of vital importance in the early stages, as this virus-induced protein is required for the synthesis of the viral RNA which in turn directs the synthesis of the viral proteins. Initiation at the replicase cistron is therefore made possible in the way described on p. 174. At a later phase of infection, however, preferential synthesis of the coat protein is required, as several molecules of this protein are used to encapsulate one viral RNA molecule. RNA synthesis has to slow down and in order to achieve this, replicase synthesis is blocked at this stage. It has been shown that coat protein can bind to R17 RNA at a position close to the replicase initiation site[62]. Binding of coat protein thus prevents ribosomes from initiating at the replicase cistron.

ELONGATION

The addition of each amino acid to the growing polypeptide chain occurs via a three-step mechanism: (1) the aminoacyl-tRNA, specified by the codon in the A-site, binds to the ribosome; (2) the amino group of this amino acid is linked in a peptide bond to the growing C-terminus of the polypeptide chain; (3) the ribosome moves along to the next codon and the different components attached to the particle are rearranged so as to make it ready to accept the next incoming aminoacyl-tRNA. Essentially the same three steps take place in prokaryotic and eukaryotic organisms. Elongation factors are necessary in both systems to carry out these reactions: for the first step elongation factors EF-Tu and EF-Ts (also called transfer factors) are required in prokaryotes and elongation factor EF-1 in eukaryotic organisms. The third step is brought about by elongation factor EF-G in the prokaryotic, EF-2 in the eukaryotic system. Two molecules of GTP are bound and hydrolysed in the course of the complete process.

First, the mechanism of these steps in prokaryotic organisms will be described. The data refer mostly to *E. coli* but are probably valid for a great number of prokaryotic organisms.

Elongation of the polypeptide chain in prokaryotes

(a) Binding of aminoacyl-tRNA to the ribosome

The aminoacyl-tRNA interacts with the ribosome in the form of a ternary complex comprised of EF-Tu, GTP and aa-tRNA. The EF-Tu·GTP complex is formed first (the reactions leading to its formation will be described later). This complex can bind rather tightly any aa-tRNA with the exception of initiator tRNA. Neither formylated nor unformylated met-$tRNA_F^{met}$ reacts with EF-Tu·GTP[63], a specificity which was noticed in early studies[64] and which prevents any mistakes in translation which might occur from misreading an internal AUG codon for an initiator codon. The tRNAs must be charged to enter the ternary complex and the amino acids they carry must not be acylated. This specificity ensures that only aa-tRNAs which can produce peptide bonds should bind to the ribosomes.

Several events occur upon interaction of the EF-Tu·GTP·aa-tRNA complex with the programmed ribosome. The anticodon of the aa-tRNA anneals to the corresponding codon, and in consequence of this interaction, the tRNA molecule undergoes a conformational change which releases its TψCG arm and allows this sequence to anneal to the 5S ribosomal RNA which is present in the A-site and which contains a complementary CGAA sequence (*see* p. 190–191). The necessity for accurate codon–anticodon interaction for the stable binding of aa-tRNA to the ribosome represents a safeguard against mis-translation: aa-tRNAs which do not correspond to the exposed codon can be rejected. With the help of this proof-reading mechanism, translation of the message is carried out with very great precision: mistakes in decoding are estimated at 10^{-4} at most (*see* also Chapter 7). The aa-tRNA is further stabilised in the A-site by interaction with ribosomal proteins, primarily with protein L16.

GTP is hydrolysed concomitantly with aa-tRNA binding. Protein L7/L12, present in the ribosomal A-site, seems to interact with EF-Tu to produce GTPase activity[83]. The resulting EF-Tu·GDP complex is released from the ribosome. GTP hydrolysis is required for the removal of the elongation factor. If GTP is replaced by the unhydrolysable $GMPP(CH_2)P$, the release of EF-Tu becomes very slow and its presence prevents further reactions on the ribosome. The EF-Tu·GDP complex is very stable, GDP binds much more tightly to EF-Tu than does GTP, with a dissociation constant two orders of magnitude lower. GDP is thus not released; the active complex is restored by recycling with the aid of the other transfer factor, EF-Ts. (In these abbreviations, T refers to the name 'transfer factor' of these two elongation factors, u and s reflect the heat-unstable and heat-stable properties of the two proteins.) EF-Ts catalyses a GDP–GTP exchange:

EF-Tu·GDP + EF-Ts → EF-Tu·EF-Ts + GDP
EF-Tu·EF-Ts + GTP → EF-Tu·GTP + EF-Ts

The EF-Tu·GTP complex is thus re-formed, and can bind another aa-tRNA.

(b) Formation of the peptide bond

The reactions we have followed so far on the *E. coli* ribosome leave us with a particle to which fmet-tRNA$_F^{met}$ is bound at the P-site and an aa-tRNA (in *figure 8.6*, val-tRNA^{val}) at the A-site. The charged CCA ends of the tRNAs are in contact

1. The A-site (shaded area) is free in the 70S initiation complex. GUC codon is present at this site

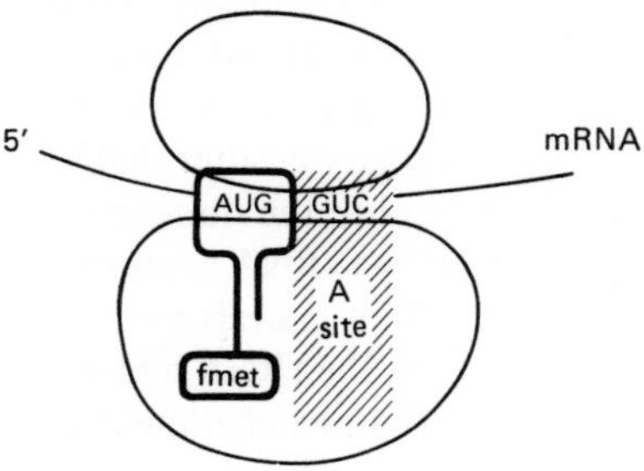

2. val–tRNA^{val} binds to A-site, by codon–anticodon annealing and interaction with ribosomal proteins. (EF–Tu involved)

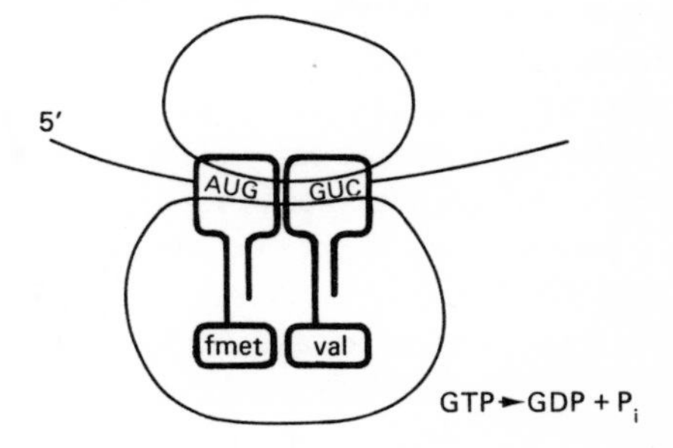

3. Peptide bond formation: peptidyl transferase produces fmet–val–tRNA^{val} (in A-site). Deacylated tRNA$_F^{met}$ remains in P-site

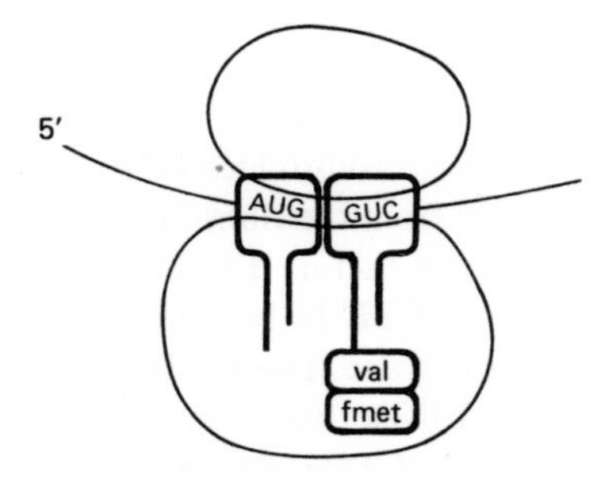

4. Translocation: deacylated initiator tRNA is replaced by peptidyl–tRNA in P-site. The A-site is free to accept the next aminoacyl–tRNA. Ribosome moved one codon further on mRNA. UUU codon is now exposed at the A-site.

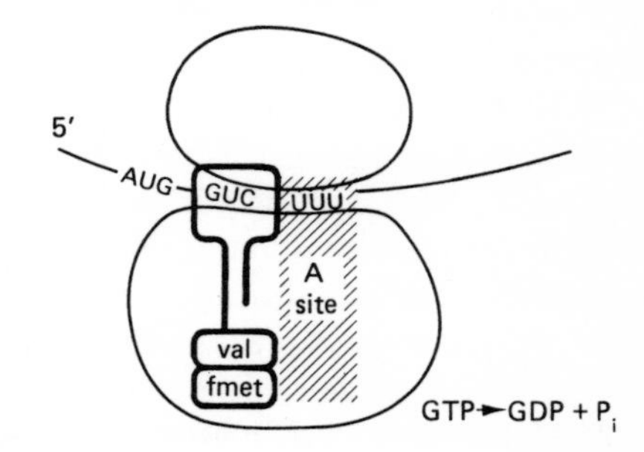

Figure 8.6 Schematic representation of the events occurring on the ribosome in the process of elongation. Steps 1 to 4 lead to the incorporation of one amino acid into the peptide chain.

with the peptidyl transferase centre, a ribosomal domain on the 50S subunit, which comprises, among others proteins L3, L11, L16, L18, L20 and a stretch of the 23S RNA[10] (*see* also p. 202–203). The tRNAs are bound to the particle in a way which ensures that the reacting ends are in the optimal stereochemical position for

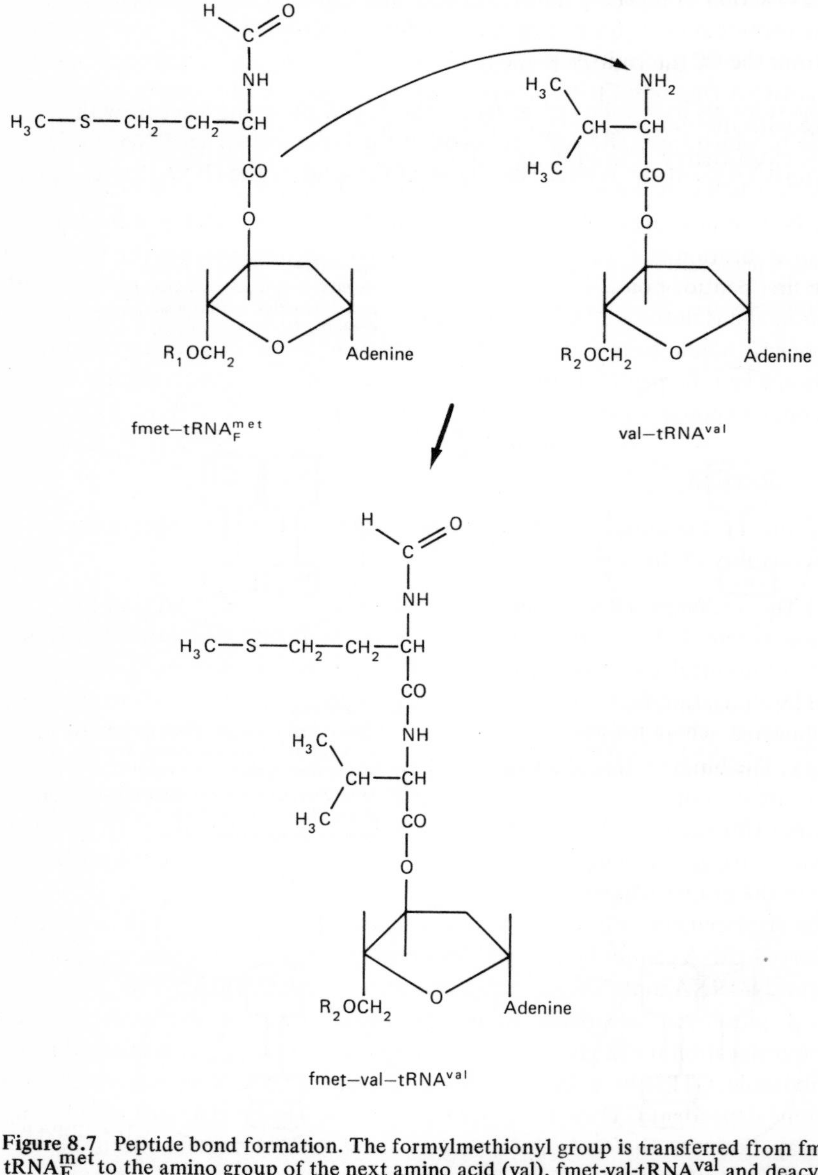

fmet–$tRNA_F^{met}$ val–$tRNA^{val}$

fmet–val–$tRNA^{val}$

Figure 8.7 Peptide bond formation. The formylmethionyl group is transferred from fmet-$tRNA_F^{met}$ to the amino group of the next amino acid (val). fmet-val-$tRNA^{val}$ and deacylated $tRNA_F^{met}$ are the reaction products. In the next step the fmet-valyl group will be transferred to the amino group of the third amino acid. R_1 = rest of $tRNA_F^{met}$ molecule; R_2 = rest of $tRNA^{val}$ molecule.

the transfer reaction. The catalytic activity which brings about peptide transfer is an inherent property of the peptidyl transferase centre, and no extra-ribosomal factors are required for it. No single ribosomal protein could be identified as possessing this enzyme activity, and the reaction may thus be the result of the co-operative action of several proteins.

The peptide bond is formed by transferring the carboxyl end of formylmethionine from the CCA terminus of the initiator tRNA to the amino group of valine in valyl-tRNA (*figure 8.7*). As the carboxyl group was already present in ester linkage with the ribose of the terminal adenosine, the transfer reaction does not require extra energy. The product of the reaction is formylmethionyl-valyl-$tRNA^{val}$, which is at this stage situated at the A-site, while the initiator tRNA in the P-site is now uncharged. The diagram in *figure 8.6* shows the individual steps in elongation and the situation on the ribosomal particle after the formation of the first peptide bond. This is very different from the state of the ribosome after initiation and is unsuitable to accept a new aa-tRNA. In the next step the different components bound to the ribosome have to be rearranged so as to recreate the situation when the peptidyl-tRNA is in the P-site, and the A-site is empty, with a new codon exposed, to allow entry of the next aa-tRNA.

(c) Translocation

This is the most complex step in elongation, in which several changes occur simultaneously on the ribosome.

(1) The uncharged tRNA, which served as donor in the peptidyl transfer reaction, is removed from the P-site and its place is taken by the peptidyl-tRNA which is thus displaced from the A-site to the P-site. These changes are brought about by elongation factor EF-G. This factor combines with GTP before attaching to the ribosome, where it binds to protein L7/L12 and to a few other proteins in its vicinity. The binding sites of EF-G and of transfer factor EF-Tu overlap, both factors are in contact with L7/L12 and with some further, different ribosomal proteins. This excludes the possibility of simultaneous binding of the two factors and may serve as a safeguard that the individual steps in elongation follow each other in the proper sequence.

The displacement of deacylated tRNA by peptidyl-tRNA occurs in one reaction. Uncharged tRNA cannot be removed independently: no reaction occurs if there is no peptidyl-tRNA in the A-site[65]. Translocation of peptidyl-tRNA can be followed by its puromycin sensitivity: while in the A-site, it cannot react with puromycin, after translocation it can give rise to a puromycin–peptide which is released from the ribosome. GTP is hydrolysed during translocation, a reaction catalysed by the ribosome-dependent GTPase activity of EF-G factor. As judged from the puromycin assay, however, this hydrolysis is not required for translocation itself, but rather for the release and cyclic re-use of the elongation factor.

(2) Concomitantly with the rearrangement of the tRNAs bound to the ribosome, the particle moves along the mRNA to the next codon, which thus becomes

exposed in the A-site. The movement of the ribosome in the 5′ → 3′ direction could be directly demonstrated by isolating and sequencing the mRNA stretch protected by the particle before translocation and after the addition of EF-G and GTP. Gupta *et al.*[66] showed that after translocation the protected stretch included three more nucleotides at its 3′ end.

The direction of ribosome movement, which is also clear from the above experimental result, has, in fact, been known for some time. Even before the individual steps in elongation had been elucidated, Thach *et al.*[67] studied the translation of synthetic hexanucleotides into dipeptides. This work was undertaken with the aim of deciphering the genetic code, but comparison of the structure of oligonucleotides with the peptides synthesised also proved unequivocally that the message is read in the 5′ → 3′ direction. The direction of synthesis of the polypeptide chain was also independently determined well before the mechanism of peptidyl transfer was known. Dintzis[68], following the amounts and kinetics of incorporation of radioactive amino acids into different parts of nascent polypeptide chains, came to the conclusion that synthesis starts at the N-terminus and proceeds in the direction towards the C-terminus.

Today, when nucleic acid sequences and protein sequences can be compared and fragments of mRNAs can be translated in *in vitro* systems, we have, of course, ample evidence which supports these early observations. The nature of the peptidyl transfer reaction and the N-formylated structure of the initiating amino acid, are in agreement only with a synthesis of proteins from the N-terminus to the C-terminus. The structure of mRNAs, and the position of initiation and termination sites, also provide unequivocal proof that ribosomes move along and read the message in the 5′ → 3′ direction. The fact that translation starts near the 5′ terminus of mRNA makes it possible for ribosomes in bacterial cells to attach to a nascent mRNA when only a small part of it has been synthesised. In transcription, the mRNA is synthesised from its 5′ terminus and as soon as the first initiation site becomes available, this can bind a ribosomal particle, and an initiation complex can be formed. As the direction of transcription and translation is the same on the mRNA, the ribosome can move along the nascent RNA molecule and translate it simultaneously with its transcription. Polysomes (p. 272) can also be formed on these nascent RNA chains, and this coupled transcription–translation process can thus be well observed in the electron microscope. *Figure 3.4* shows an electron micrograph of *E. coli* DNA in the process of transcription, with nascent RNA molecules attached and with a number of ribosomes bound to these partly synthesised RNAs.

We thus know exactly at which stage and in which direction the ribosome moves along the mRNA, but it is still not known how this movement is accomplished. It is not even clear whether an active movement of the ribosomal particle is involved or if it is rather the mRNA which plays an active role when migrating through the tunnel in the ribosome. Whichever the case, structural changes are expected to occur in several components of the system which eventually result in displacing the ribosome by exactly three nucleotides on the mRNA. Different hypotheses have been put forward attributing this movement to conformational changes in different components. Changes in the conformation of mRNA[69], tRNA[70], the ribosomal particles themselves[71] and the RNA within the ribosomal particle[72] have been assumed to be responsible for moving the ribosome forward by one codon. None of the proposed mechanisms goes beyond the stage of speculation. With our recently acquired knowledge of the tertiary structure of tRNAs, it should be possible to reinvestigate the idea suggested years ago by Woese[70] that a conformational change in the anticodon loop may 'pull' the annealed codon to a different site on the ribosome. Even if the actual structural changes involved proved to be different from those proposed in 1970, it seems probable that tRNA has a central role in controlling the step-by-step movement of mRNA relative to the ribosome. A role of tRNA in maintaining the reading frame has been substantiated by Riddle and Carbon[73], who detected a frameshift suppressor mutant which produces a $tRNA^{gly}$ species which has four instead of three G residues at its anticodon site. As frameshift mutations arise by insertion (or deletion) of a nucleotide which thus disrupts the triplet reading frame, suppression of this mutation by a four-letter anticodon suggests that in this case annealing of tetranucleotides in tRNA and mRNA results in pulling the mRNA four nucleotides further through the ribosome, thus re-establishing the original reading frame. This would imply that normally the tRNA is responsible for the three-nucleotide steps on the mRNA.

Weidner *et al.*[72] describe a possible conformational change in 5S RNA which, in connection with the above role of tRNA, would give a more complete model of the structural changes involved in translocation. They propose that 5S RNA can exist in two stable conformations (*see* p. 192–193 and *figure 6.3*), and that a switch from one to the other might be mediated by elongation factor EF-G. The main difference between the two conformations is that in one the site where tRNA anneals is in close proximity to the site where 5S RNA–23S RNA annealing occurs, while in the other these two sites are much further apart. A change in 5S RNA conformation could thus cause a movement of tRNA relative to other ribosomal components, and the movement of tRNA on the ribosome could in turn cause the migration of mRNA through the particle.

With the completion of the translocation step the ribosome is ready to start a new cycle of accepting another aa-tRNA and incorporating another amino acid into the polypeptide chain. The A-site is empty, with the next codon exposed; the P-site is occupied by peptidyl-tRNA. The same three steps occur which have been described above (*figure 8.6*): the incoming aa-tRNA is bound in the A-site, and the fmet-val residue is transferred to the amino group of the third amino acid, forming a tripeptidyl-tRNA which will in turn be translocated into the P-site so that a cyclic repetition of these steps can go on until a termination codon is reached in the message.

The nascent polypeptide chain—or at least a part of it, up to about 30 amino acids from the growing C-terminus—remains attached to ribosomal proteins and to 23S RNA. Affinity labelling experiments showed that proteins L2, L27 and L32 or L33 and L27, form a groove which accommodates the nascent peptide and that 23S RNA is also present along the site where the peptide chain is in contact with the ribosome (*see* p. 202).

When the ribosome engaged in elongation has moved far enough away from the initiation site, a new ribosome can bind to this site and can start a new polypeptide chain. Several ribosomes can translate the same message simultaneously, moving synchronously along the mRNA, at regular short distances from each other. The message can be practically covered by ribosomes, the first one completing, the last one just starting translation. These structures, which can be clearly visualised in the electron microscope, are called *polysomes. Figure 8.4* shows an electron microphotograph taken by McKnight *et al.* of polysomes in a eukaryotic system; a large number of silk gland ribosomes are simultaneously translating silk fibroin mRNA. The size of polysomes, i.e. the number of ribosomes attached to the mRNA, depends on the length of the message and on the efficiency of translation. Polysomes were first detected in the early 'sixties, both in eukaryotic and in prokaryotic organisms. In prokaryotes, where transcription and translation occur simultaneously, the formation of polysomes on the nascent RNA chains can be well observed in the electron microscope, as shown in *figure 3.4.* As the length of the RNA stretches increases the farther they are from the promoter site, so also does the number of ribosomes which are attached to them.

Elongation factors in eukaryotes

The three main steps in elongation seem to occur essentially in the same way in prokaryotic and eukaryotic organisms. The protein factors involved in the process are, however, different[3,12].

From a number of mammalian tissues, only one transfer factor, EF-1, could be isolated, but this protein was present in the cells both in monomeric form and in the form of aggregates. The aggregates (heavy form, EF-1_H) are of very high molecular weight, in the 300 000 to 1 000 000 range, and often also contain lipids. The light form, EF-1_L, isolated from different sources, has a MW of about 50 000. Both forms, and even intermediates like dimers, tetramers and hexamers of EF-1, react with GTP, but only the monomer forms a stable ternary complex with GTP and aa-tRNA. It seems that GTP may disaggregate the heavy form of the protein into monomers which then form the ternary complex:

$$\text{EF-1}\cdot\text{GTP} + \text{aa-tRNA} \rightleftarrows \text{aa-tRNA}\cdot\text{EF} - 1\cdot\text{GTP}$$

The ternary complex reacts with the ribosome, with the result that the aa-tRNA attaches to the ribosomal A-site and is fixed in this position by annealing to 5.8S RNA (*see* p. 197) and by binding to as yet unidentified ribosomal proteins. GTP is hydrolysed and EF-1·GDP released from the particle.

A further difference between EF-1 and the prokaryotic transfer factor EF-Tu is that the former has a slightly higher affinity for GTP than for GDP. This probably makes recycling of the factor unnecessary. Factors which stimulate EF-1 activity have been isolated from pig liver and from reticulocytes, but it is not yet known whether these factors have any role in the recycling of EF-1 and in restoring the EF-1·GTP complex.

The mechanism of translocation does not differ in prokaryotes and eukaryotes. The eukaryotic elongation factor, EF-2, carries out the same functions as does EF-G in *E. coli*.

TERMINATION OF TRANSLATION AND RELEASE OF THE POLYPEPTIDE CHAIN

Under normal conditions, no tRNAs exist in the cell which would react with any of the three nonsense codons, UAA, UAG, UGA. These codons serve as termination signals, one, or in some cases two of them (p. 159) following the codon for the last amino acid in the message. Their function as termination codons has been investigated by genetic techniques: mutations which convert a codon at an internal site into one of these nonsense triplets (nonsense mutations), cause early termination of translation. Nonsense suppressor mutants, which can synthesise a protein in spite of a nonsense mutation in the message, have been found to produce tRNA species which in response to a nonsense codon insert an amino acid into the polypeptide chain. These suppressor tRNAs are derived from the wild-type species, but contain a mutated base either in the anticodon which enables them to anneal to a nonsense codon, or at another part of the molecule which, in a poorly understood, indirect way, alters the interaction between the anticodon loop and the codon in the message (*see* p. 236).

Chain termination can occur only after translocation has taken place. The peptidyl-tRNA must be in the P-site and the termination codon exposed in the A-site. The actual reaction, which is probably carried out by the peptidyl transferase, is the hydrolysis of the bond between the tRNA and the carboxyl end of the polypeptide chain. Extra-ribosomal protein factors, the release factors, are required for this process; they may be responsible for affecting the proteins at the peptidyl transferase centre so as to alter their specificity, making a hydrolytic reaction instead of a transfer reaction possible.

Release factors have been isolated from prokaryotic as well as eukaryotic organisms[74]. In the former, two codon-specific factors, RF-1 and RF-2, are present: RF-1 responds to the termination codons UAA and UAG, RF-2 to UAA and UGA. A third release factor, RF-3, has been detected in *E. coli* and can stimulate the binding of RF-1 and RF-2 to ribosomes. It is not certain whether and in which way this factor participates in the process of termination. The eukaryotic factor, RF, is nonspecific, and can recognise all three termination codons. Release factors bind to the A-site of the ribosome. In suppressor mutants, they compete for this site with the suppressor tRNA. The binding and release of the factors may be accompanied by the binding and hydrolysis of GTP, but the exact details of this process are not clear. The binding site of release factors on the *E. coli* ribosome includes protein L7/L12, as do the binding sites of the elongation factors, but they also include L2 and L11. The overlapping binding sites prevent simultaneous attachment of elongation and release factors to the ribosome.

Hydrolysis of the bond between the polypeptide and tRNA releases the nascent protein which, as will be discussed below, may still undergo post-translational modifications. There is some controversy with regard to further events which take place on the ribosomal particles. The polypeptide chain itself has been in contact with ribosomal proteins and 23S RNA. It is not known how it is released from this interaction. Neither do we know how the deacylated tRNA is removed from the particle. The most controversial question is, in which form are the ribosomes released from the mRNA? Free subunits are present in the cytoplasm and they are needed for initiation of translation. Elongation and termination of a polypeptide chain are carried out by ribosomal particles in their associated form. Are the particles released in this form or is their release coupled with a dissociation process? In this latter case, free subunits would be made available for initiation and the dissociation factor would have a completely passive function: to prevent re-association of the subunits. In the former case, dissociation of ribosomes should precede the formation of an initiation complex. The dissociation factor activity of initiation factors might still play a passive role: it could shift the equilibrium towards the dissociated state[3,14]. This seems today the more probable sequence of events, although unequivocal proof for the release of undissociated ribosomes is not available.

MODIFICATIONS OF THE NASCENT PROTEIN

It follows from the mechanism of initiation of translation that nascent proteins carry a methionine residue at the N-terminus; in prokaryotes this methionine is formylated. One of the early modifications which occur in eukaryotes is the removal of the N-terminal methionine, which is cleaved off when the nascent chain is only 15–30 residues long[75]. In prokaryotes, specific deformylases remove the formyl group from the completed nascent protein. This is often, but not always, followed by cleavage of the methionine residue. Still, in *E. coli* more than 30 per cent of the proteins retain a methionine at their N-termini.

A more specialised type of post-translational modification—or, in fact, modification which may occur concurrently with translation—takes place in the synthesis of some viral proteins. Although eukaryotic mRNAs are usually monocistronic, some eukaryotic viruses produce mRNAs which, in one message, contain the information for a number of viral proteins. The message is translated into one long polypeptide chain which, on proteolytic cleavage, yields the mature viral proteins. In encephalomyocarditis virus, the viral RNA itself serves as messenger. Its MW is 2.5×10^6 which corresponds to about 8000 nucleotide residues. It is translated uninterruptedly by ribosomes, but proteolytic cleavage of the nascent polypeptide chain is already occurring during the translation process. When translation is terminated, not one but three protein molecules have been released[76].

The largest of these is subject to further proteolysis, to produce eventually a further four viral proteins. The production of polyproteins from one message also occurs in other viruses. The involvement of proteolytic cleavage in the formation of mature viral proteins is also common in bacteriophages.

Nascent proteins and the final native protein products often differ very considerably in their structure. Many proteins undergo extensive post-translational modifications before a functionally active structure is produced. These structural changes, however, do not follow a general pattern but depend on the individual properties of each protein. The study of the rather diverse field of these post-transcriptional modifications is outside the scope of this book, which was intended to deal with topics directly connected with the transfer of genetic information from DNA to protein. This process has been completed when the nascent protein, expressing the same information as had been encoded in the DNA, is released from the ribosome.

REFERENCES

1 Wahba, A. J., Chae, Y. B., Iwasaki, K., Mazunder, R., Miller, M. J., Sabol, S. and Sillero, M. A. G. *Cold Spring Harbor Symp.*, **34**, 285 (1969).
Wahba, A. J., Iwasaki, K., Miller, M. J., Sabol, S., Sillero, M. A. G. and Vasquez, C. *Cold Spring Harbor Symp.*, **34**, 291 (1969).
2 Wahba, A. J. and Miller, M. J. *Methods in Enzymology*, **30F**, 3 (1974).
Lee-Huang, S. and Ochoa, S. *Methods in Enzymology*, **30F**, 31 (1974).
3 Grunberg-Manago, M. and Gros, F. *Progr. Nucl. Acid Res. Mol. Biol.*, **20**, 209 (1977).
4 Lee-Huang, S. and Ochoa, S. *Methods in Enzymology*, **30F**, 45 (1974).
5 Brauer, D. and Wittmann-Liebold, B. *FEBS Lett.*, **79**, 269 (1977).
6 Revel, M., Herzberg, M. and Greenshpan, H. *Cold Spring Harbor Symp.*, **34**, 261 (1969).
7 Dondon, J., Godefroy-Colburn, T., Graffe, M. and Grunberg-Manago, M. *FEBS Lett.*, **45**, 82 (1974).
8 Steitz, J. A., Wahba, A. J., Laughrea, M. and Moore, P. B. *Nucl. Acid Res.*, **4**, 1 (1977).
9 Subramanian, A. R. and Davis, B. D. *Nature*, **228**, 1273 (1970).
Sabol, S., Sillero, M. A. G., Iwasaki, K. and Ochoa, S. *Nature*, **228**, 1269 (1970).
10 Stöffler, G. and Wittmann, H. G. in *Molecular Mechanisms of Protein Biosynthesis* (eds Weissbach, H. and Pestka, S.) Academic Press, New York, 117 (1977).
11 Van Duin, J., Kurland, C. G., Dondon, J., Grunberg-Manago, M., Branlant, C. and Ebel, J. P. *FEBS Lett.*, **62**, 111 (1976).
12 Revel, M. in *Molecular Mechanisms of Protein Biosynthesis* (eds Weissbach, H. and Pestka, S.) Academic Press, New York, 24 (1977).
13 Rudland, P. S., Whybrow, W. A., Marcker, K. A. and Clark, B. F. C. *Nature*, **222**, 750 (1969).
14 Godefroy-Colburn, T., Wolfe, A. D., Dondon, J., Grunberg-Manago, M., Dessen, P. and Pantaloni, D. *J. Mol. Biol.* **94**, 461 (1975).
15 Fakunding, J. L. and Hershey, J. W. B. *J. Biol. Chem.* **248**, 4206 (1973).

16 Vermeer, C., Van Alphen, W. J., Knippenberg, P. and Bosch, L. *Eur. J. Biochem.*, **40**, 295 (1973).
17 Jay, G. and Kaempfer, R. *PNAS*, **71**, 3199 (1974).
18 Benne, R., Ebes, F. and Voorma, H. O. *Eur. J. Biochem.*, **38**, 265 (1973).
19 Dahlberg, J. E., Kintner, C. and Lund, E. *PNAS*, **75**, 1071 (1978).
20 Collatz, E., Kuchler, E., Stöffler, G. and Czernilofsky, A. P. *FEBS Lett.*, **63**, 283 (1976).
21 Staehelin, T., Trachsel, H., Erni, B., Boschetti, A. and Schreier, M. H. in *Proceedings of the 10th FEBS Meeting* (eds Chapeville, F. and Grunberg-Manago, M.) North Holland/American Elsevier, New York, 309 (1975).
22 Heywood, S. M., Kennedy, D. S. and Bester, A. J. *PNAS*, **71**, 2428 (1974).
23 Marcus, A., Seal, S. N. and Weeks, D. P. *Methods in Enzymology*, **30F**, 94 (1974).
23a Benne, R. and Hershey, J. W. B. *J. Biol. Chem.*, **253**, 3078 (1978).
24 Lubsen, N. H. and Davis, B. D. *PNAS*, **69**, 353 (1972).
25 Gupta, N. K., Woodley, C. L., Chen, Y. C. and Bose, K. K. *J. Biol. Chem.*, **248**, 4500 (1973).
25a Trachsel, H. and Staehelin, T. *PNAS*, **75**, 204 (1978).
26 McKnight, S. L., Sullivan, N. L. and Miller, O. L. Jr *Progr. Nucl. Acid Res. Mol. Biol.*, **19**, 313 (1977).
27 Kozak, M. and Shatkin, A. J. *J. Mol. Biol.*, **112**, 75 (1977).
Kozak, M. *Nature*, **269**, 390 (1977).
28 Rose, J. K. *PNAS*, **74**, 3672 (1977).
29 Ilan, J. and Ilan, J. *PNAS*, **74**, 2325 (1977).
30 Shine, J. and Dalgarno, L. *Nature*, **254**, 34 (1975).
31 Sprague, K. U., Steitz, J. A., Grenley, R. M. and Stocking, C. E. *Nature*, **267**, 462 (1977).
32 Steitz, J. A. and Jakes, K. *PNAS*, **72**, 4734 (1975).
Steitz, J. A. and Bryan, R. A. *J. Mol. Biol.*, **114**, 527 (1977).
Steitz, J. A. and Steege, D. A. *J. Mol. Biol.*, **114**, 545 (1977).
33 Hagenbuchle, O., Santer, M., Steitz, J. A. and Mans, R. J. *Cell*, **13**, 551 (1978).
34 Lodish, H. F. *J. Mol. Biol.*, **50**, 689 (1970); *Nature*, **226**, 705 (1970).
35 Hall, N. D. and Arnstein, H. R. V. *Biochem. Biophys. Res. Comm.*, **54**, 1489 (1973).
36 Johnson, B. and Szekely, M. *Nature*, **267**, 550 (1977).
37 Bear, D. G., Ng, Ray, Van Derveer, D., Johnson, N. P., Thomas, G., Scleich, T. and Noller, H. F. *PNAS*, **73**, 1824 (1976).
Szer, W., Hermoso, J. M. and Boublik, M. *Biochem. Biophys. Res. Comm.*, **70**, 957 (1976).
Thomas, J. O., Kolb, A. and Szer, W. *J. Mol. Biol.*, **123**, 163 (1978).
38 Lee-Huang, S. and Ochoa, S. *Nature NB*, **234**, 236 (1971).
39 Berissi, H., Groner, Y. and Revel, M. *Nature NB*, **234**, 44 (1971).
40 Thompson, W. C., Buzash, E. A. and Heywood, S. M. *Biochemistry*, **12**, 4559 (1973).
41 Wigle, D. T. and Smith, A. E. *Nature NB*, **242**, 136 (1973).
42 Revel, M., Pollack, Y., Groner, Y., Scheps, R., Inouye, H., Berissi, H. and Zeller, H. *Biochimie*, **55**, 41 (1973).
43 Groner, Y., Pollack, Y., Berissi, H. and Revel, M. *Nature NB*, **239**, 16 (1972).
44 Suryanarayana, T. and Subramanian, A. R. *FEBS Lett.*, **79**, 264 (1977).
45 Wahba, A. J., Miller, M. J., Niveleau, A., Landers, T. A., Carmichael, G. G., Weber, K., Hawley, D. A. and Slobin, L. I. *J. Biol. Chem.*, **249**, 3314 (1974).
46 Dahlberg, A. E. *J. Biol. Chem.*, **249**, 7673 (1974).

Van Dieijen, G., Van der Laken, C. J., Van Knippenberg, P. H. and Van Duin, J. *J. Mol. Biol.*, **93**, 351 (1975).
Szer, W., Hermoso, J. M. and Leffler, S. *PNAS*, **72**, 2325 (1975).
Van Dieijen, G., Van Knippenberg, P. H., Van Duin, J., Koekman, B. and Pouwels, P. H. *Mol. Gen. Genet.*, **153**, 75 (1977).
47 Bosch, L. in *Proceedings of the 10th FEBS Meeting* (eds Chapeville, F. and Grunberg-Manago, M.) North Holland/American Elsevier, New York, 275 (1975).
48 Rourke, A. W. and Heywood, S. M. *Biochemistry*, **11**, 2061 (1972).
49 Kennedy, D. S., Bester, A. J. and Heywood, S. M. *Biochem. Biophys. Res. Comm.*, **61**, 415 (1974).
Heywood, S. M. and Kennedy, D. S. *Progr. Nucl. Acid Res. Mol. Biol.*, **19**, 477 (1976).
50 Roberts, B. E. and Paterson, B. M. *PNAS*, **70**, 2330 (1973).
51 Gross, K., Probst, E., Schaffner, W. and Birnstiel, M. *Cell*, **8**, 455 (1976).
52 Smith, A. E., Kamen, R., Mangel, W. F., Shure, H. and Wheeler, T. *Cell*, **9**, 481 (1976).
53 Andersorn, C. W., Atkins, J. F. and Dunn, J. J. *PNAS*, **73**, 2752 (1976).
Schreier, M. H., Staehelin, T., Gesteland, R. F. and Spahr, P. F. *J. Mol. Biol.*, **75**, 575 (1973).
Morrison, T. G. and Lodish, H. F. *J. Biol. Chem.*, **249**, 5860 (1974).
54 Legon, S., Model, P. and Robertson, H. D. *PNAS*, **74**, 2692 (1977).
55 Gurdon, J. B., Lane, C. D., Woodland, H. R. and Marbaix, G. *Nature*, **233**, 177 (1971).
56 Struhl, K. and Davis, R. W. *PNAS*, **74**, 5255 (1977).
Dickson, R. C. and Markin, J. S. *Cell*, **15**, 123 (1978).
Mercereau-Puijalon, O., Royal, A., Cami, B., Garapin, A., Krust, A., Gannon, F. and Kourilsky, P. *Nature*, **275**, 505 (1978).
Goeddel, D. V. *et al. PNAS*, **76**, 106 (1979).
57 Isono, K. and Isono, S. *PNAS*, **73**, 767 (1976).
Isono, S. and Isono, K. *Eur. J. Biochem.*, **56**, 15 (1975).
58 Hsu, M.-T. and Ford, J. *PNAS*, **74**, 4982 (1977).
59 Smith, A. E. in 'Gene Expression', *Proc. 11th FEBS Meeting*, **43**, Pergamon Press, 37 (1978).
Siddel, S. G. and Smith, A. E. *J. Virol.*, **27**, 427 (1978).
60 Kumar, A. and Pederson, T. *J. Mol. Biol.*, **96**, 353 (1975).
61 Blobel, G. *PNAS*, **70**, 924 (1973).
Mueller, R. U., Chow, V. and Gander, E. S. *Eur. J. Biochem.*, **77**, 287 (1977).
Sundquist, B., Persson, T. and Lindberg, U. *Nucl. Acid Res.*, **4**, 899 (1977).
62 Bernardi, A. and Spahr, P. F. *PNAS*, **69**, 3033 (1972).
63 Schulman, L. D. H., Pelka, H. and Sundari, R. M. *J. Biol. Chem.*, **249**, 7102 (1974).
64 Ono, Y., Skoultchi, A., Klein, A. and Lengyel, P. *Nature*, **220**, 1304 (1968).
65 Lucas-Lenard, J. and Haenni, A. L. *PNAS*, **63**, 93 (1969).
Modolell, J., Cabrer, B. and Vazquez, D. *PNAS*, **70**, 3561 (1973).
66 Gupta, S. L., Waterson, J., Sopori, M. L., Weissman, S. M. and Lengyel, P. *Biochemistry*, **10**, 4410 (1971).
67 Thach, R. E., Sundarajan, T. A., Dewey, K. F., Brown, J. C. and Doty, P. *Cold Spring Harbor Symp.*, **31**, 85 (1966).
68 Dintzis, H. M. *PNAS*, **47**, 247 (1961).
69 Hardesty, B., Culp, W. and McKeehan, W. *Cold Spring Harbor Symp.*, **34**, 331 (1969).
70 Woese, C. *Nature*, **226**, 817 (1970).

71 Spirin, A. S. *Cold Spring Harbor Symp.*, **34**, 197 (1969).
Chuang, D. M. and Simpson, M. V. *PNAS*, **68**, 1474 (1971).
72 Weidner, H., Yuan, R. and Crothers, D. M. *Nature*, **266**, 193 (1977).
73 Riddle, D. L. and Carbon, J. *Nature NB*, **242**, 230 (1973).
74 Caskey, C. T. in *Molecular Mechanisms of Protein Biosynthesis* (eds Weissbach, H. and Pestka, S.) Academic Press, New York, 443 (1977).
75 Wilson, D. B. and Dintzis, H. M. *PNAS*, **66**, 1282 (1970).
76 Butterworth, B. E. and Rueckert, R. R. *Virology*, **50**, 535 (1972).
77 Taniguchi, T. and Weissmann, C. *Nature*, **275**, 770 (1978).
78 Kozak, M. and Shatkin, A. J. *Cell*, **13**, 201 (1978); *J. Biol. Chem.*, **253**, 6568 (1978).
Kozak, M. *Cell*, **15**, 1109 (1978).
79 Baralle, F. E. and Brownlee, G. G. *Nature*, **274**, 84 (1978).
80 Sonenberg, N., Morgan, M. A., Merrick, W. C. and Shatkin, A. J. *PNAS*, **75**, 4843 (1978).
81 Rose, J. K., Trachsel, H., Leong, K. and Baltimore, D. *PNAS*, **75**, 2732 (1978).
82 Legon, S. personal communication.
83 Donner, D., Villems, R. and Kurland, C. G. *PNAS*, **75**, 3192 (1978).

Index